DUST PLASMA INTERACTION IN SPACE

DUST PLASMA INTERACTION IN SPACE

P. K. SHUKLA
EDITOR

Nova Science Publishers, Inc.
New York

Senior Editors: Susan Boriotti and Donna Dennis
Coordinating Editor: Tatiana Shohov
Office Manager: Annette Hellinger
Graphics: Wanda Serrano
Editorial Production: Jennifer Vogt, Matthew Kozlowski, Jonathan Rose, Ron Doda and Maya Columbus
Circulation: Ave Maria Gonzalez, Indah Becker and Vladimir Klestov
Communications and Acquisitions: Serge P. Shohov
Marketing: Cathy DeGregory

Library of Congress Cataloging-in-Publication Data
Available upon request.
ISBN 1-59033-365-9

400 Oser Ave, Suite 1600
Hauppauge, New York 11788-3619
Tele. 631-231-7269 Fax 631-231-8175
e-mail: Novascience@earthlink.net
Web Site: http://www.novapubishers.com

Printed in the United States of America

CONTENTS

Foreword

This volume on "Dust Plasma Interaction in Space" is part of the outcome of an International Team, led by Prof. Padma Kant Shukla, at the International Space Science Institute (ISSI) in Bern, Switzerland.

ISSI is an Institute where scientists from different countries can work together. Its main task is to contribute to the achievement of a deeper understanding of the results from space missions, adding value to those results through multi-disciplinary research in an atmosphere of international cooperation. One tool for fulfilling this task is its International Teams programme. Teams are composed of up to 15 scientists of different laboratories, nationalities and expertise meeting for extended periods at ISSI. Their raison d'être is the performance of a specific research project with eventual publication of the scientific outcome. Their activity is directed and organised by a team leader who is also the initiator of the proposal to ISSI. Though in close contact with the scientific staff of the Institute, the teams are largely independent in the execution of their project. Teams can be proposed and set up at any time, and are granted support by ISSI after a successful review procedure.

In a way, the "Dusty Plasma" Team was prototypical for this approach. The team was truly international, with hardly two members coming from the same country. It met at ISSI for a total of seven weeks between November 1998 and April 2002. The discussions at the meetings were very intense, lively, controversial, yet always to the point and firmly based in scholarly tradition and scientific spirit. ISSI was very happy to host this team and is pleased to having assisted in its successful conclusion with this volume. The staff and directors of ISSI thank Padma Shukla and the whole team for their intense and successful work.

Johannes Geiss, Götz Paschmann, Rudolf von Steiger
May 2002

Preface

Dust is ubiquitous in the Universe and responsible for stellar and planetary formation. Virtually all previous studies have considered the dust particulates to be a charge neutral component. Satellite missions such as Voyager revealed the extent to which charged dust plays a role in astrophysics. In most areas dust exists in the presence of a dilute plasma. In such a plasma environment the dust becomes charged due to a variety of processes, e.g. photo-ionization, collisions with electrons and ions, thermionic emission etc. Closer to home in the Earth's mesosphere, rocket measurements and radar observations conclusively demonstrate the importance of charged dust in forming complex structures responsible for Polar Mesospheric Summer Echoes.

At the same time major progress has been made in plasma processing techniques where the production of charged dust has contributed adversely to the formation of smaller etched structures. In such devices dust formes rapidly, limiting the efficiency of micro-circuitry production. Overcoming these problems produced a greater understanding of charged dust formation and their interactions with the plasma to the extent that it is now possible to create various nano-scale structures such as nanotubes. This will open up the possibility for producing new materials and coating with novel physical properties.

The discovery of charged dust crystals, formed by a strongly correlated and ordered system, and dust acoustic wave propagation characteristics increased substantially the scientific interest opening up new area of research. Of particular significance is the possibility of developing experiments and theory in the laboratory to model various space and astrophysical processes involving charged dust. This is possible because the underlying fundamental physics is the same.

This book is the outcome of a periodic collaboration between a small group of scientists meeting twice a year at the International Space Science Institute (ISSI) in Bern, Switzerland, over a period of three years. It represents only a selection of the topics covered and is by no means pretends to be an exhaustive survey of the dust plasma interactions in both space and laboratory. We hope the book will convey some of the excitement and pioneering spirit generated during the lovely debates and open discussions at ISSI.

The book begins with a description of the fundamental physical processes which characterize dusty plasmas. Starting with charging and shielding of isolated dust particles in a plasma environment, the review progresses to nonlinear dust dynamics of spherical and irregular dust. In an ensemble of a dust plasma system collective processes begin to manifest themselves in the form of new wave modes and instabilities which play an important role in the collective behavior displayed in space and astrophysical environment. These include effects of dust on instabilities in the lower ionosphere, dust grain rotation driven electrostatic and electromagnetic waves, fragmentation of molecular dust clouds and formation of structures, possibly leading to stellar and planetary formation. Possible solutions to angular momentum and magnetic flux loss problems may be resolved as a result of new tearing type instabilities in partially ionized, magnetized dust plasma system. New insights into nonlinear plasma waves in the presence of charged dust are discussed with emphasis on soliton formation, shock waves and vortices. The overriding question in space and laboratory plasmas is how the dust forms and grows. This is addressed using new insights developed to understand the role of ion drag forces in the agglomeration processes, shown to be more effective than the traditionally accepted Van der Waals forces. Finally more exotic topics, such as the Ising model for irregular dust grain alignment, classical Casimir forces and chargeons are explored.

We are grateful to the ISSI directors, Professors J. Geiss, R. von Steiger, B. Hultqvist and G. Paschmann, and to the Programme Manager Dr. V. Manno for providing support and encouragement during our stays at ISSI.

Editor P. K. Shukla and ISSI team members:

R. Bingham *D. Jovanović* *A. A. Mamun*
J. T. Mendonça *D. P. Resendes* *M. Rosenberg*
G. Sorasio *L. Stenflo* *D. D. Tskhakaya*
F. Verheest

Charging and Shielding of Dust Grains in a Dusty Plasma

D. D. Tskhakaya
Institute of Physics, Georgian Academy of Sciences,
Tbilisi 380077, Georgia

P. K. Shukla and A. A. Mamun*
Fakultät für Physik und Astronomie,
Ruhr-Universität Bochum, D-44780, Germany

L. Stenflo
Department of Plasma Physics, Umeå University,
SE-90187 Umeå, Sweden

Abstract

The charging of a dust grain in a unmagnetized plasma is examined. A description of the far-field dipole-like potential is presented for the case when the dust grain absorbs electrons and ions from the background plasma. In a Maxwellian plasma the electron and ion currents, which reach the dust grain surface, are calculated. It is shown that the trapped ions produce only an insignificant current and therefore they have little effect on the dust grain charging process. The trapped ion effects on the shielding of a dust grain in a plasma are also examined. It is shown that ions with low angular momentum cannot form a trapped cloud in a potential well. Therefore, it is necessary to examine the clouds of trapped ions with large angular momentum, which can shield the grains. The potential distribution between charged dust grains in a dusty plasma is derived analytically. The profiles of the intergrain potential distribution can have such a shape that the dust grains form a regular periodic structure (dust crystal). The charge of a given grain is not only determined by its own potential, but it also strongly depends on the potential of other dust grains that are participating in the dust crystallization. It means that although the short range interaction is provided by the Debye-Hückel potential, the long range interaction contributes to the formation of a dust crystal. For cubic symmetry and small dust grains (when the dust grain sizes are smaller than the inter-grain spacing) the calculation of the potential has been performed analytically. The

*Permanent address: Department of Physics, Jahangirnagar University, Savar, Dhaka, Bangladesh.

possibility of both longitudinal (compressional) and transverse oscillations of a cubic dust lattice is discussed.

1 Introduction

When a neutral grain is immersed in a plasma, it will usually attain a negative charge [Shukla 2001; Shukla and Mamun 2002]. The charging and interaction of grains in a plasma is a subject of central importance for the explanation of dust grain attraction in laboratory and space environments [Shukla and Mamun 2002].

There are numerous processes [Mott-Smith and Langmuir 1926; Bernstein and Rabinovitz 1959; Horanyi *et al.* 1988; Goertz 1989; Northrop 1992; Allen 1992] by which a neutral dust grain in a plasma can be charged. The charging processes include collection of the background plasma electrons and ions by the dust grain, photoionization, secondary electron emission, thermionic emission, etc. Below we consider the first process only - grain charging by electron and ion currents.

The physics of charged dust grains is usually described by the orbit limited motion (OLM) theory, and the charge on the dust grain is determined by balancing the electron and ion currents that reach the dust grain surface. The OLM theory assumes that there are no potential barriers preventing the plasma ions from reaching the positive energy points in phase space.

Recently, there has been a renewed interest in the study of the OLM theory and diverse opinions have emerged [Allen *et al.* 2000; Lampe *et al.* 2000, 2001]. While Allen *et al.* [2000] question the validity of the OLM theory for any finite sized negatively charged dust grain in a Maxwellian plasma, Lampe *et al.* [2000, 2001] re-examine the limits of validity of the OLM theory by incorporating the potential barriers that exclude some plasma ions. Lampe *et al.* [2000, 2001] reported that the OLM theory is quite exact in the limit of small grain size, and that it turns out to be quite accurate in calculating the ion current to the dust grain for typical conditions pertinent to dusty plasmas.

Since the dust grain charging involving the OLM theory is subjected to differing views at present, it is of urgent interest to re-examine the dust particle charging in an unmagnetized plasma [Tskhakaya *et al.* 2001]. Our objective is also to discuss the role of trapped ions in the shielding of the grain field [Tskhakaya *et al.* 2001].

About fifteen years ago, Ikezi [1986] theoretically predicted the possibility of a periodic distribution of the potential (dust crystallization) in dusty plasmas. The idea of Ikezi has been experimentally verified by several groups [Chu and I 1994, 1994; Chu *et al.* 1994; Thomas *et al.* 1994; Hayashi and Tachibana 1994; Morfill *et al.* 1997, 1999; Fortov *et al.* 1996, 1999; Zheng and Earnshow 1995; Pieper *et al.* 1996; Hayashi 2001; Molotkov *et al.* 2000; Zuzic *et al.* 2000], who reported the fabrication of dust crystals in radio-frequency and glow discharges. However, no satisfactory mechanisms have yet been presented for understanding the formation of periodic potential structures, although several authors have studied dust lattice oscillations [Melandsø 1996; Thomas *et al.* 1998; Farokhi *et al.* 1999; Shukla 2000]. A dust crystal is composed of an ensemble of charged macroparticles which are very much affected by the short range shielded Coulomb potential. However, in reality, the mutual inter-grain interaction is quite complex in that the charge on a specific grain is not only determined by its own surface potential, but also by the potentials of other grains which participate in the formation of the dust crystal.

Due to the repulsive character of the Debye-Hückel potential (also referred to as the Yukawa potential [Morfill *et al.* 1997 1999; Fortov *et al.* 1996, 1999; Zheng and Earnshow 1995], the symmetry of the dust crystal is determined by the form of the repulsive walls [Chu and I 1994, 1994; Chu *et al.* 1994] that surround the dust grain gas. The screened Coulomb potential approximation is backed by simulations [Zheng and Earnshow 1995] and experiments [Morfill *et al.* 1997, 1999]. For a given symmetry, one can use the methods developed in solid state physics [Kittel 1956] to investigate the properties of a dust crystal. However, contrary to solid state crystals, studies of dust crystals in plasmas are easier to perform because the form of the interaction potentials (viz. Debye-Hückel, dipole-dipole, wake, dust shadowing, etc.) is known. Accordingly, we are able to deduce more complete information on the physics of dust crystals in plasmas [Farokhi *et al.* 1999; Tskhakaya and Shukla 2001].

We have organized our present chapter in the following fashion. We describe the motion of electrons and ions in the dust grain potential in §2. We derive analytical expressions for the electron and ion current densities in §3 and the ion number density far from the grain in §4, respectively. We then estimate the dust grain potential in §5 and study the effects of the trapped ions on the shielding of the dust grains in §6. We derive an analytical periodic distribution of the potential in §7, and we study small oscillations of the dust lattice in §8. We finally present a brief discussion in §9.

2 Motion of Electrons and Ions

Let us consider a multi-component dusty plasma whose constituents are electrons, ions and extremely massive negatively charged dust grains which are discrete particles [Tskhakaya *et al.* 2001]. We assume that the dust grain size r_d is much smaller than the ion Debye radius λ_{Di}. Since micron-sized dust grains are typically billion times heavier than protons, they are supposed to be immobile on the time scales of interest here. The surfaces of the stationary dust grains are supposed to absorb the electrons and ions during the charging process. Furthermore, we assume that the field and the plasma particle (electrons and ions) distributions have radial symmetry. The finite grain size effect is incorporated by introducing a cut-off scale size r_d. The radial motion of the plasma particles is considered as one-dimensional in the field corresponding to the effective potential energy

$$U_{eff}(r;M) = U(r) + \frac{M^2}{2m_\alpha r^2}, \tag{1}$$

where the angular momentum M is an integral of motion and m_α is the mass of the species α (α equals e for electrons and i for ions). The dependence of the effective potential on the distance r to the center of force is determined by the potential energy $U(r)$ of the particles in the field that is produced by a dust grain.

It is important to determine the regions for finite and infinite particle motions, as the electron and ion distributions in those regions have different forms. Therefore, we introduce certain assumptions for the dependence of the potential energy $U(r)$ on r. At short distances (viz. r is smaller than the ion Debye radius λ_{Di}) from the grain, the Debye screening of the grain charge by the plasma particles is not significant and the potential energy $U(r)$ decreases as $1/r$, i. e. slower than $1/r^2$. When $r \sim \lambda_{Di}$, $U(r)$ decreases exponentially due to screening, i. e. faster than $1/r^2$. The behavior of $U(r)$ at large distances (viz. $r \gg \lambda_{Di}$) significantly depends on the conditions at the dust grain surface. If the dust grain surface absorbs electrons and ions, the potential energy $U(r)$ at $r \to \infty$ decays as $1/r^2$. Although this has been noted by many authors [Lampe *et al.* 2000, 2001], the limits of its validity and its physical origin have not been explained. In the following, we present a brief derivation [Alpert *et al.* 1964] of the dipole field (for a negatively charged dust grain surrounded by ion clouds).

In steady state, the kinetic equation for the ions is of the form

$$\mathbf{v} \cdot \frac{\partial f_i}{\partial \mathbf{r}} - \frac{1}{m_i}\frac{\partial U(r)}{\partial \mathbf{r}} \cdot \frac{\partial f_i}{\partial \mathbf{v}} = A(\mathbf{r}, \mathbf{v})\delta(S(\mathbf{r})), \tag{2}$$

where f_i is the ion distribution function and the value $U(r)$, defined as a singly charged ion potential energy, is negative. The right-hand side in (2) describes the absorption of particles by the grains, $A(\mathbf{r}, \mathbf{v})$ represents the rate of change of ions per unit time, and $S(\mathbf{r}) = 0$ determines the dust grain surface. Inserting $f_i = f_{i0} + \delta f_i$, where f_{i0} is a Maxwellian distribution function with temperature T_i, in (2) and Fourier decomposing in space, we obtain

$$\begin{aligned} i\mathbf{k}\cdot\mathbf{v}\delta f_{\mathbf{k}} - i\frac{U_{\mathbf{k}}}{m_i}\mathbf{k}\cdot\frac{\partial f_{i0}}{\partial \mathbf{v}} &= \frac{i}{m_i(2\pi)^3}\int d\mathbf{k}' U_{\mathbf{k}'}\mathbf{k}' \cdot \frac{\partial \delta f_{\mathbf{k}-\mathbf{k}'}}{\partial \mathbf{v}} \\ &+ \int d\mathbf{r} A(\mathbf{r}, \mathbf{v})\delta(S(\mathbf{r}))\exp(-i\mathbf{k}\cdot\mathbf{r}), \end{aligned} \tag{3}$$

where $\delta f_{\mathbf{k}}$ is the Fourier transform of δf_i, which is not assumed to be small. We are interested in finding the behavior of the perturbations at large distances ($r \to \infty$), which means that in the right-hand side of (3) we have to consider the limit $k \to 0$. The convergence of the integral on the right-hand side is ensured since we expect

$U_{\mathbf{k}}$ and $\delta f_{\mathbf{k}}$ to decay as $1/k$ at $k \to 0$. Hence, we obtain

$$i\mathbf{k} \cdot \mathbf{v} \left(\delta f_{\mathbf{k}} + \frac{U_{\mathbf{k}}}{T_i} f_{i0} \right) = I(\mathbf{v}), \tag{4}$$

where

$$I(\mathbf{v}) = \int d\mathbf{r} A(\mathbf{r}, \mathbf{v}) \delta(S(\mathbf{r})) + \frac{1}{m_i} \int d\mathbf{r} \frac{\partial U(r)}{\partial \mathbf{r}} \cdot \frac{\partial \delta f_i}{\delta \mathbf{r}}. \tag{5}$$

The function $I(\mathbf{v})$ is determined by the properties of the field in the nearest zone of the dust grain.

Let us assume that the weakly correlated electrons follow a Boltzmann distribution, so that the corresponding density is

$$n_e(\phi) = n_{e0} \exp[U(r)/T_e], \tag{6}$$

where n_{e0} is the number density of electrons when $U = 0$ and T_e is the electron temperature. Substituting (5) and (6) into Poisson's equation we readily obtain

$$\nabla^2 U = 4\pi e^2 \left(n_{e0} \exp[U(r)/T_e] - n_{i0} - \int \delta f_i d\mathbf{v} \right), \tag{7}$$

where e is the magnitude of the electron charge and $n_{i0}\,(n_{e0})$ is the unperturbed ion (electron) number density. Expanding the exponential function in powers of $U(r)/T_e$ and Fourier decomposing, we obtain from (7)

$$\left(k^2 + k_{De}^2\right) U_{\mathbf{k}} + \frac{2\pi n_{e0} e^2}{T_e^2} \frac{1}{(2\pi)^3} \int d\mathbf{k}' U_{\mathbf{k}'} U_{\mathbf{k}-\mathbf{k}'} + ... = 4\pi e^2 \int \delta f_{\mathbf{k}} d\mathbf{v}, \tag{8}$$

where $k_{De} = (4\pi n_{e0} e^2/T_e)^{1/2}$ is the inverse electron Debye radius. The linear term (with respect to $U_{\mathbf{k}}$) is of order $1/k$, while the nonlinear terms tend to a constant value at $k \to 0$. Thus, neglecting the nonlinear term and ignoring the departure from the quasi-neutrality [the k^2 term in the left-hand side of (8)], we obtain the Fourier component of the ion number density

$$\delta n_{\mathbf{k}} = \int \delta f_{\mathbf{k}} d\mathbf{v} = \frac{U_{\mathbf{k}}}{T_e} n_{e0}. \tag{9}$$

We note that (9) remains valid even in the nonlinear regime [Alpert *et al.* 1964] when $k \to 0$. Combining (9) with (4) we obtain

$$\delta n_{\mathbf{k}} \simeq -\frac{i}{1 + T_e/T_i} \int d\mathbf{v} \frac{I(\mathbf{v})}{\mathbf{k} \cdot \mathbf{v} - i\epsilon} \quad \text{at} \quad \epsilon \to +0 \tag{10}$$

from which it emerges that

$$\delta n_{\mathbf{k}} \sim \frac{1}{k} \quad \text{and} \quad U_{\mathbf{k}} \simeq \frac{1}{k} \quad \text{for} \quad k \to 0. \tag{11}$$

Consequently, in real space, we would have

$$U(r) = U_\infty \frac{r_d^2}{r^2}, \tag{12}$$

where U_∞ is the potential at the dust grain surface if the $1/r^2$ dependence would have been extrapolated over all the distances up to the grain surface. The quantity $|U_\infty|$ is always smaller than the real potential on the dust grain surface $|U(r_d)|$ [Alpert *et al.* 1964], viz.

$$|U_\infty| < |U(r_d)|, \tag{13}$$

as will also be shown later [Tskhakaya *et al.* 2001]. The expression (12) can be physically understood as follows [Alpert *et al.* 1964]. When the perturbation at large distances is caused by the absorption of plasma particles, then the quasi-neutral density perturbation (and consequently the field potential and the potential energy) is proportional to the solid angle under which the dust grain is seen from the given point r. Since the solid angle is proportional to r_d^2/r^2, we retrieve (12) on a physical ground as well. We note that (12) remains valid for distances that are smaller than the collisional mean free path $\lambda_m \gg r$. The inter-particle collisions destroy the dependence law (12).

For electrons that are repelled by a negatively charged dust grain, U_{eff} decreases monotonically because the potential energy of electrons is positive, viz. $U_e(r) = -U(r) > 0$. Thus, the electrons are reflected from a monotonically decreasing potential well. For ions $U(r) < 0$ and the character of the dependence of their effective potential energy on r significantly depends on the angular momentum M. It then follows from (1) that the extreme values of U_{eff} are determined from the condition

$$\frac{dU_{eff}}{dr} = -\frac{1}{r^3}\left[\frac{M^2}{m_i} - r^3\frac{dU(r)}{dr}\right] = 0. \tag{14}$$

According to the above discussions of the behavior of the potential energy $U(r)$ for absorbing grains, the qualitative dependence of the term $r^3 dU/dr$ on r must have the form depicted in figure 1. The maximum value of $r^3 dU/dr$ is reached at a certain point $r_k \geq \lambda_{Di}$. By using the Debye-Hückel shielded potential, we obtain $r_k \approx 1.61\lambda_{Di}$. From equation (12) and figure 1 we determine

$$M_0^2 = m_i\left(r^3\frac{dU}{dr}\right)_{r\to\infty} = 2m_i r_d^2 |U_\infty| \tag{15}$$

and

$$M_k^2 = m_i\left(r^3\frac{dU}{dr}\right)_{r=r_k}. \tag{16}$$

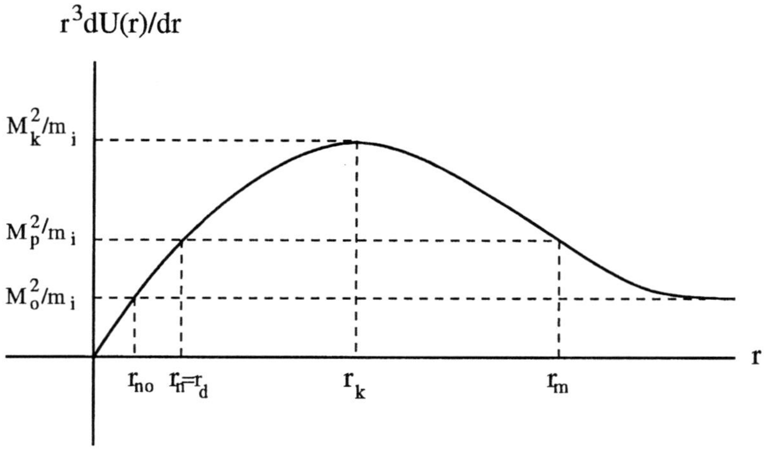

Figure 1: Qualitative dependence of the auxiliary function $r^3 dU/dr$ on r.

At $M \leq M_0$, the effective potential energy U_{eff} has only one extreme value that is minimum. If the angular momentum is in the range

$$M_0 < M < M_k, \tag{17}$$

then $U_{eff}(r)$ will have both maximum and minimum at points r_m and r_n, respectively. The values of these extreme points depend on the angular momentum M, but the condition $r_n \leq r_k \leq r_m$ is always satisfied. At $M = M_k$ the extreme points coincide, viz. $r_n = r_m$, and at this point the function $U_{eff}(r)$ has an inflection point. If $M > M_k$, the effective potential energy $U_{eff}(r)$ decreases monotonously with the increase of r.

Let us determine the critical value of the angular momentum M_p from the condition of equality of the maximum value $U_{eff}(r_m; M)$ and an effective potential energy on the surface of the grain, which is

$$U_{eff}(r_d; M_p) = U_{eff}(r_m; M_p). \tag{18}$$

Equation (18) yields

$$M_p^2 = 2m_i r_d^2 \left[|U(r_d)| - |U(r_m)| \right] \left(1 - r_d^2/r_m^2\right)^{-1}. \tag{19}$$

Since $r_m \geq r_k \gg r_d$ and $|U(r_d)| \gg |U(r_m)|$, we obtain the approximate value of M_p, namely

$$M_p^2 \simeq 2m_i r_d^2 |U(r_d)|. \tag{20}$$

It is obvious from (13), (15) and (20) that $M_p > M_0$. By means of (1) and (18) we can make sure that $U_{eff}(r_d; M) > U_{eff}(r_m; M)$ at $M > M_p$, while $U_{eff}(r_d; M) <$

$U_{eff}(r_m; M)$ at $M < M_p$. The value of the zero point r_s of the effective potential energy $U_{eff}(r_s; M)$ is determined from the condition

$$U_{eff}(r_s; M) = 0 \qquad \text{at} \qquad M \leq M_0. \tag{21}$$

According to (1) and (13) r_s is always less that the radius of the grain, viz. $r_s < r_d$. Accounting for the above mentioned properties of the function $U_{eff}(r; M)$, we can

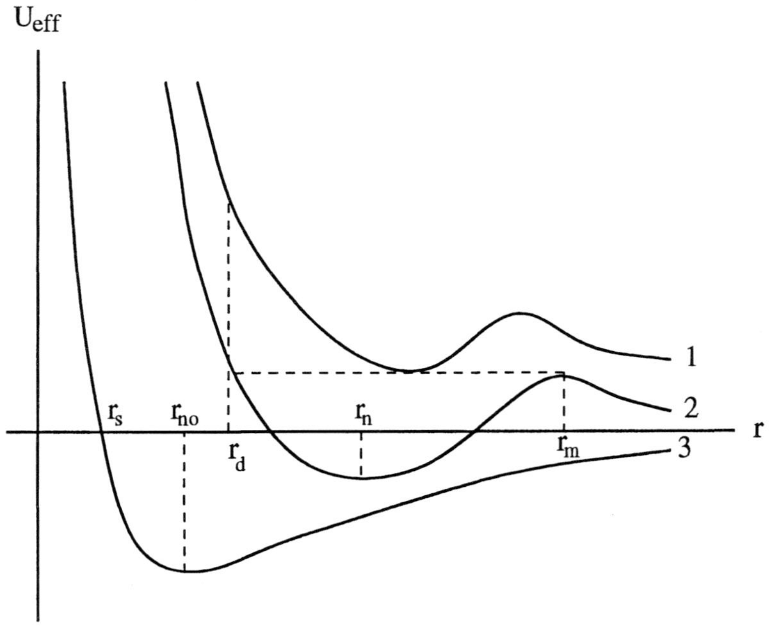

Figure 2: Qualitative dependences of $U_{eff}(r; M)$ on r for various values of the angular momentum: $M_p < M < M_k$ (curve 1), $M = M_p$ (curve 2) and $M \leq M_0$ (curve 3)

determine the surface that separates regions of infinite and finite particle motions in the velocity space (v_r, v_θ), where v_r and v_θ are the components of the velocity along and across the radial direction. We are interested in calculating the current associated with those particles which reach the dust grain surface. According to figure 2, they must have the energy

$$E(v_r, v_\theta, r) = \frac{m_i v_r^2}{2} + U_{eff}(r; M) \geq U_{eff}(r_d; M). \tag{22}$$

The usual definition of the angular momentum reads

$$M = m_i v_\theta r. \tag{23}$$

Three cases are apparent. First, if $M \leq M_0$, the particles with the positive energy

$$E(v_r, v_\theta, r) \geq 0 \tag{24}$$

will move infinitely and all such particles will reach the grain surface since $r_s < r_d$.

Second, for $M_0 < M \leq M_p$ all the free ions attach onto the grain surface since for such values of angular momentum, as is mentioned above, the condition $U_{eff}(r_d; M) < U_{eff}(r_m; M)$ is fulfilled. The region of infinite ion motion is determined by the condition

$$E(v_r, v_\theta, r) \geq U_{eff}(r_m; M). \tag{25}$$

Third, if $M > M_p$, then $U_{eff}(r_m; M) < U_{eff}(r_d; M)$, and only those particles whose energy satisfies (22) are able to reach the dust grain surface.

3 Current Densities

In plasmas, the electron thermal speed is much larger than the ion thermal speed. Accordingly, the electrons will reach the dust grain surface faster than the ions. The grain surface will thus acquire a huge negative charge. When it has a critical capacity of absorbing electrons, it will start repelling electrons and attracting ions which form a sheath around the grain.

In the following, we discuss the form of the electron and ion currents that reach a dust grain surface in a plasma with Maxwellian particle distributions

$$f_\alpha(v_r, v_\theta, r) = n_{\alpha 0}\left(\frac{m_\alpha}{2\pi T_\alpha}\right)^{3/2} \exp\left[-\frac{m_\alpha(v_r^2 + v_\theta^2)}{2T_\alpha} \mp \frac{|U(r)|}{T_\alpha}\right], \tag{26}$$

where $\alpha = e, i$ and the $-(+)$ sign in the argument of the exponential function corresponds to electrons (ions). The current of the plasma particles per unit area with external unit normal vector $\mathbf{n}$ located at the point r will be determined by the expression

$$J_\alpha = -2\pi q_\alpha \int (\mathbf{n}\cdot\mathbf{r}) f_\alpha(v_r, v_\theta, r) v_\theta dv_\theta dv_r, \tag{27}$$

where $q_e = -e$ and $q_i = e$ and the integration in (27) has to be performed over velocities for which $\mathbf{n}\cdot\mathbf{v} < 0$, i. e. $v_r < 0$. In (26) we have denoted $|U(r)| = e|\phi(r)|$, where $\phi(r) < 0$ is the electric potential at the point r.

For the electron current, we have

$$J_e = -en_{e0}V_{Te}\exp\left[-\frac{e|\phi(r)|}{T_e}\right], \tag{28}$$

where $V_{Te} = (T_e/2\pi m_e)^{1/2}$.

The boundaries confining the region in velocity space from which the ions can get onto the dust grain surface significantly depend on the value of M. For this reason, the determination of the current of infinite ions should be made separately for each of the three characteristic cases of the angular momentum regimes, as described in the previous section. We obtain the following expressions for the currents of free ions. First, for the region $M \leq M_0$, we have

$$J_{i1} = en_{i0}V_{Ti}\frac{r_d^2}{r^2}\frac{|U_\infty|}{T_i}, \tag{29}$$

where $V_{Ti} = (T_i/2\pi m_i)^{1/2}$. Second, in the region $M_0 < M \leq M_p$, we have

$$\begin{aligned} J_{i2} &= en_{i0}V_{Ti}\frac{r_m^2}{r_d^2}\left[\exp\left(-\frac{r_d^2}{r_m^2}\frac{|U_\infty|}{T_i}\right) - \exp\left(-\frac{r_d^2}{r_m^2}\frac{e|\phi(r_d)|}{T_i}\right)\right] \\ &\times \exp\left(\frac{e|\phi(r_m)|}{T_i}\right). \end{aligned} \tag{30}$$

Third, in the region $M > M_p$ we have

$$J_{i3} = en_{i0}V_{Ti}\exp\left[\frac{e|\phi(r_d)|}{T_i} - \frac{M_p^2}{2m_i r_d^2 T_i}\right]. \tag{31}$$

The total current of ions with an infinite motion onto the grain surface is the sum of (29)-(31), which differs from the expression deduced by Alpert *et al.* [1964]. The reason for the discrepancy is that Alpert *et al.* [1964] did not introduce the critical value for the angular momentum M_p and that the ions were not classified by the values of the ratio M/M_p. If the potential energy of the ions at a distance r_m is much less than the thermal energy, viz. $|U(r_m)| \ll T_i$ and also if $r_d \ll r_m$, then by means of (20) and (31) we obtain

$$J_i = en_{i0}V_{Ti}\left(1 + \frac{e|\phi(r_d)|}{T_i}\right), \tag{32}$$

which agrees with Mott-Smith and Langmuir [1926]. This is expected since near the grain the field decreases according to Coulomb's law. It is important to note that in (32) the smallness of the ratio $e|\phi(r_d)|/T_i$ is not assumed.

The determination of the current of finite ions (viz those trapped by the potential well; see figure 2) is difficult, since it is necessary to determine their distribution function. If the ions are not absorbed by the grain, then it is appropriate to use the Boltzmann distribution for the ions in those regions where the motion is finite. However, when the grain absorbs ions, the determination of the distribution function of trapped ions is a complicated independent problem [Alpert *et al.* 1964; Gurevich 1968]. But in the case of a rarefied plasma when the mean free path of the ions exceeds the ion Debye radius and the grain absorbs the ions, we can estimate the current of the trapped ions.

In steady state, the decrease of the number of ions in the region of finite motion (due to absorption on the dust grain surface) is compensated by the flow of ions from the region of infinite motion into the region of finite motion. This flow is caused by collisions of ions with each other and with the neutral particles. The particles which are trapped this way may then be absorbed by the dust grain. Consequently, in the stationary case, the flow of trapped ions on the surface of the grain cannot exceed the flow from the region of infinite motion into the region of finite motion.

The integral of ion collisions with each other and with neutrals can be represented in the form of a divergence in phase space [Silin 1971], and the kinetic equation can be written in the form

$$\mathbf{v} \cdot \frac{\partial f_i}{\partial \mathbf{r}} - \frac{\partial U(r)}{\partial \mathbf{r}} \cdot \frac{\partial f_i}{\partial \mathbf{p}} = -\frac{\partial}{\partial \mathbf{p}} \cdot \mathbf{S}(f_i, f_\beta), \tag{33}$$

where $\mathbf{p} = m_i \mathbf{v}$, $S(f_i, f_\beta)$ is the flow density in the momentum space and $\beta = i, n$. The expression for S can be derived for ion-ion collisions from the Landau collision integral and for ion-neutral collisions from the Boltzmann collision integral. However, in the case of rare collisions, one can use, to a first approximation, the Maxwell-Boltzmann distribution for both ions and neutrals. We have also to consider the inhomogeneous character of those distribution functions, otherwise the collision integrals would vanish. Let us determine the current density of the trapped ions in the form

$$J_i^t(r) = e \int v_r f_i d\mathbf{v}, \tag{34}$$

where the integration is carried out over velocities from the region of finite motion. Then from (33) we obtain

$$\frac{\partial}{\partial r} J_i^t(r) \simeq -\frac{e}{m_i} \oint_\sigma \mathbf{S} \cdot d\sigma, \tag{35}$$

where σ is the surface (in the velocity space) enclosing the region of finite motion. The right-hand side of (35) is caused by the ion flow through this surface. The current on the surface of the grain, $r \approx r_d$, is of interest to us. Since $r_m \gg r_d$ the equation for the surface in the first two cases of the previous section has the form

$$v_r^2 + v_\theta^2 \simeq \frac{2}{m_i} |U(r_d)|. \tag{36}$$

As a result, for the current of trapped ions on the dust grain surface, we obtain the following estimates. For ion-ion collisions, we have

$$J_i^t \approx e n_{i0} V_{Ti} \frac{\lambda_{Di}}{\lambda_{ii} \Lambda} \frac{|U(r_d)|}{T_i}, \tag{37}$$

where λ_{ii} is the mean free path for ion-ion collisions and Λ is the usual Coulomb logarithm. For ion-neutral collisions we have

$$J_i^t \approx e n_{i0} V_{Ti} \frac{d}{\lambda_{in}} \frac{|U(r_d)|}{T_i}, \tag{38}$$

where d is the radius of interaction of ions and neutrals and λ_{in} is the mean free path for these collisions. Since in our case $\lambda_{ii} \gg \lambda_{Di}$ and $\lambda_{in} \gg d$, a comparison of (37) and (38) with (32) reveals that

$$J_i^t \ll J_i. \tag{39}$$

Hence, in a rarefied plasma the ion current on the dust grain surface is mainly determined by the ions which execute an infinite motion.

4 Particle Number Densities

Let us now determine the electron and ion number densities far from the grain (viz. $r \gg \lambda_{Di}$). The general definition of the particle number density is

$$n_\alpha = 2\pi \int f_\alpha(v_r, v_\theta, r) v_\theta dv_\theta dv_r, \tag{40}$$

where the function $f_\alpha(v_r, v_\theta, r)$ is defined by (26). The limits of integration must be determined according to the character of the particle motion. For example, for the ions one has to follow the conclusions about the ion motion that are indicated in figure 2. The grain absorbs only those electrons which have the energy

$$E_e = \frac{m_e}{2}(v_r^2 + v_\theta^2) + |U(r)| \geq \frac{M_e^2}{2m_e r_d^2} + |U(r_d)| \tag{41}$$

for $v_r < 0$. Here, we have introduced $M_e = m_e v_\theta r$. Far from the grain (viz. $r \gg \lambda_{Di}$), when $|U(r)|/T_e \ll 1$, after integrating (40) outside the area (41), we obtain

$$\begin{aligned}\frac{n_e}{n_{e0}} &= 1 - \frac{|U(r)|}{T_e} + \frac{r_d^2}{2r^2}\left(\frac{|U(r_d)|}{T_e} - \frac{1}{2}\right)\left(1 - \operatorname{erf}\left(\sqrt{\frac{|U(r_d)|}{T_e}}\right)\right) \\ &\quad - \frac{r_d^2}{2\sqrt{\pi}r^2}\sqrt{\frac{|U(r_d)|}{T_e}} \exp\left(-|U(r_d)|/T_e\right),\end{aligned} \tag{42}$$

where

$$\operatorname{erf}(x) = \frac{2}{\sqrt{\pi}} \int_0^x dt \exp(-t^2) \tag{43}$$

is the error function. For the ions we have to consider different ranges of their angular momentum, as described in §2.

For $M \leq M_0$ all the free ions impact the grain surface and are absorbed. According to the results given in the previous section the number of the trapped ions is negligibly small. Then the necessary area of integration in the velocity space is determined by the conditions

$$v_r^2 + v_\theta^2 \geq \frac{2}{m_i}|U(r)|, \tag{44}$$

$$0 \leq v_\theta \leq \frac{M_0}{m_i r} \qquad \text{and} \qquad v_r < 0. \tag{45}$$

Far from the grain, when $|U(r)|/T_i \ll 1$, we obtain

$$n_i = \frac{n_{i0}}{2}\frac{|U(r)|}{T_i}. \tag{46}$$

For the angular momentum in the region $M_0 < M \leq M_p$ the area in the velocity space for ions impinging on the grains is determined by the relations

$$v_r^2 + v_\theta^2 \geq v_\theta^2 \frac{r^2}{r_m^2} + \frac{2}{m_i}\left[|U(r)| - |U(r_m)|\right], \tag{47}$$

$$\frac{M_0}{m_i r} < v_\theta \leq \frac{M_p}{m_i r}, \qquad \text{and} \qquad v_r \leq 0. \tag{48}$$

The quantities M_0 and M_p are defined by (15) and (20). Performing the integration outside of this area we find the ion number density far from the grain (viz. $r \gg \lambda_{Di}$ and $r_m \geq \lambda_{Di}$)

$$n_i \simeq n_{i0} \frac{2}{\sqrt{\pi}} \frac{r_d^3}{r^3} \left(\frac{|U(r_d)|}{T_i} \right)^{3/2}, \tag{49}$$

which is negligible because $n_i \propto 1/r^3$. We note that the number density of free ions which are reflected by the potential hump, $|U(r_m)|$is also small (cf. curve 2 in figure 2). For the angular momentum $M > M_p$ (cf. curve 1 in figure 2) only the ions from the area

$$v_r^2 + v_\theta^2 \geq v_\theta^2 \frac{r^2}{r_d^2} - \frac{2}{m_i}\left(|U(r_d)| - |U(r)|\right), \tag{50}$$

and $v_r < 0$ are absorbed by the grain. Those free ions which have velocity satisfying the opposite inequality in (50) can move both in the positive and negative radial directions, $v_r < 0$ and $v_r > 0$. Far from the grain ($r \gg \lambda_{Di}$) the area of integration in velocity space is presented in figure 3 (the shaded regions). The curve in the region $v_r > 0$ in figure 3 is determined by (50) when the left- and right-hand sides

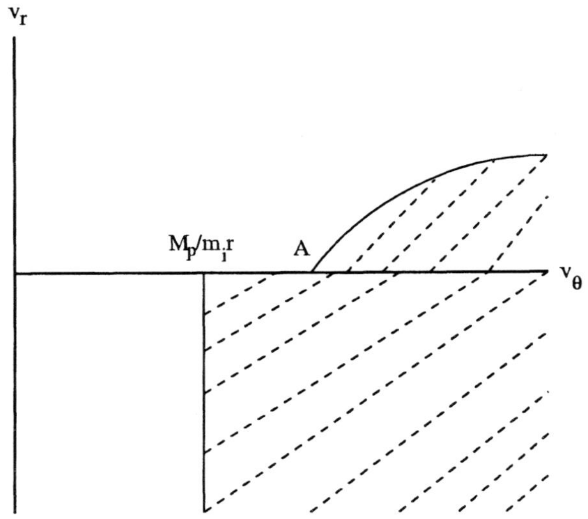

Figure 3: The area (shaded parts) of integration for finding the ion number density far from the grain at $M > M_p$. The point A corresponds to $v_\theta = [(2/m_i)(|U(r_d)| - |U(r)|)]^{1/2} (r^2/r_d^2 - 1)^{-1/2} > M_p/m_i r$.

are equal. Performing the corresponding integration for the ion number density we obtain

$$n_i = n_{i0}\left[1 + \frac{|U(r)|}{T_i} - \frac{r_d^2}{r^2}\left(\frac{|U(r_d)|}{T_i} + \frac{1}{4}\right)\right]. \tag{51}$$

By adding (46) and (51) we obtain the total ion number density far from the grain

$$n_i \simeq n_{i0}\left[1 + \frac{3}{2}\frac{|U(r)|}{T_i} - \frac{r_d^2}{r^2}\left(\frac{|U(r_d)|}{T_i} + \frac{1}{4}\right)\right]. \tag{52}$$

The quasi-neutrality condition $n_i = n_e$ must hold far from the grain. Using the relation (12) and equating (42) and (52) we obtain

$$\begin{aligned} 3\frac{|U_\infty| - |U(r_d)|}{T_i} &= -\frac{|U(r_d)|}{T_i} - \operatorname{erf}\left(\sqrt{\frac{|U(r_d)}{T_e}}\right)\left(\frac{|U(r_d)|}{T_e} - \frac{1}{2}\right) \\ &\quad - \frac{1}{\sqrt{\pi}}\sqrt{\frac{|U(r_d)|}{T_e}}\exp\left(-|U(r_d)|/T_e\right]. \end{aligned} \tag{53}$$

In the next section [e.g. Eq. (56)] it will be shown that $|U(r_d)|/T_e > 1$. Hence it follows from (53) that $|U_\infty| < |U(r_d)|$, which was used in §2 [cf. Eq. (13)].

5 Dust Grain Potential

In steady state the potential on the dust grain surface is determined by putting to zero the sum of the electron and ion currents at $r = r_d$. It means that

$$J_i(r_d) + J_e(r_d) = 0. \tag{54}$$

For $n_{e0} = n_{i0}$ we find that

$$\left[1 + \frac{e|\phi(r_d)|}{T_i}\right] \exp\left[\frac{e|\phi(r_d)|}{T_e}\right] = \frac{V_{Te}}{V_{Ti}}. \tag{55}$$

Equation (55) has a solution

$$\frac{e|\phi(r_d)|}{T_e} = \frac{|U(r_d)|}{T_e} > \frac{1}{2}, \tag{56}$$

if $1 \leq T_e/T_i < m_i/m_e$. The relation (56) has been used in the preceding section for the analysis of (53). Taking $V_{Te}/V_{Ti} \simeq 200$ for an isothermal plasma with $T_e \approx T_i = T$, we obtain from (55)

$$|\phi(r_d)| \simeq 3.75T/e. \tag{57}$$

The electric field strength near the grain is

$$E_0 = \frac{|\phi(r_d)|}{r_d}. \tag{58}$$

For $r_d \simeq 10^{-2}\mu m$ and $T \sim 0.01$ eV, the dust grain potential and the electric field strength turn out to be 6.5×10^{-2} V and 6.5×10^4 V/cm, respectively. Strong electric fields can cause dust grain disruption and produce additional ionization of neutral grains.

6 Trapped Ion Effects on Shielding of Dust Grains

Recently, Lampe *et al.* [2001a,2001b] investigated the effects of trapped ions on the shielding of a dust grain in a plasma. Specifically, they employed the OLM theory for the ion motion in the electrostatic field around a single dust grain, and showed that ion-neutral charge exchange can give rise to a buildup of negative energy trapped ions which may dominate the shielding cloud of the dust grain field. In the following, we examine the validity of the trapped particle effects on the shielding of a dust grain in an unmagnetized plasma [Tskhakaya *et al.* 2001].

Lampe *et al.* [2001b] have accounted for the first interval for the ion angular momentum, $M \leq M_0$ (cf. §2), and used a corresponding restriction on the total energy of the ions, $E \leq 0$ (cf. Eq. (2a) of Lampe *et al.* [2001b]). This case corresponds to the curve 1 in figure 2. According to (13) and the remark made just after (53),

$$\frac{|U_\infty|}{|U(r_d)|} < 1. \tag{59}$$

The dependence of $|U_\infty|$ on the grain surface potential is depicted on page 317 in Alpert *et al.* [1964] for the case when the electron and ion temperatures are equal, (viz. $T_e \approx T_i$) and it is shown that always $|U_\infty| \,/\, |U(r_d)| < 1/3$. With increasing $|U(r_d)|$ this ratio decreases. Obviously (12) is valid only for large r. However, if one extrapolates that equation to $r \to r_d$, one finds, in view of (59), that the dependence in (12) on r does not compensate a sharp decrease of the potential energy $U(r)$ at the distances $r > \lambda_{Di}$. From (59) it also follows that the zero point r_s of the effective potential energy $U_{eff}(r; M)$ at $M \leq M_0$ is always close to the center ($r = 0$) in comparison with the grain surface. That means $r_s < r_d$. Actually, from (1) one finds that $U(r; M)$ at $r = r_d$ for $M \leq M_0$ is negative (see the curve 1 in figure 2), i.e.

$$U_{eff}(r_d; M) \leq |U_\infty| - |U(r_d)| < 0, \quad \text{for } M \leq M_0. \tag{60}$$

Moreover, it turns out that at $M \leq M_0$ also the distance for the minimum point of $U_{eff}(r; M)$ from the center $(r = 0)$ can be less than the grain radius (see the curve 1 in figure 2). For the distances $r \leq \lambda_{Di}, r_k$ we can choose the test function

$$U(r) \simeq -\frac{r_d}{r} |U(r_d)| \exp(-r/\lambda_{Di}) \tag{61}$$

for the potential energy. Then from (14) we obtain for $M \leq M_0$

$$r_{\min} \equiv r_{n0} \simeq 2r_d \frac{|U_\infty|}{|U(r_d)|}. \tag{62}$$

For $|U_\infty| < (1/2)\,|U(r_d)|$ we have $r_{\min} < r_d$. This result is physically expected, as the ions with small angular momentum impact the dust grain surface and are absorbed.

At this point, it is worth stressing that the free ions fall into the potential well within a time $\tau = \lambda_{mfp}/V_i$, where λ_{mfp} is the ion mean free path due to ion-neutral and ion-ion collisions and V_i is the relative speed between the colliding particles; this speed can be larger than the ion thermal speed V_{Ti} (see the end of this section). The time scale τ is much larger than the characteristic time $\tau_w (= \lambda_{Di}/V_i)$ for ions to traverse the potential well, because usually $\lambda_{mfp} \gg \lambda_{Di}$ [Shukla 2001; Shukla and Mamun 2002; Lampe *et al.* 2001b]. It turns out that ions, falling into the potential well very quickly stick onto the grain surface and are lost due to absorption. It means that the ions do not have enough time to be accumulated in the well and to form a stationary trapped cloud. So we can conclude that for ions, with the angular momentum M smaller than a certain critical value M_0, the dust grain surface is within the well and the ions falling into the potential well are immediately lost due to absorption onto the dust grain surface. Hence, the formation of ion clouds, that can shield the grain, is possible only for ions with $M > M_0$. For the latter case, we have to properly calculate the distribution function and the corresponding density for the ions which are trapped in a potential well. This leads to a lengthy analysis because a correct calculation requires that the trapped ions satisfy the boundary condition on the surface in the velocity space, which separates the region of infinite and finite motion [Lifshitz and Pitaevskii 1981].

Dusty plasmas typically have $T_i \ll T_e$ and $e\,|\phi(r_d)| \sim T_e$. Accordingly, we have

$$\frac{|U(r_d)|}{T_i} = \frac{e\,|\phi(r_d)|}{T_i} \gg 1. \tag{63}$$

Lampe *et al.* [2001b] assumed that the ion temperature is equal to the temperature of the neutrals and that the ion-neutral charge exchange collision frequency ν does not depend on energy. However, from (63) it follows that the velocity of the ions in the electric field exceeds their thermal speed. Consequently, the relative velocity of the colliding species in ion-neutral charge-exchange collisions differs essentially from their thermal speed. In this case, charge exchange collisions can be described to a good approximation by a constant cross section [McDaniel and Mason 1973]. It means that for problems connected with the ion motion in the grain field at $T_i \ll T_e$, the constancy of the mean free path at charge-exchange collisions $(\lambda_{mfp} = \text{const})$ is more applicable than the constancy of the collision frequency $(\nu = \text{const})$.

7 Periodic Potential Distribution of Potential

In this section, we construct a theory for the three-dimensional (3D) periodic distribution of the electric potential in a dusty plasma. Specifically, we show that charged macroparticles (dust grains) can organize themselves in the form of regular periodic structures if the external boundary conditions are appropriately considered [Tskhakaya and Shukla 2001]. For this purpose, we calculate the spatial distribution of electrostatic potentials for 3D dust lattices accounting for interactions between all the dust grains that form the dust crystal. Furthermore, we have also investigated oscillations of a dust lattice and found that transverse lattice oscillations have anisotropic behavior relative to the direction of the lattice wave propagation. We note that our investigation of 3D dusty crystals and associated oscillations is superior to those existing in solid state physics where there does not exist a self-consistent theory for 3D crystals including the interaction forces.

Let us consider again a multi-component dusty plasma whose constituents are electrons, ions and extremely massive negatively charged dust grains which are discrete particles. The plasma potential $\phi(\mathbf{r})$ in such a system is determined from Poisson's equation

$$\nabla^2\phi(\mathbf{r}) - 4\pi e(n_e - n_i) = -4\pi \sum_{\mathbf{n}} q_{\mathbf{n}}(\mathbf{r}), \tag{64}$$

where $n_e\,(\phi)$ and $n_i\,(\phi)$ are the electron and ion number densities and the summation on the right-hand side of (64) is taken over all grains and $\mathbf{n} \equiv (n_1, n_2, n_3)$. Here, n_1, n_2 and n_3 enumerate the grain position along the x, y and z axes, respectively. The charge of the grain $q_{\mathbf{n}}(\mathbf{r})$ can be represented in the form

$$q_{\mathbf{n}} = \frac{Q_{\mathbf{n}}}{S_{\mathbf{n}}} \int_{S_{\mathbf{n}}} dS_{\mathbf{n}}(\mathbf{r}')\delta\left(\mathbf{r} - (\mathbf{r_n} + \mathbf{r}')\right) \tag{65}$$

by assuming that the grains are conductors. Here, $\delta(\mathbf{r})$ is the Dirac delta function. The integral in (65) has been taken over the dust grain surface $S_{\mathbf{n}}$. The center of mass of the grain has been denoted by $r_{\mathbf{n}} \equiv (x_{n1}, y_{n2}, z_{n3}) \equiv \left(x_{n1}^{(1)}, x_{n2}^{(2)}, x_{n3}^{(3)}\right)$. In the following, the superscripts (1), (2) and (3) shall denote the x, y and z components, respectively, i.e. $A^{(1)} = A_x, A^{(2)} = A_y, A^{(3)} = A_z$ and $x^{(1)} = x, x^{(2)} = y, x^{(3)} = z$. We stress that the charge of the n-th grain, $Q_{\mathbf{n}}$, is not only determined by its own potential, but it also depends on the potential of other grains [Landau and Lifshitz 1960]. Thus, we have

$$Q_{\mathbf{n}} = \sum_{\mathbf{n}'} C_{\mathbf{nn}'}(\mathbf{r_n}, \mathbf{r_{n'}})\phi_{\mathbf{n}'}(\mathbf{r_n}), \tag{66}$$

where $C_{\mathbf{nn}}$ and $C_{\mathbf{nn}'}$ $(n \neq n')$ are the capacity and electrostatic induction coefficients, respectively. These coefficients depend on the shape and the relative position of conducting grains.

Equation (64) has to be solved with appropriate boundary conditions on the grain surfaces which read: (i) the grain surface must be equipotential. It means that regardless of the direction of approach to any point on the grain surface, the solution of (64) must satisfy the condition

$$\phi\big|_{S_{\mathbf{n}}} = \phi_{\mathbf{n}} = \text{constant}. \tag{67}$$

(ii) the derivative of the potential (viz. the electric field) on the grain surface has a jump. Thus, substituting (65) and (66) into (64) and integrating the resultant equation over the grain volume we obtain the second boundary condition

$$\int_{S_{\mathbf{n}}} \nabla\phi dS_{\mathbf{n}} = -4\pi \sum_{\mathbf{n}'} C_{\mathbf{nn}'}\phi_{\mathbf{n}'}(\mathbf{r_n}) = -4\pi Q_{\mathbf{n}}. \tag{68}$$

Let us assume that weakly correlated electrons and ions follow Boltzmann distributions, so that the corresponding densities are, respectively,

$$n_e(\phi) = n_{e0}\exp(e\phi/T_e) \qquad \text{and} \qquad n_i = n_{i0}\exp(-e\phi/T_i), \tag{69}$$

where n_{e0} and n_{i0} are as usual the number densities of the electrons and ions when $\phi = 0$, and $T_e > T_i$. Substituting (65), (66) and (69) into (64) for small potentials we readily obtain

$$\nabla^2\varphi - 2k_D^2\varphi = -\sum_{\mathbf{nn}'}\frac{C_{\mathbf{nn}'}}{S_n}(\varphi_{\mathbf{n}'} + \hat{\varphi})\int_{S_\mathbf{n}} dS_\mathbf{n}(\mathbf{r}')\delta(\mathbf{r} - (\mathbf{r}_\mathbf{n} + \mathbf{r}')). \tag{70}$$

Here, $\varphi = -e\phi/T_i (\ll 1)$ and $k_D^2 = 2\pi e^2(n_{e0}/T_e + n_{i0}/T_i)$. Equation (70) exhibits that non-equal electron and ion number densities (at $\phi = 0$) simply shift the potential by $\hat{\varphi}$, where $\hat{\varphi} = (n_{e0} - n_{i0})/T_i(n_{e0}/T_e + n_{i0}/T_i)$ is a constant. Thus, in the following, without loss of generality, we shall restrict ourselves to the case when the electron and ion densities at $\phi = 0$ are equal, i.e. $n_{e0} = n_{i0}$.

The general method for describing the dust grain system in the plasma is as follows: The solution of (70) has to be found by solving it in a volume outside the grains [this means we have to neglect the right-hand side of equation (64) and solve the homogeneous equation] and match the solutions on the surfaces of each dust grain by using the boundary conditions (67) and (68). We consider the dust grain size to be much smaller than the inter-grain spacing, viz. $a_\mathbf{n} \ll d$. (here for convenience the radius of the nth grain is denoted by $a_\mathbf{n}$, different from §1, where the grains radius is denoted by r_d). Then, we have $C_{\mathbf{nn}} \simeq a_\mathbf{n}$ and $C_{\mathbf{nn}'} \ll C_{\mathbf{nn}}$, where $n \neq n'$. Hence, equation (70) reduces to the Schrödinger equation

$$\nabla^2\varphi(\mathbf{r}) + \left[-2k_D^2\varphi(\mathbf{r}) - V(\mathbf{r})\right]\varphi(\mathbf{r}) = 0, \tag{71}$$

where the potential energy is

$$V(\mathbf{r}) = -\sum_n a_n\delta(\mathbf{r} - \mathbf{r}_n). \tag{72}$$

As an application of the above mentioned method, we shall consider a simple case. Keeping in mind the cubic symmetry, we present the solution of the homogeneous equation (71) by the method of separation of variables. That means, we introduce $\varphi(r) = \prod_{i=1}^{3}\varphi_i\left(x^{(i)}\right)$ and obtain

$$\begin{aligned}\varphi\left(x^{(1)}, x^{(2)}, x^{(3)}\right) &= \varphi_\mathbf{n}\prod_{i=1}^{3}\left\{\alpha_{\mathbf{n}_i}^{(i)}\sinh\left[s_i\left(x^{(i)} - x_{\mathbf{n}_i}^{(i)}\right)\right]\right. \\ &\left.+\cosh\left[s_i\left(x^{(i)} - x_{\mathbf{n}_i}^{(i)}\right)\right]\right\},\end{aligned} \tag{73}$$

which is defined in the rectangular volume $x_{n_i}^{(i)} < x^{(i)} < x_{n_i+1}^{(i)}$. Here, $\varphi_\mathbf{n}$ is the potential of the grain with coordinates $x_{n_1}, y_{n_2}, z_{n_3}$ and $s_1^2 + s_2^2 + s_3^2 = 2k_D^2$. Obviously the quantities $1/s_1$, $1/s_2$ and $1/s_3$ represent the characteristic scales of potential screening along the x, y and z axes, respectively. The boundary condition (67) is then reduced to the requirement of the potential continuity on the three planes $x^{(i)} = x^{(i)}_{ni}$, where $i = 1, 2, 3$. This allows us to deduce

$$\alpha^{(i)}_{\mathbf{n}_i} = \left\{ \frac{\varphi_{\mathbf{n}+\mathbf{I}_i}}{\varphi_{\mathbf{n}}} \sinh^{-1}\left[s_i \left(x^{(i)}_{\mathbf{n}_i+1} - x^{(i)}_{\mathbf{n}_i} \right) \right] - \coth\left[s_i \left(x^{(i)}_{\mathbf{n}_i+1} - x^{(i)}_{\mathbf{n}_i} \right) \right] \right\} \sinh^{-1}\left[s_i \left(x^{(i)}_{\mathbf{n}_i+1} - x^{(i)}_{\mathbf{n}_i} \right) \right], \tag{74}$$

where $\mathbf{I}_i = \delta_{ik}$ is the Kronecker symbol and the notation $\varphi_{\mathbf{n}+\mathbf{I}_i}$ means $\varphi_{n_1+\delta_{i1}, n_2}$ $+\delta_{i2}, n_3 + \delta_{i3}$. Substituting (74) into (73), we obtain the distribution of the potential

$$\varphi\left(x^{(1)}, x^{(2)}, x^{(3)}\right) = \frac{1}{\varphi^2_{\mathbf{n}}} \prod_{i=1}^{3} \sinh^{-1}\left[s_i \left(x^{(i)}_{\mathbf{n}_i+1} - x^{(i)}_{\mathbf{n}_i} \right) \right] \left\{ \varphi_{\mathbf{n}+\mathbf{I}_i} \sinh\left[s_i \left(x^{(i)} - x^{(i)}_{\mathbf{n}_i} \right) \right] + \varphi_{\mathbf{n}} \sinh\left[s_i \left(x^{(i)}_{\mathbf{n}(i)+1} - x^{(i)} \right) \right] \right\}. \tag{75}$$

The second boundary condition, given by (68), yields

$$\varphi_{\mathbf{n}} \left\{ \sum_{i=1}^{3} s_i \left[\tanh[\frac{s_i}{2} \left(x^{(i)}_{n_i+1} - x^{(i)}_{n_i} \right)] + \tanh[\frac{s_i}{2} \left(x^{(i)}_{n_i} - x^{(i)}_{n_i-1} \right)] \right] - \frac{1}{a_{\mathbf{n}}} \right\} \simeq \sum_{i=1}^{3} \left\{ \frac{\varphi_{\mathbf{n}+\mathbf{I}_i} - \varphi_{\mathbf{n}}}{\sinh\left[s_i \left(x^{(i)}_{n(i)+1} - x^{(i)} \right) \right]} + \frac{\varphi_{\mathbf{n}-\mathbf{I}_i} - \varphi_{\mathbf{n}}}{\sinh\left[s_i \left(x^{(i)}_{\mathbf{n}(i)} - x^{(i)}_{\mathbf{n}(i)-1} \right) \right]} \right\}. \tag{76}$$

In obtaining (76), without loosing the essential physics, we have supposed that in every cage the grains have the form of a small cube with the site length $2a_{\mathbf{n}}$ and the surface integral has been performed over such cubic surfaces. Also used is the condition $s_i a_{\mathbf{n}} < 1$.

When the grains have identical potentials, viz. $\varphi_{\mathbf{n}} = \varphi_0 =$ constant, we obtain $x^{(i)}_{n_i+1} - x^{(i)}_{ni} = d^{(i)}$, which implies that along every axis the intergrain distances are equal, and the following relation is satisfied

$$\sum_{i=1}^{3} s_i \tanh\left(\frac{s_i d^{(i)}}{2} \right) \simeq \frac{1}{2a}, \tag{77}$$

where $a_{\mathbf{n}} \simeq a$. On the other hand, if the dust grains are separated with equal distances and the relation (77) is satisfied the grain surface potentials turn out to be equal as well. It follows that for the existence of a regular periodic dust crystal the plasma parameters have to satisfy some definite condition, given by (77).

In the general case when the grains potentials are not equal, the relation (76) connects the potential of different grains to each other, i.e. $\varphi_{\mathbf{n}} = \sum_{\mathbf{m}\neq 0} D_{\mathbf{n},\mathbf{m}} \varphi_{n_1}$

$+m_1, n_2+m_2, n_3+m_3$, where $D_{\mathbf{n},\mathbf{m}}$ can be expressed as a combination of the coefficients of $\varphi_{\mathbf{n}}$ in (76). Consequently, the potential (75) can be seen to arise from the effect of the whole grain system in the dust crystal.

8 Small Oscillations of a Dust Lattice

Let us now consider small oscillations of the grains around their equilibrium position. For simplicity we consider cubic symmetry and take uniform charge and mass distributions for the dust grains. Thus, $Q_{\mathbf{n}} = Q_0$, $m_{\mathbf{n}} = m_0$, $s_1 = s_2 = s_3 = s$ and $d^{(1)} = d^{(2)} = d^{(3)} = d$. Then, for the total electrostatic interaction energy for the grains we have

$$U = Q_0\varphi_0 \sum_{\mathbf{n}} \prod_{i=1}^{3} \frac{\cosh\left[\frac{s_i}{2}\left(x^{(i)}_{n_i+1} + x^{(i)}_{n_i-1} - 2x^{(i)}_{n_i}\right)\right]}{\cosh\left[\frac{s_i}{2}\left(x^{(i)}_{n_i+1} - x^{(i)}_{n_i-1}\right)\right]}, \tag{78}$$

where from the sum in (78) we have excluded the interaction of the charged grain with its own field. By using (78) we can construct the equation of motion for the dust grain displacement $\xi^{(i)}_{n_i}$. The dust grains execute small oscillations around their equilibrium position $x^{(i)}_{0n_i}$ with the displacement $\xi^{(i)}_{n_i} = x^{(i)}_{n_i} - x^{(i)}_{0n_i}$. The equation for the n-th grain displacement is

$$\frac{\partial^2 \xi^{(i)}_{n_i}}{\partial t^2} = -\frac{s^2}{4}\frac{Q_0\varphi_0}{m_0}\cosh^{-3}(sd)\left(\xi^{(i)}_{n_i+1} + \xi^{(i)}_{n_i-1} - 2\xi^{(i)}_{n_i}\right), \tag{79}$$

Seeking a wave-train solution of the form $\xi^{(i)}_{n_i} = A^{(i)}\exp(i\mathbf{k}\cdot\mathbf{n}d - i\omega t)$, where $\mathbf{k}$ is the wave vector and ω is the frequency, we obtain from (79)

$$\omega_i^2 = \frac{s^2}{4}\frac{Q_0\varphi_0}{m_0}\frac{k_i^2 d^2}{\cosh^3(sd)}\left[\frac{\sin(k_i d/2)}{k_i d/2}\right]^2. \tag{80}$$

We emphasize that (80) determines the frequencies of both longitudinal (compressional) and transverse oscillations. If the wave vector is aligned along one of the main lattice axes, there appear only longitudinal waves as $\mathbf{k}\times\mathbf{A} = 0$. On the other hand, if the wave vector forms an angle with the main axis of a cubic cell, then there also develop transversal components of the oscillations as $\mathbf{k}\times\mathbf{A} \neq 0$. Physically, the longitudinal and transverse particle dynamics can be understood as follows. The minimum of the potential energy is located along the main axis of a cubic crystal lattice, which provides channels for guiding the dust grains. When the wave propagates along one of these axes, it pushes the particles along the channels and the direction of the particle displacement coincides with the wave vector, and only longitudinal (compressional) waves appear. On the other hand, if the wave vector forms an angle with the crystal axis, then the displacement of the dust grains, which move along the channels, would have both longitudinal and transverse components with respect to the wave vector.

We have considered a physically realistic situation in which the charge of a specific dust grain is not only determined by its own surface potential, but it also incorporates the effects of other dust grain potentials. Thus, we have analytically deduced an expression for the electrostatic potential when electrons and ions are Boltzmann distributed, and when dust grains are strongly interacting not only with their neighbours but also with other grains that are participating in the formation of the periodic structure. The results, given above, show that a regular crystal structure appears only when the plasma parameters satisfy a specific condition that is expressed by equation (77). It should be mentioned that the formalism developed here can be applied to any symmetry (depending on the symmetry of the bounding walls confining the dusty plasma).

9 Discussion

We have considered a realistic dusty plasma system, and have, particularly, focused on various basic features of dusty plasmas, namely the motion of the electrons and the ions in the dust grain potential, the electron and ion current densities, the ion number density far from the dust grain, the dust grain potential, effects of the trapped ions on the shielding of the dust grains, the distribution of the periodic potential, the oscillations of dust lattices, etc.

We have first studied the dust grain charging in a plasma accounting for the absorption of electrons and ions by a dust grain. It is found that besides the shielded Debye-Hückel potential there also appears a dipole-like potential at large distances (in comparison with the ion Debye radius). The effective potential of the dust grain is then drastically modified. We have analyzed the dynamics of the plasma particles in the effective potential, and discussed the charging currents that are essential for the determination of the potential of a dust grain. We have found that the current due to the trapped ions on the dust grain surface is insignificant, and that the dust grain charging is solely determined by the currents of the free electrons and ions that reach the dust grain surface.

We have then examined the effects of the trapped ions on the shielding of a dust grain and clarified the underlying physics. We have shown that the shapes of the effective potential and the potential well strongly depend on the ion angular momentum M. Therefore, the ion trapping in the potential well is controlled by M. When M is smaller than a certain critical value M_0, the dust grain surface is within the well and the ions falling into the potential well are thus immediately lost due to absorption onto the dust grain surface. Hence, the formation of ion clouds that shield the grain is possible only for $M > M_0$. This will be addressed in more detail in chapter 2.

We have also presented a theory for a dust crystal and its oscillations. We have considered a physically realistic situation in which the charge of a specific dust grain is not only determined by its own surface potential, but also incorporates the effects of other dust grain potentials. We have thus been able to deduce an analytic expression for the electrostatic potential when the electrons and ions are Boltzmann distributed, and when the dust grains are strongly interacting not only with their neighbours, but also with other grains that are participating in the dust crystal assembly. Clearly, short and long range interactions between dust grains are self-consistently included in our theory for dust crystal. Our results show that a regular crystal structure appears only when the plasma parameters satisfy a specific condition that is given by equation (77). We have finally presented a transparent treatment for longitudinal and transverse oscillations of our dust crystal. In particular, we have derived analytical expressions for the frequencies of the dust lattice oscillations and explained the underlying physics of those modes.

Acknowledgements

The authors acknowledge the support of the International Space Science Institute at Bern (Switzerland) for the team "Complex Plasmas in Space: Applications to the Solar System and Near Earth Space".

References

Allen J. E. 1992, *Phys. Scripta* **45**, 497.

Allen J. E. , Annaratone B. M. and Angelis U. de 2000, *J. Plasma Phys.* **63**, 299.

Alpert Ya. L., Gurevich A. V. and Pitaevskii A. P. 1964, *The Satellites in the Rarefied Plasma* (Nauka, Moscow).

Bernstein I. B. and Rabinovitz I. N. 1959, *Phys. Fluids* **2**, 112.

Chu J. H. and I Lin. 1994, *Physica A* **205**, 183.

Chu J. H. and I Lin. 1994, *Phys. Rev. Lett.* **72**, 4009.

Chu J. H. , Du J. B. and I Lin. 1994, *J. Phys. D: Appl. Phys.* **27**, 296.

Farokhi B., Shukla P. K., Tsintsadze N. L. and Tskhakaya D. D. 1999, *Phys. Plasmas* **7**, 814.

Fortov V. E. *et al.* 1996, *JETP Lett.* **64**, 92.

Fortov V. E., Nefedov A. P. and O. F. Petrov O. F. 1999, *Phys. Plasmas* **6**, 1759.

Goertz C. K. 1989, *Rev. Geophys.* **270**, 271.

Gurevich A. V. 1968, *JETP* **26**, 575.

Hayashi Y. and Tachibana K. 1994, *Jpn. J. Appl. Phys.* **33**, 804.

Hayashi Y. 1999, *Phys. Rev. Lett.* **83**, 4764.

Hayashi Y. 2001, *Physica Scripta* **T89**, 112.

Horanyi M., Houpis H. L. F. and Mendis D. A. 1988, *Astrophys. Space Sci.* **144**, 215.

Ikezi H. 1986, *Phys. Fluids* **29**, 1764.

Kittel C. 1956, *Introduction to Solid State Physics* (John Wiley and Sons, Inc, New York), Chapter 11.

Lampe M., Joyce G., Ganguli G. and Gavrishchaka V. 2000, *Phys Plasmas* **7**, 3851.

Lampe M., Joyce G. and Ganguli G. 2001a, *Phys. Scripta* **T89**, 106.

Lampe M., Gavrishchaka V., Ganguli G. and Joyce G. 2001b, *Phys. Rev. Lett.* **86**, 5278.

Landau L. D. and Lifshitz E. M. 1960, *Electrodynamics of Continuous Media* (Pergamon Press, Oxford), p. 4.

Lifshitz E. M. and Pitaevskii L. P. 1981 *Physical Kinetics* (Pergamon, Oxford), pp. 146-149.

McDaniel E. W. and Mason E. A. 1973, *The Mobility and Diffusion in Gases* (Wiley, New York), Chap. 5-7.

Melandsø F. 1996, *Phys. Plasmas* **3**, 3890.

Molotkov *et al.* 2000, *JETP Lett.* **71**, 102.

Morfill G. E., Thomas H. M. and Zuzic M. 1997, in *Advances in Dusty Plasmas*, edited by P. K. Shukla *et al.* (World Scientific, Singapore), pp. 99-142.

Morfill G. E., Thomas H. M., Konopka U. and Zuzic M. 1999, *Phys. Plasmas* **6**, 1769 .

Mott-Smith Jr. H. and Langmuir I. 1926, *Phys. Rev.* **28**, 27.

Northrop T. G. 1992, *Phys. Scripta* **45**, 497.

Pieper J. B., Goree J. and Quinn R. A. 1996, *Phys. Rev.* **E 54**, 5636.

Shukla P. K. 2000, *Phys. Rev. Lett.* **84**, 5328.

Shukla P. K. 2001, *Phys. Plasmas* **8**, 1791.

Shukla P. K. and Mamun A. A., *Introduction to Dusty Plasma Physics* (Institute of Physics Publ. Ltd, Bristol, 2002).

Silin V. P. 1971, *Introduction to the Kinetic Gas Theory* (Nauka, Moscow).

Thomas H. M. *et al.* 1994, *Phys. Rev. Lett.* **73**, 652.

Thomas H. M., Jokipii J. R., Morfill G. E. and Zuzic M. 1998, in *Strongly Coupled Coulomb Systems*, edited by G. Kalman *et al.* (Plenum Press, New York), pp. 187-192.

Tskhakaya D. D., Tsintsadze N. L., Shukla P. K. and Stenflo L. 2001, *Phys. Scripta* **64**, 366.

Tskhakaya D. D. and Shukla P. K. 2001, *Phys. Lett.* **A 286**, 277.

Tskhakaya D. D., Shukla P. K. and Stenflo L. 2001, *Phys. Plasmas* **8**, 5333.

Zheng X. H. and J. C. Earnshaw J. C. 1995, *Phys. Rev. Lett.* **75**, 4214.

Zuzic M. *et al.* 2000, *Phys. Rev. Lett.* **85**, 4064.

Electrostatic Shielding of Dust Grains by Orbiting Ions

D. Jovanović
Institute of Physics, P. O. Box 57, Yu–11001 Belgrade, Yugoslavia
P. K. Shukla
Institut für Theoretische Physik IV,
Ruhr–Universität Bochum, D–44780 Bochum, Germany

Abstract

The electrostatic shielding of a charged dust grain in a collisional plasma is calculated self-consistently by solving the kinetic equation for ions coupled with Boltzmann distributed electrons and Poisson's equation. The ion velocity distribution, in the form of a spherically symmetric ion hole, is found to be anisotropic in the velocity space in the presence of charge-exchange collisions. The number of trapped ions and their spatial distribution are determined by the conditions for localization of the potential and the trapping and de-trapping rates. Our results confirm the existence of a strong Debye shielding of the dust charge, which in some cases yields the over-population of the trapped ion distribution in the form of an ion hump.

1 Introduction

Charged dust grains are omnipresent in the dusty universe [*Spitzer* 1982] comprising astrophysical environments [*Evans* 1993], comets and planetary rings [*Verheest* 2000]. In the near-Earth plasmas, they appear in the micrometeorites' showers and noctilucent clouds in the upper ionosphere, rocket exhausts, thermonuclear fireballs, etc. In laboratory plasmas, they are encountered in plasma processing (computer microchips) [*Bouchoule* 1999] as well as in magnetically confined fusion devices [*Shukla and Mamun* 2002]. The book by *Shukla and Mamun* [2002] contains an up-to-date review of the main topics, methods and results in dusty plasmas.

One of the most important and still unresolved problems in the physics of dusty plasmas is the determination of the dust charge, whose first systematic measurements were conducted only recently [*Horányi et al.* 1995, *Walch et al.* 1995]. As the grains are not directly accessible, the data include also the effects of the surrounding electron and ion clouds, and their interpretation depends on the theoretical model of the electron and ion dynamics inside the Debye sphere. The standard model for the charge collection by a grain is the orbit-motion-limited (OML) theory, see e. g. [*Bernstein and Rabinovich* 1959, *Al'pert et al.* 1965, *Laframboise and Parker*

1973, *Dougherty et al.* 1992, *Whipple et al.* 1985, *Lampe* 2001]. The latter regards the steady state in which collisionless electrons and ions move along ballistic trajectories which are determined by the electrostatic potential originating from the space charge distribution including that of the dust and of the surrounding electrons and ions.

A major shortcoming of the OML theory comes from neglecting the low energy ions which are orbiting the grain on closed trajectories. It was noted that trapped ions may emerge due to energy loss by collisions [*Al'pert et al.* 1965, *Bernstein and Rabinovich* 1959]. The effects of trapped ions on the grain charging were recently studied by Monte-Carlo simulations [*Goree* 1992], and a simple analytical model [*Lampe et al.* 2001] has also been proposed. The simulations by *Goree* [1992] revealed that the ion trapping is significant for laboratory plasmas where it produces a strong Debye shielding. Due to the oversimplified model, based on the linear plasma response and on the locality of collisional effects, *Goree* [1992] came to a surprising conclusion that the number of trapped ions, determined by the balance between the collisional trapping and de-trapping, is independent of the effective collision frequency. *Lampe et al.* [2001] attempted to analytically calculate the correction to the OML charging due to the presence of trapped ions. They used a plausible model collision integral for the creation and destruction of trapped ions by charge exchange between ions and neutrals, and treated Poisson's equation self-consistently, but did not solve the full kinetic equation for the ions. Instead, both for the free and trapped ions, the distribution function is adopted in the form of a Boltzmannean corrected by a geometrical factor consistent with the OML theory, accounting for the ion loss to the dust grain. While such an approximation is fully justified for the free ions [*Lampe* 2001], the distribution of the trapped ions considerably departs from the Boltzmannean and needs to be determined self-consistently.

The self-consistent numerical simulations [*Zobnin et al.* 2000] considered the nonlinear ion dynamics in the presence of a single dust grain, in the electrostatic field resulting from the grain charging and its shielding by electron and ion clouds. A charge-exchange mechanism similar to that of *Lampe et al.* [2001] was included. The results revealed an efficient Debye-Hückel type shielding of the dust grain by trapped ions, at a distance much shorter (few %) than the Debye radius. Thus, in the regime when the ion mean free path is much larger than the Debye radius, the OML theory is inappropriate for the determination of the dust charge.

In this article we present the first analytical, fully self-consistent, calculations of the electrostatic shielding of a charged dust grain in a collisional plasma. In the regime when the mean free path for the ion-dust collisions is larger than that for the ion-neutral collisions, we solve the kinetic equation for the ions, coupled with Boltzmann distributed electrons and Poisson's equation, in the form of a spherically symmetric ion hole. The ion velocity distribution is found to be anisotropic in the presence of charge-exchange collisions. The number of trapped ions and their spatial distribution, related to the angular anisotropy of the ion distribution function, are self-consistently determined from the mean free paths for the trapping/de-trapping. Our results confirm the existence of a strong Debye shielding of the dust charge, which in some cases may produce the over-population of the trapped ion distribution, manifesting as the ion hump.

2 Governing Equations

We consider an unmagnetized, spatially homogeneous, dusty plasma comprising electrons, ions, dust particulates and neutral atoms. Their unperturbed number densities are $n_e^{(0)}$, $n_i^{(0)}$, n_d, and n_n, and the charges are $-e$, $z_i e$, $-z_d e$, and 0, respectively, which satisfy the charge neutrality condition $z_d n_d + n_e^{(0)} = z_i n_i^{(0)}$. The electron, ion and neutral temperatures are T_e, T_i, and $T_n = T_i$, respectively. The plasma is weakly ionized, viz. $n_n \gg n_i^{(0)} \sim n_e^{(0)} \sim z_d n_d$. The dust grains are spherical, with the radius r_d which is much smaller than the Debye radius, $r_d \ll \lambda_D$ (where $\lambda_D^{-2} = e^2 n_e^{(0)}/\epsilon_0 T_e + z_i^2 e^2 n_i^{(0)}/\epsilon_0 T_i$), and the plasma is weakly collisional, $\lambda_{mfp}^{(n)} \gg \lambda_D$ ($\lambda_{mfp}^{(n)}$ is the mean free path for ion-neutral collisions). The grains are also assumed to be small enough so that ion-dust collisions are less frequent than ion-neutral collisions. Expressing the mean free paths through the collision cross sections, $\lambda_{mfp}^{(d)} = n_d \pi r_d^2$, $\lambda_{mfp}^{(n)} = n_n^{(0)} \sigma$, the condition $\lambda_{mfp}^{(d)} \gg \lambda_{mfp}^{(n)}$ may be rewritten as

$$\frac{\lambda_{mfp}^{(d)}}{\lambda_{mfp}^{(n)}} = \frac{n_n}{n_i^{(0)}} \frac{n_i^{(0)}}{z_d n_d} \frac{\sigma}{\pi \delta^2} \frac{e\phi\left(r_d\right)}{T_e} \frac{\delta^2}{r_d^2} \frac{4\pi\epsilon_0 r_d T_e}{e}, \tag{1}$$

where δ is the inter-atom distance in the grain and $\phi\left(r_d\right)$ is the electrostatic potential at its surface. For a typical dust potential $e\phi\left(r_d\right)/T_e \sim 2.5$ (cf. *Shukla and Mamun* [2002], p. 36), a hot electron glow discharge with $T_e \geq 10$ eV, and sub-micron grains ($r_d \sim 10^{-7} m$) made of carbon, we have from Eq. (1) $\lambda_{mfp}^{(d)}/\lambda_{mfp}^{(n)} \sim (n_n/n_i^{(0)})(n_i^{(0)}/z_d n_d)(\sigma/\pi\delta^2)$. This is a very large number ($> 10^3$) in a weakly ionized plasma with $n_n/n_i^{(0)} \sim 10^3 - 10^4$ and the dust charging collisions may be neglected on the ion-neutral collision scale.

The ions are described by the kinetic equation

$$\frac{\partial f_i}{\partial t} + \vec{v} \cdot \nabla f_i - \frac{q_i}{m_i} \nabla \phi \cdot \frac{\partial f_i}{\partial \vec{v}} = \left(\frac{\partial f_i}{\partial t}\right)_{coll}. \tag{2}$$

Following *Goree* [1992], *Lampe et al.* [2001], *Zobnin et al.* [2000] we assume that the ions are subjected only to charge-exchange collisions in which an incoming ion with the velocity $\vec{v}_{in}$ exchanges its charge with a neutral whose velocity is $\vec{v}_{out}$. In a stationary state close to the thermodynamic equilibrium the main contribution to the collision integral $(\partial f_i/\partial t)_{coll}$ in Eq. (2) comes from the asymmetric events in which

i) an energetic (free) ion exchanges its charge with a slow neutral and disappears, while the newly born ion is trapped in the potential minimum (trapping), or
ii) a trapped ion collides with a fast neutral, and the newly born ion escapes from the well (detrapping).

3 Effects of Ion-Neutral Collisions

The collision integral may be written as the sum of the appropriate sources and sinks for the free and trapped ions denoted by the superscripts F and T, respectively,

$$\left(\partial f_i^F(\vec{v})/\partial t\right)_{coll} = \int d^3\vec{v}'\,\sigma\left(|\vec{v}-\vec{v}'|\right)\left[f_i^T(\vec{v}')\,|\vec{v}'|\,f_n(\vec{v}) - f_i^F(\vec{v})\,|\vec{v}|\,f_n(\vec{v}')\right], \quad (3)$$

$$\left(\partial f_i^T(\vec{v})/\partial t\right)_{coll} = \int d^3\vec{v}'\,\sigma\left(|\vec{v}-\vec{v}'|\right)\left[f_i^F(\vec{v}')\,|\vec{v}'|\,f_n(\vec{v}) - f_i^T(\vec{v})\,|\vec{v}|\,f_n(\vec{v}')\right]. \quad (4)$$

In a weakly ionized plasma, $n_n \gg n_i$, the distribution function of the neutrals, f_n, is Maxwellian $f_n(\vec{v}) = n_n/(2\pi v_{T,i}^2)^{3/2}\exp(-v^2/2v_{T,i}^2)$, as its perturbation by collisions with ions is a small quantity of higher order. In the thermal range of velocities, the collisional cross section σ depends on the relative velocity as $\sigma(|\vec{v}-\vec{v}'|) = \sigma_0[\log(b/|\vec{v}-\vec{v}'|)]^2$, where σ_0 and b are constants [*Zobnin et al.* 2000]. For simplicity, we neglect this weak logarithmic dependence and adopt $\sigma = $ const.

In a weakly collisional case, $(\partial f_i/\partial t)_{coll} \ll \vec{v}\cdot\nabla f_i$, we expand the ion distribution function as $f_i = f_{i,0} + \delta f_i$, where the leading component $f_{i,0}$ is found from the stationary Vlasov equation

$$\vec{v}\cdot\nabla f_{i,0} - \frac{z_i e}{m_i}\nabla\phi_0\cdot\frac{\partial f_{i,0}}{\partial\vec{v}} = 0, \quad (5)$$

while the perturbation δf_i obeys

$$\vec{v}\cdot\nabla\delta f_i - \frac{z_i e}{m_i}\left(\nabla\phi_0\cdot\frac{\partial\delta f_i}{\partial\vec{v}} + \nabla\delta\phi\cdot\frac{\partial f_{i,0}}{\partial\vec{v}}\right) = \left(\frac{\partial f_{i,0}}{\partial t}\right)_{coll}. \quad (6)$$

Here the potential ϕ_0 and its perturbation $\delta\phi$ are calculated self-consistently from Poisson's equation.

It will be shown later that in the regime studied in this article the dust grains are strongly shielded by orbiting ions (see also *Lampe et al.* [2001]), and we regard them as fully isolated from each other, surrounded by spherically symmetric electrostatic potential. Using $\nabla = (\vec{r}/r)(\partial/\partial r)$, where $\vec{r}$ is the radius vector, one easily verifies that the Vlasov equation (5) has two conserved quantities, the energy W_0 and the angular momentum $\vec{L}$

$$W_0 = \frac{m_i v^2}{2v_{T,i}^2} + \frac{e z_i \phi_0(r)}{T_i}, \qquad \vec{L} = \frac{\vec{v}\times\vec{r}}{v_{T,i}\lambda_D}, \quad (7)$$

and the leading order ion distribution function can be expressed as a function of W_0 and $\vec{L}$, i. e. $f_{i,0} = f_{i,0}(W_0, \vec{L})$. The free and trapped particles correspond to $W_0 > 0$ and $W_0 < 0$, respectively. In the $(\vec{v}, \vec{r})$ phase space they lie outside and inside the separatrix $W_0 = 0$, which is the locus of the branch points for the velocity $\vec{v}$.

4 Modeling the Ion Distribution

It was shown by *Lampe* [2001] that the effects of the angular momentum on the free ion distribution, which manifest as the particle trapping by the centrifugal potential barrier, reduce the shielding by $\sim 10^{-2}(T_e/T_i)(r_d/\lambda_D)$, and that the deviation from the OML is negligible if $(T_e/T_i)(r_d/\lambda_D) \ll 30$. As in the $r_d \to 0$ limit the OML result reduces to the Boltzmann distribution, the distribution of the free particles is adopted in the form

$$f_{i,0}^F(\vec{v}) = n_i^{(0)} \left(2\pi v_{T,i}^2\right)^{-3/2} e^{-W_0} \quad \text{for} \quad W_0 > 0. \tag{8}$$

For the trapped particles we use Schamel's distribution [*Schamel* 2000] $f_{Sch}^T \sim \exp(-\beta W_0)$, where the parameter β accounts for the number of trapped particles that are created during collisions. As the potential $\phi(r)$ is strongly shielded, the density of the trapped particles is expected to increase sharply close to the grain. Then, the creation and destruction rates of the trapped particles, Eq. (4), vary strongly between $r \ll \lambda_D$ and $r \sim \lambda_D$. Consequently, the trapped ion population on elongated orbits is different from that on circular orbits with the same energy. When the mean free path is much longer than the circumference of the orbit, the numbers of particles approaching to, and receding from the grain are the same. Such features of the trapped particle distribution are modelled by multiplying Schamel's distribution by the form factor $f_{form}^T \sim \exp(\gamma L\sqrt{-W_0})$, yielding for $W_0 < 0$

$$f_{i,0}^T(\vec{v}) = n_i^{(0)} \left(2\pi v_{T,i}^2\right)^{-3/2} e^{-\beta W_0} \exp\left(\gamma L\sqrt{-W_0}\right). \tag{9}$$

Using $\int d^3\vec{v} = 4\pi \int_0^\infty v^2\, dv \int_0^{\pi/2} \sin\theta\, d\theta$ and $L \equiv |\vec{L}| = (v\, r/v_{T,i}\lambda_D)\sin\theta$, where θ is the angle between $\vec{r}$ and $\vec{v}$, we integrate Eqs. (8) and (9) and obtain the leading-order densities of the free and trapped ions in the form

$$n_{i,0}^F = n_i^{(0)} e^{-W_m}\left[1 + 2\, e^{W_m}\sqrt{-W_m/\pi} - \operatorname{erf}\left(\sqrt{-W_m}\right)\right], \tag{10}$$

$$n_{i,0}^T \approx \frac{n_i^{(0)} e^{-\beta W_m}}{2\beta^{3/2}}\, \pi\left[I_1(-\xi) - L_{-1}(\xi)\right]\left[2\, e^{\beta W_m}\sqrt{-\beta W_m/\pi} - \operatorname{erf}\left(\sqrt{-\beta W_m}\right)\right], \tag{11}$$

where W_m $(= z_i\, e\, \phi_0(r)/T_i)$ is the energy of the the zero-velocity ions, $\operatorname{erf}(\zeta) = (2/\sqrt{\pi})\int_0^\zeta \exp(-t^2)\, dt$, I_1 and L_{-1} are the modified Bessel and Struve functions of the order 1 and -1, respectively, and $\xi = \gamma\, r\, W_m/\sqrt{2}\lambda_D$.

5 Nonlinear Plasma Response

Using Eqs. (10), (11) and the Boltzmann electron distribution, $n_{e,0} = n_e^{(0)} \exp [-e\phi_0(r)/T_e]$, Poisson's equation becomes

$$\frac{\epsilon_0}{e}\left(\frac{\partial^2}{\partial r^2}+\frac{2}{r}\frac{\partial}{\partial r}\right)\phi_0(r) = z_d n_d + n_{e,0} - z_i\left(n_{i,0}^F + n_{i,0}^T\right). \tag{12}$$

We proceed by deriving an analogous equation for the potential $\delta\phi$. Without loss of generality, the perturbation of the ion distribution function, δf_i, is sought as a function of r, W_0 and $\vec{L}$. For the solution which is vanishing at $r \to \infty$, the kinetic equation (6) is solved as

$$\delta f_i = z_i e \frac{\delta\phi}{T_i}\frac{\partial f_{i,0}}{\partial W_0} - \int_r^\infty \frac{dr'}{V_r(r')}\left(\frac{\partial f_{i,0}}{\partial t}\right)_{coll}, \tag{13}$$

where the integration may be performed if the kernel of the integral is expressed as a function of r', W_0 and $\vec{L}$. The collisional integrals (3) and (4) are easily cast in such a form by using the expressions (8) and (9) for the distribution function with the ion velocity V and its radial component, $V_r = \vec{v}\cdot\vec{r}/r$, expressed as

$$V(r) = \sqrt{2}\, v_{T,i}\left[W_0 - z_i e\,\phi_0(r)/T_i\right]^{1/2}, \tag{14}$$

and

$$V_r(r) = \sqrt{2}\, v_{T,i}\left[W_0 - z_i e\,\phi_0(r)/T_i - \lambda_D^2 L^2/2r^2\right]^{1/2}. \tag{15}$$

Likewise, distribution of the neutrals takes the form $f_n[V(r)] = n_n/(2\pi v_{T,i}^2)^{3/2} \exp[-V^2(r)/2v_{T,i}^2]$. Integrating Eq. (13) in the velocity space, the perturbation of the ion number density, δn_i, may be written as

$$\delta n_i = z_i e\frac{\delta\phi}{T_i}\left(N_h - n_{i,0}^F - \beta n_{i,0}^T\right) - \int_r^\infty \frac{dr'}{\lambda_{mfp}^{(n)}} N_{coll}, \tag{16}$$

where

$$N_h \approx \frac{n_i^{(0)} e^{-\beta W_m}}{2\beta^{3/2}}\left[2\,e^{\beta W_m}\sqrt{-\frac{\beta W_m}{\pi}} - \mathrm{erf}\left(\sqrt{-\beta W_m}\right)\right]\times$$
$$\frac{\pi\gamma r}{\sqrt{2}\lambda_D}\left[\frac{I_1(-\xi)}{\xi} - I_2(\xi) + \frac{2\xi}{3\pi}\,{}_1F_2\left(2,\left\{\frac{3}{2},\frac{5}{2}\right\},\frac{\xi^2}{4}\right)\right], \tag{17}$$

and

$$N_{coll} = \int_{W_0<0}\frac{d^3\vec{v}}{n_n}\frac{f_n[V(r')]}{V_r(r')}\int_{W_0>0} d^3\vec{v}'\,v' f_{i,0}^F(v') +$$
$$\int_{W_0>0}\frac{d^3\vec{v}}{n_n}\frac{f_n[V(r')]}{V_r(r')}\int_{W_0<0} d^3\vec{v}'\,v' f_{i,0}^T(v') - \int_{W_0=W_m}^\infty d^3\vec{v}'\,\frac{V(r')}{V_r(r')}\,f_{i,0}(v'). \tag{18}$$

Here $\lambda_{mfp}^{(n)} = 1/\sigma n_n$ is the mean free path for -ion-neutral collisions, I_2 is the modified Bessel function of the order 2, and ${}_1F_2$ is the generalized hyper-geometric function. The five integrals in Eq. (18) can be evaluated analytically[1] using the symbolic algebra software "Mathematica", but the results are too lengthy to be displayed here. It should be noted that collisional effects enter Eq. (16) in an essentially nonlocal way.

Poisson's equation for $\delta\phi$ is then readily written

$$\frac{\epsilon_0}{e}\left(\frac{\partial^2}{\partial r^2} + \frac{2}{r}\frac{\partial}{\partial r}\right)\delta\phi(r) = \frac{en_{e,0}}{T_e}\,\delta\phi(r) - z_i\delta n_i. \tag{19}$$

The coupled nonlinear equations (12) and (19) are solved numerically using at $r = r_d$ the boundary values determined by the charging process, viz. the dust floating potential $\phi_0(r_d)$ (e. g. *Shukla and Mamun* [2002]) and the charge coefficient $S = [\partial\phi_0(r_d)/\partial r_d][r_d/\phi_0(r_d)]$ (where $1.01 \leq S \leq 1.15$) found in the numerical simulations [*Zobnin et al.* 2000]. The parameters of the trapped particle distribution function, β and γ, are fitted so as to produce a localized potential at $r \to \infty$. The calculations are performed for a nonisothermal ($T_e/T_i = 10$) hydrogen plasma, with 20% of the negative charge on the dust, viz. $z_d n_d/n_i^{(0)} = 0.2$. The dust radius is adopted as $r_d/\lambda_D = 0.01$ As expected, the solution is strongly dependent on the value of the charging parameter S. A localized solution does not exist for the unshielded Coulomb potential, $S = 1$. For the minimum value determined by *Zobnin et al.* [2000], $S = 1.011$, we find a relatively small population of the trapped ions, $\beta = -1.5$. The effect of the ion hole is further enhanced by its radial broadening due to the strong anisotropy, $\gamma = -5$. For a larger charge parameter, $S = 1.15$, an overpopulation of the trapped ions is found, corresponding to an ion hump with $\beta = 0.577$, $\gamma = 3$. The shielding of the potential in both cases is almost the same (less than 1% difference). The ion distribution functions and the radial profile of the potential are displayed in Figs. 1 and 2, respectively.

6 Conclusions

In conclusion, our results confirm the existence of a strong Debye shielding of the dust charge, and provide a theoretical explanation for such behavior of the electron and ion clouds. We have self-consistently calculated the electrostatic shielding of a charged dust grain in a collisional plasma by solving the kinetic equation for the ions, coupled with Boltzmann distributed electrons and Poisson's equation. The ion velocity distribution, in the form of a spherically symmetric ion hole, is found to be anisotropic in the velocity space in the presence of charge-exchange collisions. The number of trapped ions and their spatial distribution are determined from the

[1]Some of the integrals involving trapped ions are evaluated approximately, by averaging the form factors f_{form}^T over the negative ion energies.

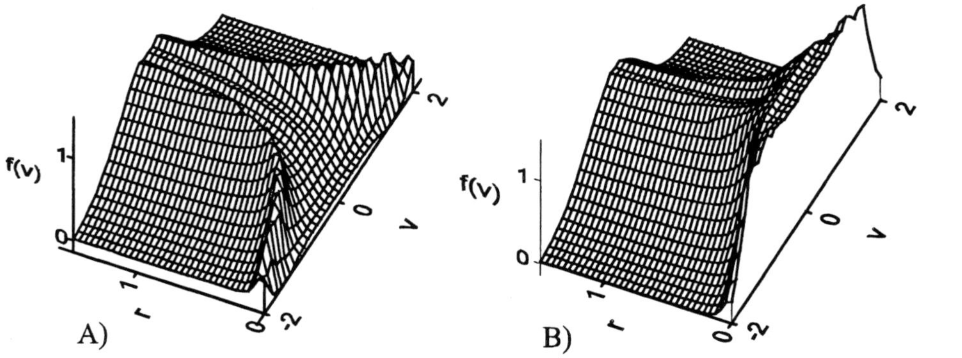

Figure 1: Ion distribution function (in arbitrary units), as a function of the particle velocity and distance from the dust grain, in the case of a hole (A) and hump (B). The velocity is normalized by $\sqrt{2}\,v_{T,i}$, and the distance by the ion Debye radius, $\lambda_{D,i} = [e^2 n_i^{(0)}/\epsilon_0 T_i]^{-1/2}$.

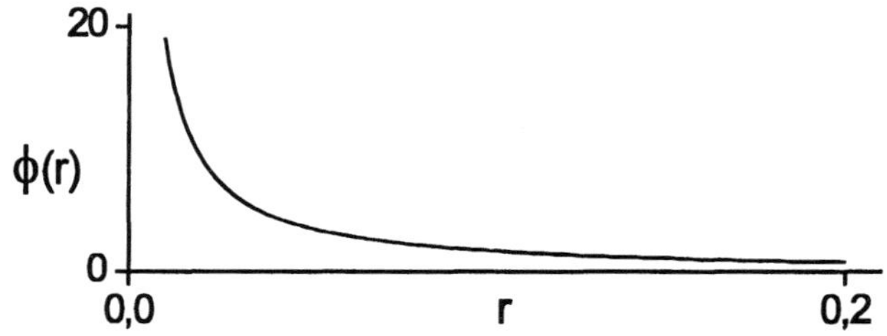

Figure 2: Radial profile of the dimensionless electrostatic potential $e\phi(r)/T_i$. The distance is normalized by $\lambda_{D,i}$. The grain size is $0.01\lambda_{D,i}$.

interplay between the collective plasma interaction and the trapping/de-trapping process. Physically, the stationary solution in the presence of collisions results from the self-tuning of the trapped ion density profile by the feedback based on the nonlocality of the collisional terms, Eqs. (16) and (18), and on the ion mixing in the radial direction along elongated orbits. The collisional corrections to the collisionless distribution function are small quantities, whose role is to determine the value of the parameters β and γ for which both nonlinear Poisson's equations (11) and (19) are satisfied for the given charging parameter S. In some cases, we demonstrated the over-population of the trapped ion distribution, that manifests as the ion hump. The earlier model of *Lampe et al.* [2001] corresponds to the limit $\beta = 1$, $\gamma = 0$, and it produces a distribution function with a gentle maximum (i. e. neither the hole, nor the overpopulation of the resonant particles were allowed). Our results are in a good agreement with the simulations reported by *Zobnin et al.* [2000], and also with the recent laboratory experiments by *Kingrey et al.* [2001], where the total charge of trapped ions of the order of 20-30% of the dust charge was found, corresponding to $\sim 10^5$ trapped particles.

Acknowledgements

The authors acknowledge the support of the International Institute for Space Science of Bern (Switzerland). The research was partially supported also by the European Commission (Brussels) through the contract No. HPRN-CT-2000-00140. One of the authors (D. J.) acknowledges the financial support of the Deutsche Forschungsgemeinschaft for a three months' stay at the University of Bayreuth, F. R. Germany.

References

Al'pert Ya. L., Gurevich A. V. and Pitaevskii L. P. 1965, *Space Physics with Artificial Satellites*, (Consultants Bureau, New York).

Bernstein I. B. and Rabinovich I. N. 1959, Phys. Fluids **2**, 112.

See, many articles in 1999, *Dusty Plasmas: Physics, Chemistry and Technological Impacts in Plasma Processing*, edited A. Bouchoule (John Wiley & Sons, Chichester).

Dougherty J. E. *et al.* 1992, Phys. Plasmas **5**, 3934.

Evans A. 1993, *The Dusty Universe* (Ellis Harwood, London).

Goree J. 1992, Phys. Rev. Lett. **69**, 277.

Horányi M, Robertson S. and Walch B. 1995, Geophys. Res. Lett. **22**, 2079.

Kingrey D., Robertson S. and Sternovsky Z. 2001, Bull. Am. Phys. Soc. **46**, 120.

Laframboise J. G. and Parker L. W. 1973, Phys. Fluids **16**, 629.

Lampe M. 2001, J. Plasma Phys. **72**, 171.

Lampe M., Gavrishchaka V, Ganguli G. and Joyce G. 2001, Phys. Rev. Lett. **86**, 5278.

Schamel H. 2000, Phys. Plasmas, **7**, 4373.

Shukla P. K. and Mamun A. A. 2002, *Introduction to Dusty Plasma Physics* (Institute of Physics Publ., Bristol).

Spitzer L. 1982, *Physical Processes in the Interstellar Medium* (John Wiley & Sons, New York).

Verheest F. 2000, *Waves in Dusty Space Plasmas* (Kluwer Academic Publishers, Dordrecht).

Walch B., Horányi M. and Robertson S. 1995, Phys. Rev. Lett. **75**, 838.

Whipple E. C., Northrop T. G. and Mendis D. A. 1985, J. Geophys. Res. **90**, 7405.

Zobnin A. V., Nefedov A. P., Sinel'shchikov V. A. and Fortov. V. E. 2000, Zh. Eksp. Teor. Fiz. **118**, 554 [JETP 2000, **91**, 483].

Dust Grain Oscillation in Plasma Sheaths under Low Pressures

G. Sorasio, R. A. Fonseca, D. P. Resendes
GOLP, Instituto Superior Técnico, 1096 Lisboa Codex, Portugal

P. K. Shukla,
Theoretische Physik IV, Fakultät für Physik & Astronomie,
Ruhr-Universität Bochum, D-44780 Bochum, Germany

Abstract

We analyzed the dynamics of a dust particle in a plasma sheath, at very low pressures, under the influence of various forces and presented analytical and numerical results. The main forces acting on a levitating dust grain have been considered: ion drag and gravity acting downwards and the electrostatic force acting upwards. The influence on the grain dynamics of an externally applied force as well as of an electrode potential modulation have been analytically and numerically explored. For the first time, the effects of different excitation techniques on the plasma sheath environment have been theoretically analyzed, and the grain motion resulting from the sheath modification induced by the external driver have been studied. After a brief introduction to the fluctuation models presented by previous authors, a new mechanism for the excitation of self-excited oscillations has been proposed. Comparison with experimental data inferred from published investigations is also presented.

1 Introduction

During the past few years, the physics of dusty plasmas has been mostly concerned with the effects of dust on waves, instabilities, and dust crystals phase transitions, namely on collective behavior, both in laboratory and space environments [Shukla *et al* 1996, Shukla and Mamun 2002, Verheest 2000, Sorasio and Rosenberg 2001c]. On the other hand, several theoretical and experimental studies concerning the dynamics of single dust grains in space plasmas (e.g. charging of infalling meteoroids [Sorasio *et al* 2001a]), and in laboratory environments (i.e. nonlinear behavior [Ivlev *et al* 2000, Schollmeyer *et al* 1999, Zafiu *et al* 2001, Sorasio *et al* 2002] and self excited oscillations [Nunomura *et al* 1999, Takamura *et al* 2001, Sorasio *et al* 2001b, Resendes *et al* 2002, Takamura *et al* 2002] have recently attracted the attention of the complex plasmas community. In this manuscript we have focused our attention on the dynamics of dust grains levitating in the sheath region of a radio frequency (rf) or a dc plasma discharge above a negatively biased electrode. Due to the high mobility of electrons, dust particles are usually negatively charged. Levitation results from balance between the main forces acting on the grain: gravity and ion drag acting vertically downwards and the electrostatic force acting vertically upwards.

At the very lowest pressures, principally in the 1-40 mtorr range, large amplitude vertical oscillations have been observed in one (linear chains) and two (sheets) dimensional dust crystals [Nunomura *et al* 1999, Takamura *et al* 2001, Samarian *et al* 2001, Misawa *et al* 2000, Takamura *et al* 2000]. Such dust oscillations are observed when the chamber background gas pressure, or alternatively the plasma number density, are below a threshold. Under such low background pressures, these dust oscillations have been reported for time scales of the order of 10 s. Neighboring dust grains do not necessarily oscillate coherently, some do not oscillate while others drop onto the negative electrode [Takamura *et al* 2001]. In the first experimental investigation of the single layer case [Nunomura *et al* 1999], the spontaneous vertical oscillations were attributed to a finite dust grain charging time; ion and electron collection currents flowing onto the grain are not instantaneous and the dust grain surface reaches its equilibrium potential with a time delay. Recent papers have demonstrated that such a mechanism cannot explain an energy gain large enough to sustain the observed large amplitude grain oscillations. Samarian *et al.* [Samarian *et al* 2001], for example, have demonstrated that, in the 30-40 mtorr pressure range, the charge delay mechanism disagrees with the experimental results as the energy gain is only a negligible portion (less than 1%) of the energy loss by neutral gas damping. Furthermore, theoretical arguments have shown that charge delay, in the presence of ion and electron collection currents, is always dissipative [Resendes *et al* 2002, Takamura *et al* 2002]. For the time scales of relevance to the grain oscillation dynamics, the time and velocity dependence of the grain charge may be neglected; only the position dependence is of importance. Either new charging currents, i.e. suprathermal electrons [Takamura *et al* 2001], or fluctuations in the plasma sheath parameters [Sorasio *et al* 2001b] need to be considered. While stochastic heating may be important in low pressure capacitive rf discharges, secondary emission heating does not play much of a role in low pressure discharges[Lieberman and Lichtenberg 1994]. However, self excited large amplitude oscillations have been observed in a variety of discharges suggesting that plasma fluctuations are more relevant.

In complex plasmas, the role of stochastic variations in dust dynamics has been previously considered by several authors. Vaulina *et al.* [Vaulina *et al* 1999] have explored the effect of charge fluctuations on the dynamics of dust particles in the electrode region of a gaseous discharge. The fluctuating component was determined by considering stochastic fluctuations of charge on macroscopic particles neglecting the change in the charge as the particle moves, that is, neglecting the position dependence of the charge. They conclude that stochastic fluctuations of the grain charge may heat a grain and that the effective temperature of a system of macroscopic particles is always somewhat higher than thermal level. Quinn and Goree [Quinn and Goree 2000a] analyzed, both theoretically and experimentally, the role of electrostatic field and grain charge fluctuations using a single-particle Langevin treatment. Two main types of electrostatic heating, induced by ion density fluctuations experimentally detected above the sheath edge, are discussed; heating due to the interaction of electric field fluctuations with the mean particle charge and heating due to the interaction of charge fluctuations with a dc electric field. Their main result is that the calculated fluctuations in charge or field do not contribute to the observed dust grain temperature.

By considering a constant mean charge or electric field, these previous works consider the small amplitude grain oscillation limit as opposed to the present paper which is principally concerned with large amplitude oscillations[Nunomura *et al* 1999, Takamura *et al* 2001, Samarian *et al* 2001, Misawa *et al* 2000, Takamura *et al* 2000] where the trapping potential well is highly nonlinear [Ivlev *et al* 2000, Zafiu *et al* 2001, Ohno *et al* 2002] and the grain charge varies strongly with position [Zafiu *et al* 2001, Sorasio *et al* 2002]. In the present theory, we analyzed the stochastic forcing due to plasma sheath fluctuations; that is, fluctuations in the plasma sheath variables of state, namely the ion and electron densities, ion flow velocity and sheath electric field. These sheath fluctuations then induce charge fluctuations as well as fluctuating forces on the grain in a self-consistent way that includes, in particular, the couplings between the various fluctuating quantities. If we represent the net force on a grain in a steady and instantaneously (compared to the grain dynamics time scale) fluctuating plasma sheath by an external potential, the grain, whose charge is position dependent and also fluctuating, is subject to a fluctuating external potential. This approach is thus different from the previous analyses in which the fluctuating part was introduced as a stochastic time dependent force acting on a dust particle trapped in a time independent parabolic potential well with fixed equilibrium dust charge and electric field.

A heating mechanism involving the fluctuations of the external potential well in which dust grains find themselves is investigated in this paper. This is accomplished via plasma density variation [Sorasio *et al* 2001b], and will be seen to provide an appropriate model for self-excited large amplitude oscillations. When gas pressures or plasma densities are below a threshold, the external potential energy, which acts as a source for the dust grain energy, becomes very sensitive to plasma density so that small changes are associated with large energy inputs. Small variations or fluctuations in these quantities, which are always present in plasmas [Quinn and Goree 2000a], are able to create and sustain the dust grain oscillations indefinitely.

For sufficiently high gas pressures and plasma density, dust grains suspended over the negatively biased electrode can be considered as trapped in a potential well with a parabolic profile [Tomme *et al* 2000]. The situation changes dramatically when experiments are performed in low pressure regimes [Ivlev *et al* 2000, Zafiu *et al* 2001, Sorasio *et al* 2002]; here the potential well strongly deviates from the parabolic profile and the dust grain dynamics shows typical features of a nonlinear oscillator. In recent experiments, dust grain oscillations were driven either by an external sinusoidal voltage applied to a wire above a negatively biased electrode [Ivlev *et al* 2000], or by a low frequency modulation of the rf voltage [Zafiu *et al* 2001, Schollmeyer *et al* 1999]. While these methods are easy to apply, it is not known how they influence the sheath environment and how the grain is affected. What has been previously considered [Ivlev *et al* 2000, Zafiu *et al* 2001] is the careful analysis of the grain motion resulting from the application of these driving mechanisms, using the standard theory of anharmonic oscillations [Landau and Lifshitz 1960], without understanding the physical effects that were driving the oscillations. In this manuscript, for the first time, we explain theoretically the physical mechanisms of grain motion under the influence of the different mentioned techniques and analyze the resulting grain dynamics.

In this study, nonlinear resonances are driven by two different means in regimes where the the pressure and the plasma density are above the threshold which yields self excited oscillations. Firstly, we impose a sinusoidal (external) force of definite strength. Nonlinear resonances are observed. Secondly, the grain oscillations are driven by varying sinusoidally the electrode voltage around a fixed value ϕ_{wf}. The potential well in which the dust grain is trapped oscillates and changes its shape with the same frequency of electrode potential, giving rise to nonlinear grain oscillations. Both primary and sub-harmonic nonlinear resonance regimes are investigated. The results show that a small amplitude modulation of the electrode potential, at the natural frequency of the well, may produce large nonlinear grain oscillations. The results show also that the amplitude of the potential modulation, near the sub-harmonic frequency, has to be above a threshold in order to induce large grain oscillations (parametric resonance). These have only very recently been identified[Ohno *et al* 2002].

It should be noticed that, in experiments, the sinusoidal voltage applied to a wire may act not only as an external force but may also induce plasma density modulation. The reader can find detailed information about the effects of plasma density modulation in [Sorasio *et al* 2002]. In the present theoretical analysis, attention is focused on the influence of the external force even if they may coexist. The main difference between the present model and the previous analysis is that the potential well in which the dust grains are trapped is not an "a priori" assumption, but comes "a posteriori" from the balance between the main forces namely, gravity and ion drag (acting vertically downwards) and the electric force (acting upwards). Thus, the full potential and grain charge is taken into account. To use the language of the previous analysis, rather than expanding the potential and the charge and keeping one or two nonlinear, time independent terms, this approach amounts to keeping time dependent terms to all orders in both the potential and the grain charge. As a result, not only is the stable equilibrium accounted for, but so is the unstable one closer to the electrode. This allows exploration of the dynamics to saturation regimes as well as explains why the grain sometimes falls onto the electrode. These latter phenomenology are absent in previous works.

The nonlinear dynamics of a single dust grain in a plasma sheath is analyzed in this work. While a single grain is considered, the results are expected to be applicable for linear chains and sheets. An order of magnitude estimate of the repulsive nearest neighbor shielded horizontal interaction energy, $U_{int} = (Q^2/R)\exp(-R/\lambda_{De})$, in one and two-dimensional dust crystals, assuming an electron Debye radius λ_{De} of $\simeq$ 400 μ m (equal to the intergrain separation distance R) and 10^4 elementary charges per dust particle, yields an interaction energy of 10^{-10} ergs. This is smaller than the external vertical interaction energy by roughly two orders of magnitude, as will become evident in this analysis, allowing one to conclude that the horizontal grain-grain coupling is weak, comparatively speaking, in a single layer crystal. Additionally, it has been experimentally observed recently that the frequency and amplitude of the dust grain oscillations are independent of the particle number in a monolayer structure [Samarian *et al* 2001]. A theoretical model for the dust grain dynamics in a plasma discharge sheath which predicts a broad range of low pressure nonlinear oscillation phenomena is presented.

The manuscript is organized as follows. Section §2 is dedicated to the analysis of the sheath model. The weakly collisional model used in the analysis of the argon plasma discharge is analytically and numerically explored. In §3 the main forces acting on a dust grain as well as the charging processes are presented. The self excited oscillations observed for the first time by Nunomura *et al.* [Nunomura *et al* 1999] and rigorously detected in several experiments [Takamura *et al* 2001, Samarian *et al* 2001] are analytically and numerically investigated in §4. A new mechanism for excitation of dust grain large amplitude oscillation is proposed. The energy source which excites the oscillations is due to stochastic plasma sheath fluctuations which are always present in a plasma [Quinn 2000b]. Comparison with previous fluctuation models [Vaulina *et al* 1999, Quinn and Goree 2000a] as well as with experimental observations [Nunomura *et al* 1999, Takamura *et al* 2001] is presented. In §5 the dynamics of dust particles under the influence of an external driver is examined. The nonlinear oscillation regimes driven by an externally applied force are investigated rigorously and comparison with existing [Ivlev *et al* 2000] experimental data are presented. Nonlinear and parametric resonance regimes driven by a modulation of the electrode potential are analyzed in detail. Comparison with published experimental observations [Ohno *et al* 2002] is presented. A brief summary of these studies is presented in §6 .

2 Plasma Sheaths

When a plasma is in contact with a negatively biased surface, as an electrode, a strong electric field arises between the plasma and the planar wall. This region is

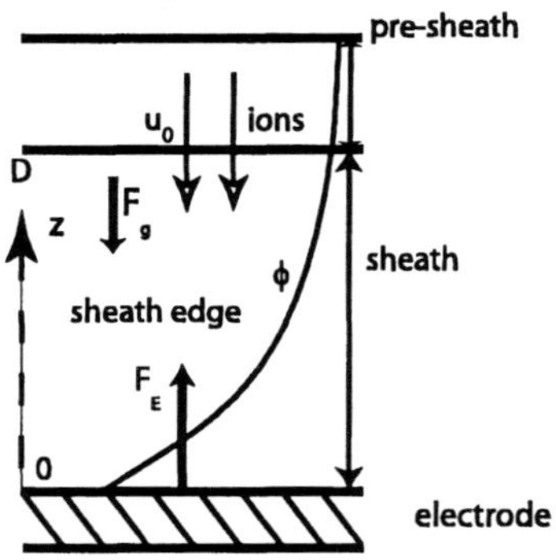

Figure 1: Plasma sheath configuration

called a sheath (see Fig. 1). In the present model, we assume that the electrode is perfectly absorbing with negligible secondary electrons emission and that the plasma in unmagnetized. We assume also that no suprathermal electrons are produced inside the chamber. Under low pressures limit the amount of ionization in the sheath can be neglected, due to the low neutral density, and in the present analysis a source free model will be used. This assumption implies an ion mean free path ℓ much larger than the sheath thickness, which can be as large as hundreds of Debye lengths λ_{Di}. In the energy range of interest ($\sim$ eV) the ion-neutral cross section can be considered to be constant and is taken to be $\sigma_s = 5 \times 10^{-15}\ cm^2$. With a neutral density $n_n \sim 10^{14}\ cm^{-3}$ (10 mtorr) the resulting ion mean free path is $\ell = \sigma_s/n_n \approx 10\ cm$. On the other hand, the ion Debye length for a ion number density $n_i = 1 \times 10^8\ cm^{-3}$ and ion temperature $T_i = 0.05\ eV$ is $\lambda_{Di} = \sqrt{(k_B Ti)/(4\pi e^2 n_i)} \approx 0.001\ cm$ and the resulting sheath thickness would be less than a centimeter and thus much less than the ion mean free path ℓ.

In a weakly ionized plasma the energy which maintains the plasma is generally heating of electrons by the source, while the ions are in thermal equilibrium with the neutrals (the equilibrium is due to the way the plasma is created and not to the ion-neutral collisions). The ion can be considered as a cold beam of positive carriers accelerated through the sheath, due to the electric potential, while the electrons density decreases according to Boltzmann factor.

In the plasma both ion and electrons (n_i and n_e respectively) have the same density equal to the mean plasma density n_0. As clarified above, in the sheath, the electrons are thermalized and the corresponding number density is $n_e = n_0 exp(e\phi_s/k_B T_e)$; here e is the magnitude of the electron charge, k_B the Boltzmann constant, T_e is the electron temperature, and ϕ_s is the local electrostatic potential. The plasma may be considered in steady state with regard to either neutral or plasma density variations. This is appropriate since the neutral collision frequency at 1 mtorr is roughly 10 KHz, the plasma density modulations analyzed are around 10-70 Hz, whereas the slowest discharge timescale, the ion plasma frequency, is roughly a thousand times faster under the discharge conditions of interest. Since single grain dynamics is considered and the dust particle does not contribute to the total space charge this is an "empty sheath" model[Nitter *et al* 1994]. In steady state, the ion continuity equation is

$$\nabla \cdot (n_i \mathbf{v_i}) = 0, \tag{1}$$

where v_i is the ion velocity. Equation 1 implies not only that in the sheath the ionization is negligible but also that plasma density and gas background pressure are two independent variables. The latter implication is not quite correct even under the low pressure regimes investigated in the manuscript but is a reasonable approximation of the real situation. The authors intend to implement a more complete sheath model in the future. Accelerated sheath ions of mass m_i experience a drag force from collisions with neutrals of density n_n with the cross-section σ_s given by $\mathbf{F_c} = m_i n_n \sigma_s {v_i}^2$. This term has to be included in the analysis because even if it is true that the probability of a collision between an ion and a neutral particle within a Debye length is quite small ($\lambda_{Di}/\ell \ll 1$) it is also true that, at least close to the sheath edge, the acceleration due to the electrostatic force and the ion drag are of the same order. The cross-section σ_s is independent of the ion velocity and constant for the energies of interest. The steady state ion equation of motion is $m_i (\mathbf{v}_i \cdot \nabla \mathbf{v}_i) = -e\nabla\phi_s - \mathbf{F}_c$ [Sheridan and Goree 1991]. The properties of the steady state plasma sheath are then governed by the ion continuity and the set of differential equations

$$v_i \frac{dv_i}{dz} = -\frac{e}{m_i}\frac{d\phi_s}{dz} - n_n \sigma_s v_i^2 - 3\frac{T_i}{m_i n_0^2} n_i \frac{dn_i}{dz}, \tag{2}$$

$$\frac{d^2\phi_s}{dz^2} = -4\pi e n_0 \left[\frac{v_{i0}}{v_i} - exp\left(\frac{e\phi_s}{k_B T_e}\right)\right], \tag{3}$$

where T_i is the ion temperature. An Argon plasma with $T_e = 1$ eV, $T_i = 0.05$ eV and $T_n = 0.05$ eV is considered. The ion velocity at the sheath edge, v_{i0}, is the ion sound velocity (Bohm sheath criterion). In the ion momentum equation (2) we have also considered the effect of the ion pressure gradient.

In the present configuration (Fig. 1), the sheath edge is at $z = D$ (where z is the vertical axis), while the electrode is at $z = 0$. As boundary conditions, we take the electric potential at the sheath edge to be zero, viz. $\phi_s(D) = 0$, where the ion acceleration is also zero [Nitter 1996]. The latter condition ensures that ions accelerate monotonically towards the sheath and are not decelerated at the sheath edge (giving rise to an anomalous region of ion concentration). This leads to the condition $d\phi/dz|_{z=D} = -m_i n_n \sigma_s v_i^2/e$. The sheath thickness D is determined by the requirement that the potential take on a given value at the electrode $\phi_s(z = 0) = \phi_w$. In order to express the sheath equations (2,3) in term of the fundamental plasma quantities the following normalization, have been introduced

$$\eta = -\frac{e\phi}{k_b T_e} \qquad u_i = \frac{v_i}{v_{thi}}, \tag{4}$$

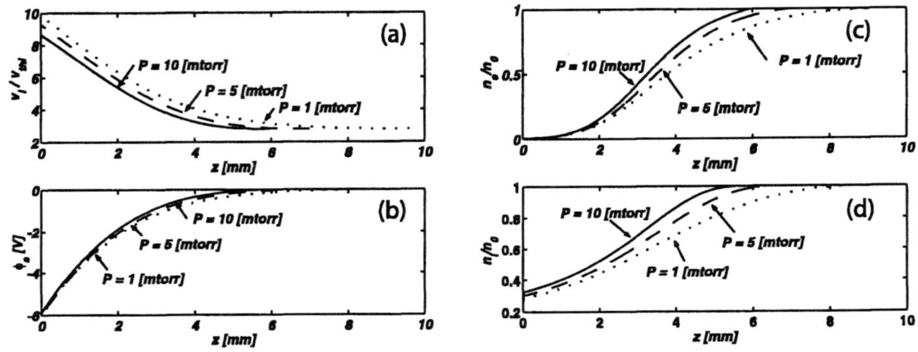

Figure 2: (a) normalized ion velocity, (b) electric potential ϕ_s, (c) normalized electron density and (d) normalized ion density as a function of the distance from the electrode for there different pressures of the background Argon gas: P = 1 mtorr (dotted), P = 5 mtorr (dashed), P = 10 mtorr (solid). The parameters are: $T_e = 1\ eV$, $T_i = T_n = 0.05\ eV$, $n_0 = 1 \times 10^8\ cm^{-3}$. The grain radius is $R = 2.5\ \mu m$ while the mass density is $\rho_d = 1.5\ g\,cm^{-3}$.

$$\zeta = \frac{z}{\lambda_{Di0}} \qquad u_0 = \frac{v_{i0}}{v_{thi}}, \tag{5}$$

where $v_{thi} = \sqrt{8k_BT_i/\pi m_i}$ is the mean ion thermal velocity and $\lambda_{Di0} = \sqrt{(k_BTi)/(4\pi e^2 n_0)}$ is the ion Debye length in the plasma. The system of equation can thus be written in the form:

$$u_i \frac{du_i}{d\zeta}\left(1 - 3\frac{\pi}{8}\frac{u_0^2}{u_i^4}\right) = \frac{\pi}{8}\frac{T_e}{T_i}\frac{d\eta}{d\zeta} - \alpha u_i^2, \tag{6}$$

$$\frac{d^2\eta}{d\zeta^2} = \frac{T_i}{T_e}\left[\frac{u_0}{u_i} - exp(-\eta)\right], \tag{7}$$

where $\alpha = n_n \sigma_s \lambda_{Di0}$ expresses the number of collisions within a Debye length (degree of collisionality). The term expressing the ion pressure gradient (left hand side of equation 6) is negligible since the ion velocity is monotonically increasing towards the electrode and is always greater than the normalized ion sound velocity ($u_i \geq u_0$, $\forall\, z$).

Figure 2 shows (a) ion velocity v_i normalized to the mean ion thermal velocity v_{thi}, (b) electric potential ϕ_s, electron (c) and ion (d) densities n_e and n_i normalized to plasma density n_0 as a function of distance from the electrode for three different pressures. The parameters are: $T_e = 1\ eV$, $T_i = T_n = 0.05\ eV$, $n_0 = 1 \times 10^8\ cm^{-3}$. The grain radius is $R = 2.5\ \mu m$ while the mass density is $\rho_d = 1.5\ g\,cm^{-3}$. The background gas is Argon.

Figure 2 shows that the lower is the pressure, the thicker is the plasma sheath; the plasma shields the electrode potential more effectively at high pressures. The pressure influence appears in the ion drag term of equation 6. The higher is the pressure the higher is the ion drag. The ions are accelerated more effectively by the electric field under low pressures. In the collisionless case, the ion velocity (and thus the ion density) would depend only on electric field. In the collisional case the ion drag term decreases the influence of the electric field on the ion acceleration without affecting the electron density dependence on the electric field. The higher is the pressures the larger is the difference between the ion and electron density dependence on the electric field. This increase in the difference between the ion and electron density dependence on the electric field at higher pressures leads to a more effective sheath. From (a) it can be seen that the ion velocity increases more rapidly at low pressures, as expected, since the number of collisions is proportional to the number of neutral particles. Because of ion continuity (1) the ion density n_i is inversely proportional to the ion velocity v_i thus leading to a larger density, at higher pressures, for the same distance from the electrode. The larger lack of equilibrium between ion and electron densities leads to a larger field and thus the electrode is shielded with a thicker sheath under lower pressures.

From equations 6 it could be inferred that the lower the ion temperature the lower the effect of ion collisions. On the other hand, the ion temperature (i.e. eq. 7) strongly affects the space derivative of the electric field. From our calculations, the ion temperature does not influence significantly the plasma sheath. Once the sheath potential ϕ_s, ion density n_i and ion velocity v_i have been calculated all along the sheath, the grain dynamics can be examined in detail.

3 Grain Forces

The potential difference between a dust particle and the plasma (normalized by the electron thermal energy), $y(z,v) \equiv e(\phi_g - \phi_s)/k_B T_e$, is obtained from a simple charging equation that is based on the "orbit limited current (OLM)" approach [Walch*et al* 1995], which assumes plasma equilibrium for the electrons. For sake of clarity we did not consider the influence of suprathermal electrons that may be produced in the chamber or the secondary electrons currents due to gamma electrons emitted by the biased electrode. The OLM theory is applicable for dust grain radii much smaller than the Debye radius, and provided that the region of influence of the dust grain, i.e. its Debye radius ($> 100\ \mu m$) is smaller than the neutral mean free path ($\ell \approx 10$ cm). For the ion current, we replace the thermal velocity and energy terms in the OLM expression by the mean speed $v_s = \sqrt{v_{thi}^2 + \bar{v}_i^2}$ with which ions approach a dust grain. Here, $\bar{v}_i = (v_i - v)$ is the directed ion speed relative to a stationary dust grain while v is the grain velocity. The ion current thus represents either a mono-energetic uni-directional current which charges a dust grain, if the directed ion speed is large in comparison with the ion thermal speed, or a spherically symmetric current in the opposite limit [Barnes 1992]. By equating the sum of the electron and ion currents to zero, we obtain an implicit equation for $y(z,v)$, which depends on the position of the dust particle in the plasma-sheath. Depending on the sign of $y(z,v)$, we have

$$\left(\frac{T_e}{T_i}\right)^{1/2}\left(\frac{m_i}{m_e}\right)^{1/2}\frac{n_e}{n_i}\exp(y)=\tilde{v}_s\left[1-\pi(T_e/T_i)y/4\tilde{v}_s^2\right], \tag{8}$$

for $y<0$, and

$$\left(\frac{T_e}{T_i}\right)^{1/2}\left(\frac{m_i}{m_e}\right)^{1/2}\frac{n_e}{n_i}(1+y)=\tilde{v}_s\exp[-\pi(T_e/T_i)y/4\tilde{v}_s^2], \tag{9}$$

for $y\geq 0$. Here, $\tilde{v}_s=\left(1+\bar{v}_i^2/v_{th_i}^2\right)^{1/2}$ and v_{th_i} denotes the mean ion thermal speed. Assuming a conducting dust grain, the particle charge can be calculated from

$$Q(z,v)=k_BT_eRy(z,v)/e. \tag{10}$$

Given the electric field E and the charge Q on a dust grain, the electric force is simply $F_e(z,v)=Q(z,v)E(z)$. The grain charge dependence on particle velocity v is quite weak since the ion speed, all along the sheath, is larger than the ion sound velocity $c_s=v_{i0}\approx 1\times 10^4$ cm/s, while the grain speed is of the order of few cm/s.

The main forces on a dust grain are the electrostatic force acting vertically upwards, and gravity and ion drag acting vertically downwards (positive z-direction). These forces are responsible for the dust particle levitation under equilibrium in a dusty plasma-sheath. We consider a dust particle of size R and bulk density ρ_d. The ion drag results from momentum transfer from the positive ion current, which is driven by the electric field and consists of two parts: the collection current related to the ions directly collected by the dust particle and the Coulomb drag [Barnes 1992, Winske and Jones 1994]. Because the collection radius (equal to the grain radius, R) is much smaller than the impact radius, $(b_\pi/2)$, corresponding to a 90 degree deflection, the collection force is negligible. We included its contribution in the numerical calculations for sake of completeness. The ion drag force [Barnes 1992] is then

$$F_{io}(z,v)=4\pi n_iv_{th_i}\tilde{v}_sm_i\bar{v}_ib_{\frac{\pi}{2}}^2\ln\Gamma, \tag{11}$$

where $b_{\frac{\pi}{2}}^2=e^2Q^4/m_i^2\bar{v}_i^4$ and $\ln(\Gamma)=(1/2)\ln\left(1+b_{max}^2/b_{min}^2\right)$. The maximum and minimum impact parameters, corresponding to maximum and minimum deflection angles, are $b_{max}=\bar{\lambda}=\left(2n_e\lambda_{D_e}+n_i\lambda_{D_i}\right)/(n_e+n_i)$ and $b_{min}=r_p$, respectively. Here, the ion and electron Debye radii are $\lambda_{D_i}=\left[4\pi e^2n_i/((1/2)m_i\bar{v}_i^2+k_BT_i)\right]^{-1/2}$ and $\lambda_{D_e}=\left(4\pi e^2n_e/k_BT_e\right)^{-1/2}$, respectively. Neutral drag on a dust particle provides friction. The drag force due to neutral collisional coupling[Sorasio *et al* 2001a, Barnes 1992, Baines 1965], is

$$F_{n_d}(v)=-(8/3)\sqrt{2\pi}R^2n_nk_BT_n(v/v_{th_n}). \tag{12}$$

where $v_{th_n}=\sqrt{T_n/m_n}$ is the neutral thermal speed. The total force acting on a dust grain is then $F_t=F_e+F_g+F_{io}+F_{n_d}$. The total force on a dust particle depends critically on the steady discharge, as this affects the external electric field, ion flows to a dust grain (and consequently the electric charge), the plasma number density as well as friction. At equilibrium, i.e. at zero dust particle speed, the total force is conservative. In this case, the potential energy is $U_t(z) = -\int_0^z dz' F_t(z')$ relative to the electrode.

Having specified the forces, the dust grain equations of motion are

$$dz/dt = v \tag{13}$$

and

$$dv/dt = \tau_0 F_t / m_p v_{thi}, \tag{14}$$

where the displacement is normalized by the ion Debye radius λ_{Di} in the plasma, the time by $\tau_0 = (\lambda_{Di}/v_{thi})$, and the velocity by v_{thi}.

4 Self Excited Oscillation

Before exploring the dynamics of the dust grain we discuss the equilibrium situation when the electric force balances ion drag and gravity and the grain has zero velocity. From the dust grain equilibrium position (distance from the electrode $z = 0$)it is in fact possible to understand how the grain would behave by varying background pressure or plasma density. At equilibrium, for a zero grain speed, the total force is conservative and and the trapping potential energy can be expressed as $U_t(z) = -\int_0^z dz' F_t(z')$.

Figure 3 (a) shows electric force F_e, ion drag F_{id} and gravity F_g as a function of grain position z, while (b) shows the grain trapping potential $U(z)$. Three different cases are shown: $n_0 = 2 \times 10^8 cm^{-3}$ (solid line), $n_0 = 3 \times 10^8 cm^{-3}$ (dashed line) and $n_0 = 4 \times 10^8 cm^{-3}$ (dotted line). The pressure $P = 4.6\ mtorr$ is held fixed while $T_e = 1\ eV$, $T_i = T_n = 0.05\ eV$, $R = 2.5\ \mu m$, $\rho_d = 1.5\ g\,cm^{-3}$ and $\phi_w = 6\ V$. Figure 3 (b) shows that the grain equilibrium position moves towards the electrode by increasing plasma density n_0, as experimentally observed by Takamura *et al.* [Takamura *et al* 2000]. The potential well changes its shape and becomes less deep and wider as n_0 decreases. It should be noticed that the sheath edge is at different distances from the electrode for different densities. The sheath edge is a theoretical point defined by the boundary conditions and chosen as zero potential reference.

As can be seen in (a) the electric force crosses gravity twice thus leading to two equilibrium points, one stable and one unstable. The ion drag decreases with the plasma density since it depends on the ion density. The electric forces, corresponding to different n_0, do not cross one each other at the same point but in two different points A and B. The crossing point between the electric forces, calculated for different plasma densities, moves away from the electrode decreasing plasma density. This means that there would be a value of the plasma density for which the trend

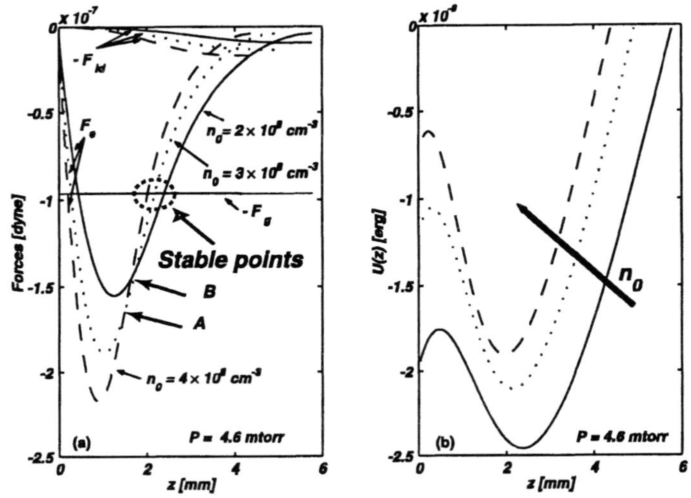

Figure 3: (a) Forces and (b) trapping potential profiles $U(z)$ as function of distance from the electrode for: $n_0 = 2\times10^8 cm^{-3}$ (solid line), $n_0 = 3\times10^8 cm^{-3}$ (dashed line), $n_0 = 4 \times 10^8 cm^{-3}$ (dotted line). The parameters are: P = 4.6 mtorr, $T_e = 1\ eV$, $T_i = T_n = 0.05\ eV$, $R = 2.5\ \mu m$, $\rho_d = 1.5\ g\,cm^{-3}$, $\phi_w = 6\ V$.

of the equilibrium position will flip. Cartoon 4 clarifies the behavior of the stable equilibrium point as a function of the plasma density. When the crossing points (i.e. A,B) between different electric forces (calculated for different plasma densities) are below gravity the stable dust grain equilibrium position moves away from the electrode with decreasing n_0. When the crossing point is above gravity (i.e. C) the trend inverts and the stable equilibrium point moves towards the electrode with decreasing n_0 until the plasma density reaches the condition for which no levitation is possible (line 1).

When the trend of the stable equilibrium position to move away from the electrode flips, the system becomes extremely sensitive to little changes in n_0.

Figure 5 shows the position of the stable (a) and unstable (b) dust grain equilibrium points as a function of plasma density n_0 for different pressures: $P = 0.5\ mtorr$ (solid line); $P = 1\ mtorr$ (dashed line); $P = 3\ mtorr$ (dotted line); $P = 5\ mtorr$ (dash-dot line) and $P = 20\ mtorr$ (dotted with point marks). The parameters are the same as figure 3. For large plasma densities ($n_0 > 1.5\ cm^{-3}$), the stable equilibrium position moves away from the electrode with decreasing n_0. No significant difference can be noticed between different pressures except for $P = 20\ mtorr$. When $P = 20\ mtorr$ the dust grain equilibrium position is shifted close towards the electrode but the shape remains the same. When n_0 is below a threshold ($n_0 < 1.5\ cm^{-3}$), the curves diverge and the position of the stable equilibrium points moves towards the electrode with different slopes under different pressures. In the case $P = 20\ mtorr$ the stable position moves towards the electrode smoothly

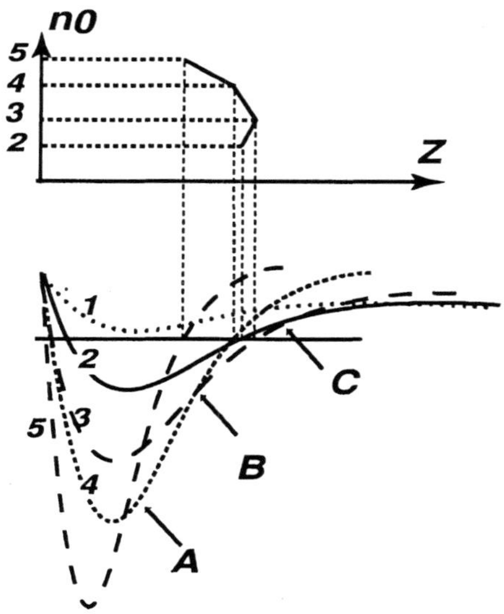

Figure 4: Forces cartoon

as n_0 decreases. When the pressure is low enough $P \leq 5\ mtorr$ a little plasma density decrease reflects in a large shift of the stable equilibrium position towards the electrode. Even if only empirically, the cause of the large amplitude oscillations, detected under low pressures, can already be inferred: when the background pressure is low ($P \leq 5\ mtorr$ for the parameters explored in the present analysis) there is a threshold value of n_0 (different for different gas pressure) below which the grain equilibrium position varies strongly as a function of n_0: little fluctuations (that are always present in the plasma [Quinn and Goree 2000a]) of the plasma density may induce large grain oscillations. The unstable equilibrium position is plotted in 5 (b). As explained above, when $n_0 < 1.5\ cm^{-3}$ a small plasma fluctuation can have a large effect on the grain dynamics. On the other hand, under these low plasma densities, the stable and unstable equilibrium points are one close to the other excluding the possibility of large amplitude oscillations: when the grain reaches the unstable equilibrium point falls onto the electrode. In order to observe large amplitude dust grain oscillations the plasma density should be less than a threshold value but, in order to create large oscillation, cannot be too small. When the stable and unstable equilibrium points coincide, the trapping potential disappears.

In figure 6 the stable (a) and unstable (b) dust grain equilibrium positions are plotted as a function of pressure for different plasma densities: $n_0 = 0.7 \times 10^8\ cm^{-3}$ (solid line); $n_0 = 0.707 \times 10^8\ cm^{-3}$ (dashed line); $n_0 = 0.8 \times 10^8\ cm^{-3}$ (dotted line); $n_0 = 0.9 \times 10^8\ cm^{-3}$ (dashed pointed line); $n_0 = 1 \times 10^8\ cm^{-3}$ (dotted with asterisk line). The position of the stable equilibrium point (a) moves towards the electrode with decreasing pressure. As observed in the experiments performed by Takamura *et al.* [Takamura *et al* 2000], the grain shifts towards the electrode very slowly with

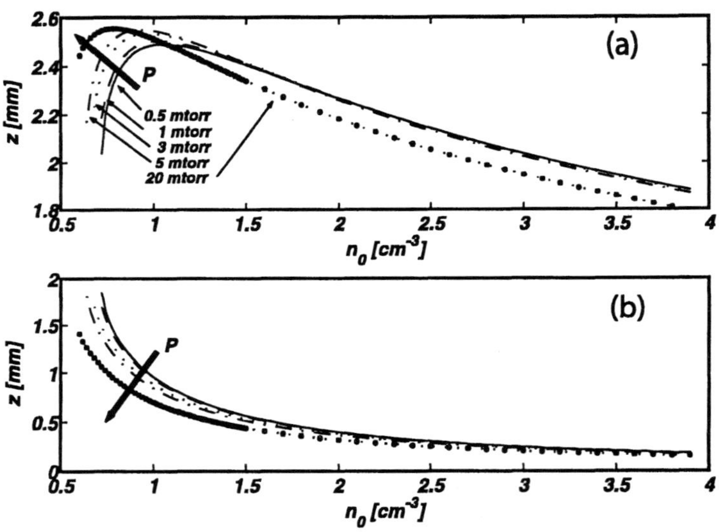

Figure 5: Dust grain stable (a) and unstable (b) equilibrium position as a function of plasma density n_0 for different pressures. The parameters are the same as figure 3.

pressure. When the plasma density is low ($n_0 < 0.9 \times 10^8\ cm^{-3}$), the stable (a) dust grain equilibrium position curves, plotted for different plasma densities, diverge by decreasing the pressure.

Consider the solid and the dotted lines corresponding to $n_0 = 0.7 \times 10^8\ cm^{-3}$ and $n_0 = 0.707 \times 10^8\ cm^{-3}$, respectively. We should pinpoint that the difference in plasma density between the two curves is as low as 1% n_0 and that such small plasma density fluctuations are always present in a plasma. The magnitude of the ion density fluctuations just above the visible sheath edge have been experimentally calculated by R. Quinn [Quinn 2000b] and found to be around 0.6 % n_i and independent of pressure; since, in the plasma, ion and electron densities can be considered to be equal to the plasma density n_0, the experimental results obtained by [Quinn 2000b] are used as reference in the present analysis. When the pressure is above a threshold ($P > 3\ mtorr$) the equilibrium position of the dust grains is not influenced by the plasma density fluctuations. When the pressure is below a threshold ($P < 3\ mtorr$) the grain becomes extremely sensible to plasma density: the curves representing the stable equilibrium position for a 1% difference in plasma density diverges and the effect is enhanced by lowering the pressure. It should be noticed that several conditions have to be satisfied in order to observe the instability and both pressure and plasma density have to be below a threshold. This feature was also observed in the experiment which first detected large amplitude dust grain oscillations [Nunomura *et al* 1999]; there the amplitude of the oscillations was proportional to both n_0 and P.

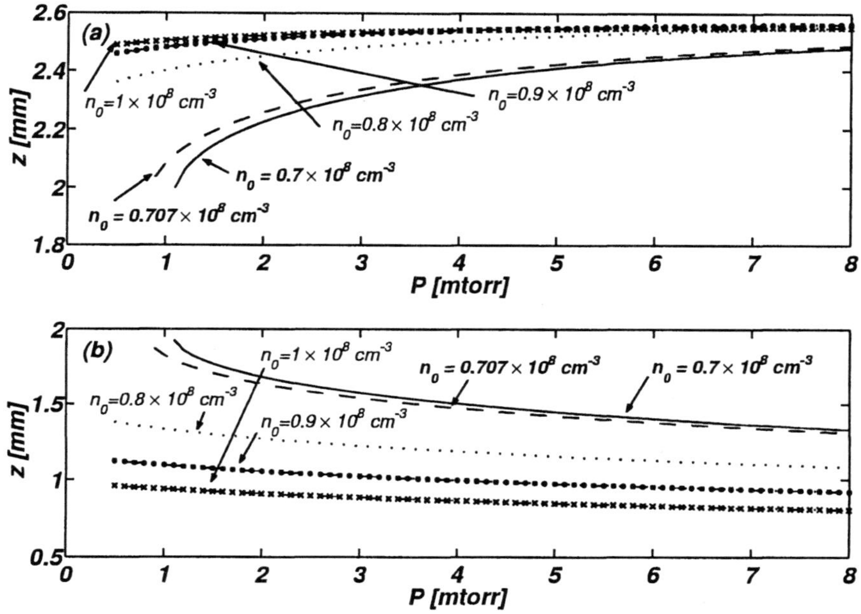

Figure 6: Dust grain stable (a) and unstable (b) equilibrium position as a function of pressure P for different plasma densities: $n_0 = 0.7 \times 10^8\ cm^{-3}$ (solid line); $n_0 = 0.707 \times 10^8\ cm^{-3}$ (dashed line); $n_0 = 0.8 \times 10^8\ cm^{-3}$ (dotted line); $n_0 = 0.9 \times 10^8\ cm^{-3}$ (dashed pointed line); $n_0 = 1 \times 10^8\ cm^{-3}$ (dotted with asterisk line). The parameters are the same as figure 3.

By looking at figure 7 the concept of threshold pressure becomes more clear. The curves represent the difference in the dust grain equilibrium positions induced by a 1% plasma density fluctuation around the plasma densities: $n_0 = 1 \times 10^8\ cm^{-3}$ (dash-dotted), $n_0 = 0.8 \times 10^8\ cm^{-3}$ (dotted), $n_0 = 0.75 \times 10^8\ cm^{-3}$ (dashed), $n_0 = 0.7 \times 10^8\ cm^{-3}$ (solid). The curves are function of pressure P. It can be seen that, when the plasma plasma density is below a threshold, the dust grain equilibrium position shift induced by a 1% plasma density fluctuation varies strongly with pressure. It has to be stressed that the equilibrium position shift induced by a density fluctuation does not coincide with the dust grain oscillation amplitude; on the other hand, it is a useful tool since, where the difference is large, the grain dynamics would be unstable. In the present sheath model, the plasma density and the gas background pressure are two independent variables. In a real experiment, the plasma density n_0 is proportional to the gas background density n_n [Vladimirov and Cramer 1999]. This means that when the pressure decreases the plasma density also decreases and the threshold behavior presented in figure 7 would be greatly enhanced.

The mechanism of energy gain is presented in the cartoon 8. Let us suppose that the grain is at rest in A (n_{0_1}) and a plasma density fluctuation, instantaneous with

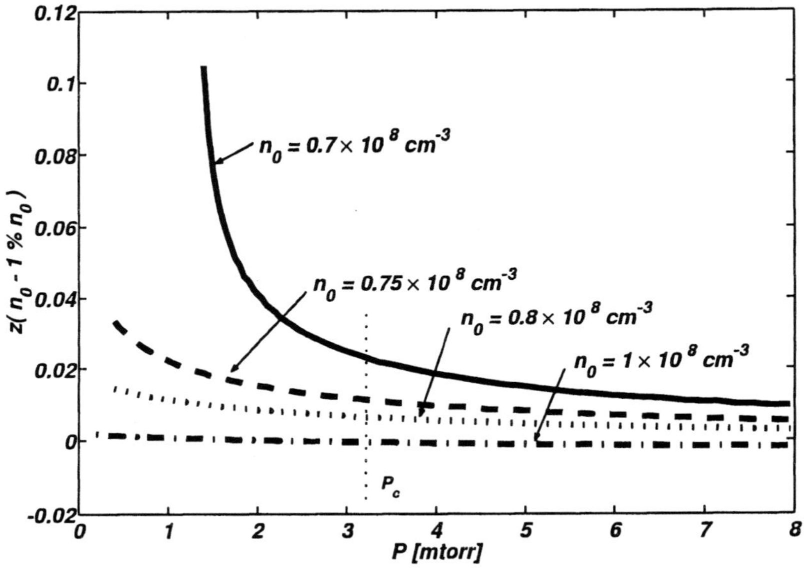

Figure 7: Shift in stable dust grain equilibrium position induced by a 1% plasma density fluctuation as a function of pressure for different plasma densities: $n_0 = 1 \times 10^8 \ cm^{-3}$ (dash-dotted), $n_0 = 0.8 \times 10^8 \ cm^{-3}$ (dotted), $n_0 = 0.75 \times 10^8 \ cm^{-3}$ (dashed), $n_0 = 0.7 \times 10^8 \ cm^{-3}$ (solid). The parameters are the same as figure 3.

respect to the grain dynamics time scale, occurs. The dust grain would suddenly find itself out of equilibrium (D) with respect to the new potential well and would start to oscillate. In this case, the grain would actually gain an amount of energy ΔE_{DE} from the fluctuation. Suppose now the grain is initially at C, and thus with a potential energy ΔE_{CA} , and suppose also that it is oscillating with velocity v_C , when a plasma density fluctuation, from n_{0_1} to n_{0_2}, occurs. The dust grain would find itself in E loosing all the potential energy it had but conserving its velocity v_C. It is clear from this simple example, that the grain does not gain energy from the different levels of potential for n_{0_1} and n_{0_2}, but gains energy from the change in the position of the equilibrium point (i.e. from A to E when the plasma fluctuates from n_{0_1} to n_{0_2}) and from the change in the shape of the trapping potential.

A large shift of the stable equilibrium position alone does not correspond to a large energy gain, but it arises from a large change in the grain potential height with respect to the trapping potential bottom of the well. Looking at figure 7 one could, in fact, argue that the lower is n_0 the larger is the grain oscillations induced by a 1% fluctuation in plasma density. On the other hand, lower values of n_0 exclude large oscillations because of the small distance between stable and unstable equilibrium points. From 8 it can be seen that in order to have a large energy gain, when a plasma density fluctuation occurs (e.g. from n_{01} to n_{02}), the grain needs to find itself out of equilibrium (stable equilibrium position shift) and at different height with respect to the bottom of the trapping potential. If n_0 is too low ($n_0 < 0.8 \times 10^{-8}\ cm^{-3}$), the trapping potential would be extremely weak (which implies low dust grain oscillation frequencies) thus leading to a small energy gain even with a relatively large dust grain equilibrium shift. At higher n_0 (e.g. $n_0 \sim 1 \times 10^8\ cm^{-3}$) and low pressures, the equilibrium position shift resulting from a plasma fluctuation is much smaller than at lower n_0 but may result in a much larger energy gain. When n_0 is too large (i.e. $n_0 > 2 \times 10^8\ cm^{-3}$) and the pressure too hight the equilibrium position shift is too small and the grain gains no energy at all from the plasma density fluctuations.

The grain does not follow closely the change in equilibrium position. If this was the case, the oscillation amplitude would coincide with the shift in dust grain equilibrium position thus excluding the observed large amplitude oscillations. As will be discussed in the following, the power of the fluctuation spectrum gives its contribution mostly at frequencies larger than the grain oscillation frequency (but not higher than 60 Hz). The grain oscillates in a time varying trapping potential and its energy comes from the average hight (over an oscillation) above the oscillating (in position) bottom of the potential well.

In the model presented in this work, the fluctuations are completely stochastic. The grain may gain or loose energy. Depending on when the plasma density fluctuations occur, the grains in the sheath may oscillate or may fall. The particle may gain enough energy to escape from the trapping potential only if there are many lucky cases, otherwise it just keeps oscillating. The influence that a plasma density fluctuation may have on the dust grain dynamics depends on its time rate of probability $\zeta(t)$, on its amplitude $\chi(P, n_0)$ and on the type of process considered $\xi(a)$.

The equations of grain motion (13 and 14) have been solved numerically using the plasma parameters given above taking particular care with the fluctuations induced. Plasma density fluctuations coherent with the grain oscillation can greatly enhance the maximum oscillation amplitude, as demonstrated by [Sorasio *et al* 2002], since they may act as a forcing term. In this case, all the suspended dust grains should show the same dynamical evolution. The self excitation experimental observations have revealed that dust grains do not oscillate coherently: some oscillate, some fall onto the electrode and others stand still. In order to eliminate any coherency, the time rate probability is chosen to be constant, when $t \in (\varsigma_0\ \varsigma_t)$, and 0 otherwise, with ς_t being the maximum time interval between two different fluctuations. A fluctuation has the same probability of occurring anytime between ς_0 and ς_t. It should be pointed out that the fluctuations considered in this article are not white noise or thermodynamics deviations [Landau and Lifshitz 1989]. The origin of these fluctuations is undetermined; their low frequency, far below the ion plasma frequency, suggests that they might be due to disturbances in the ionization balance. Substantial peaks at 60 Hz and its harmonics were experimentally detected [Quinn 2000b]. The 60 Hz modulations were found to be global and homogeneous in the plasma;

their contribution have to be neglected since they could play no role in moving one particle differently from another. In the single-particle Langevin model presented by Quinn *et al.* [Quinn and Goree 2000a] the heating of the dust particle was theoretically found to be due, in primis, to plasma density fluctuations with frequency $\omega < \gamma$, where γ is the damping frequency due to neutral drag. Since only a small fraction of the plasma density fluctuation power is contained in frequencies $\omega < \gamma$ no particle heating was found [Quinn 2000b].

In the present analysis, the plasma density fluctuation spectrum is quite broad. Since the data indicates that the total power of the fluctuation spectrum is contained in the higher frequency range we used a constant spectrum, where frequency above 60 Hz and below $2\Omega_0$ were excluded. The temperature like particle motion cannot be due to plasma collective effects ($\omega \approx 60\ Hz$) and dust grain resonance kind of behavior have to be excluded. With these assumptions, one can estimate both ς_0 and ς_t. In the proposed numerical simulation $\varsigma_0 = 1/60\ s$ and $\varsigma_t = 0.06{-}0.02\ s \sim 2\pi/3\Omega_0$ where Ω_0 is the natural frequency of the trapping well. The mechanism proposed for the energy gain is based on the fact that little plasma density fluctuations may substantially change the equilibrium position of the dust particle and its height with respect to the trapping potential bottom. The fluctuations are not kicking the grain but are making it move in a time fluctuating potential well.

As described above, a fluctuation occurring at different pressures or plasma densities has different effects on the confining potential (see figure 7). In order to be as general as possible and to eliminate any possible source of coherency, the plasma fluctuation amplitude is chosen randomly between 0 and ς_p, where ς_p is the maximum allowed fluctuation. For the present choice of parameters, according to the experimental data presented by [Quinn 2000b], the maximum plasma density step is $\varsigma_p = 1\%\ n_0$. The system is solved with a relative tolerance of 10^{-10}. It should be noted that the larger is ς_p, the higher is the pressure level where the grain dynamics would become unstable. The importance of fluctuations is due to the large sensitivity of the trapping potential to the plasma density, under low pressure conditions, as has already been emphasized.

In order to compare the magnitude of the fluctuations considered in this paper with previous fluctuation theories, we estimated the influence of a plasma density fluctuation, with a conservative maximum amplitude ς_p, on the grain charge. The mean fluctuation in dust grain charge has been previously investigated by several authors. Morfill. *et al.* [Morfill *et al* 1980] suggested that the number of electronic charges Z residing on a dust particle should obey Poisson statistics and thus the mean charge fluctuation δQ should be $\delta Q/e = \sqrt{Z}$. The same approximation was also used by Vaulina *et al.* [Vaulina *et al* 1999]. Cui and Goree [Cui and Goree 1994] calculated numerically the mean square charge fluctuation; the fluctuation

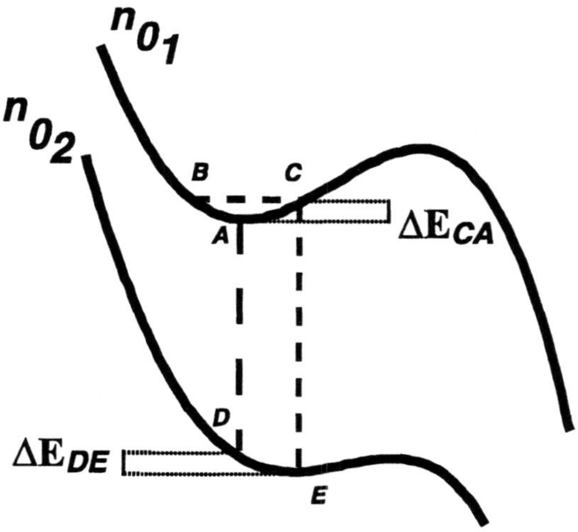

Figure 8: Cartoon representing the energy gain mechanism induced by the present fluctuation model

level results revealed the simple power-law relation

$$\frac{\delta Q}{Q} = \frac{\sqrt{Z}}{2Z}. \tag{15}$$

In the present model the grain charge varies as a function of pressure or plasma density. A plasma density fluctuation of $\varsigma_p = 1\% n_0$ performed at $P = 1.5$ mtorr and $n_0 = 0.7 \times 10^8\ cm^{-3}$ induces a charge variation of about $\delta Q/Q_{P_{ref}} \approx 0.37\%$. The maximum charge fluctuation, i.e calculated with the maximum allowed plasma density fluctuation ς_p, considered in the present analysis, is consistent with the previous results of Cui and Goree [Cui and Goree 1994] which give $\sqrt{Z}/2Z \approx 0.8\%$. Since the plasma density fluctuation probability is constant over $(0, \varsigma_p)$, the charge fluctuations of the present simulation are always lower than the value predicted by previous fluctuation theories [Morfill *et al* 1980, Cui and Goree 1994]. In order to stress the differences between the present fluctuation model and previous ones and to understand its implications, it is necessary to examine in detail the dynamics of the dust grain. Let us suppose that ion drag is negligible (as performed by [Vaulina *et al* 1999, Quinn and Goree 2000a]) and that the equilibrium position is given by the balance between gravity and the electric force. If we expand spatially the electric force around the equilibrium position z_{eq}, we can write

$$F_E(z) = F_E(z_{eq}) + \left.\frac{dF_E}{dz}\right|_{z_{eq}} (z - z_{eq}) + O\left((z - z_{eq})^2\right). \tag{16}$$

This is equivalent to

$$F_E(z) = E(z_{eq})Q(z_{eq}) + Q(z_{eq}) \left.\frac{dE}{dz}\right|_{z_{eq}} (z - z_{eq})$$
$$+E(z_{eq}) \left.\frac{dQ}{dz}\right|_{z_{eq}} (z - z_{eq}) + O\left((z - z_{eq})^2\right). \tag{17}$$

At the stable equilibrium point $F_e(z_{eq}) = m_d g$. The dust grain equation of motion can be rewritten

$$m_d \frac{d^2(z)}{dt^2} = Q(z_{eq}) \left.\frac{dE}{dz}\right|_{z_{eq}} (z - z_{eq})$$
$$+E(z_{eq}) \left.\frac{dQ}{dz}\right|_{z_{eq}} (z - z_{eq}) + O\left((z - z_{eq})^2\right). \tag{18}$$

If the oscillations are so small that the electric field can be considered linear (i.e. $\left.\frac{dE}{dz}\right|_{z_{eq}}$ $= constant$) the charge constant (i.e. $\left.\frac{dQ}{dz}\right|_{z_{eq}} = 0$), and $\frac{dz_{eq}}{dt} = 0$ the dust grain equation of motion can be written as

$$\frac{d^2(z - z_{eq})}{dt^2} + \gamma \frac{d(z - z_{eq})}{dt} = -\Omega_0^2 (z - z_{eq}) + \zeta(t), \tag{19}$$

where $\gamma \frac{d(z-z_{eq})}{dt}$ is damping due to neutral drag. In the limits considered, electric and charge fluctuations will act on the grain as a stochastic forcing term $\zeta(t)$, as considered by [Vaulina *et al* 1999, Quinn and Goree 2000a]. In the present analysis, the electrostatic field is highly nonlinear, the charge varies strongly with position, and the equilibrium position z_{eq} varies with time. If both pressure and plasma density are below the threshold for instability, the dust grain equilibrium position varies strongly for little fluctuations of n_0 and thus z_{eq} cannot be considered constant with time. On the other hand, when both pressure and plasma density are above the threshold for instability, when the ion drag is negligible, and if the oscillations considered have small amplitude, the fluctuation model presented agrees with [Vaulina *et al* 1999, Quinn and Goree 2000a].

As observed through the analysis, the dust grain equilibrium and thus the dust grain dynamics are strongly influenced by the plasma density. The frequency of the grain oscillation is also strongly affected by n_0. In figure 3 we observed that the lower is n_0 the weaker is the trapping potential well. The shape of the potential is directly proportional to dust grain oscillation frequency; larger plasma density would give narrow potential well with consequently higher oscillation frequencies. As observed in experiments [Takamura *et al* 2001], the oscillation frequency increases with plasma density.

In figure 9, the dynamics of a dust grain under the influence of plasma density fluctuations is presented. The pressure is held constant at 0.9 *mtorr* while the mean

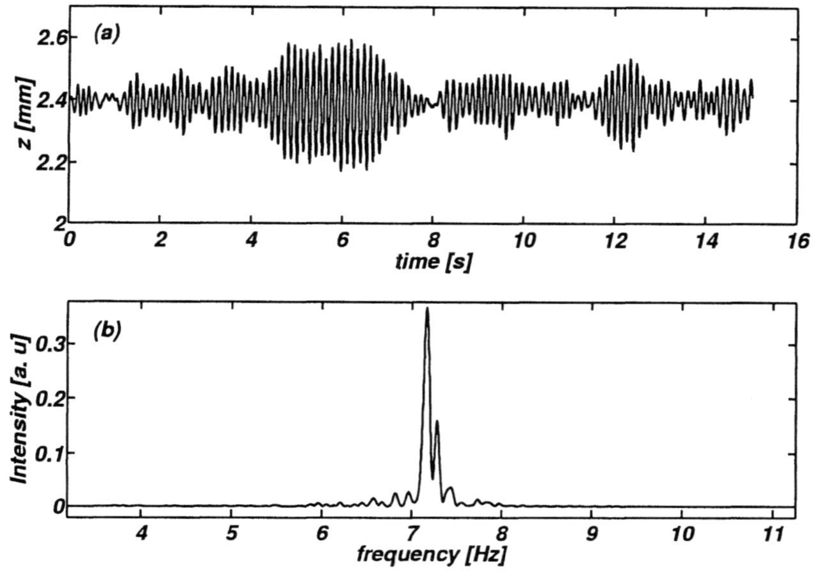

Figure 9: Dust grain oscillations induced by a 1% fluctuation in plasma density. The simulation parameters are: $P = 0.9\ mtorr$, $n_0 = 0.8 \times 10^8\ cm^{-3}$, $T_e = 1\ eV$, $T_i = T_n = 0.05\ eV$, $R = 2.5\ \mu m$, $\rho_d = 1.5\ g\,cm^{-3}$, $\phi_w = 6\ V$, $\varsigma_t = 0.06$, $\varsigma_p = 1\% n_0$

plasma density is $n_0 = 0.8 \times 10^8\ cm^{-3}$. The maximum allowed fluctuation amplitude is $\varsigma_p = 1\% n_0$. The grain is initially at equilibrium. The simulation parameters are $T_e = 1\ eV$, $T_i = T_n = 0.05\ eV$, $R = 2.5\ \mu m$, $\rho_d = 1.5\ g\,cm^{-3}$, $\phi_w = 6\ V$, $\varsigma_t = 0.06$. In (a) the position of the dust grain is plotted as a function of time while in (b) the power spectrum of the grain oscillation is presented. It should be noticed that the frequency spectrum is broad: the trapping potential is nonlinear and the frequency of the oscillations depend on the oscillation amplitude [Landau and Lifshitz 1960]. In the experiments, it have been observed that the larger the oscillations the higher the frequency [Nunomura *et al* 1999]. In the numerical simulation, the oscillation amplitude varies with time as observed by [Takamura *et al* 2001] and since the process is stochastic, different particles would have different oscillations.

Figure 10 shows the grain dynamics for the same parameters used in the previous simulation but under a higher pressure ($P = 8\ mtorr$) and a higher plasma density ($n_0 = 2.5 \times 10^8\ cm^{-3}$). The grain is initially out of equilibrium with an initial velocity of 3 cm/s. The neutral drag dampens rapidly the oscillations such that after 5 s they are reduced to about 30 – 50 μm. The frequency peak (16.3 Hz) is higher than the previous case because of the increase in plasma density (from $n_0 = 0.8 \times 10^8\ cm^{-3}$ to $n_0 = 2.5 \times 10^8\ cm^{-3}$).

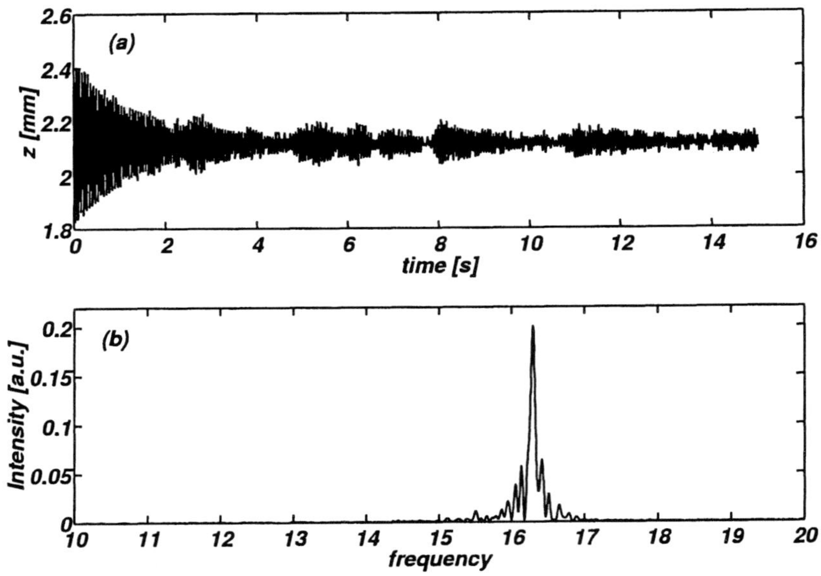

Figure 10: Dust grain oscillations induced by a 1% fluctuation in plasma density at $P = 8$ *mtorr* with $n_0 = 2.5 \times 10^8$ cm^{-3}. The parameters are the same as 9; (a) dust particle position respect to the electrode; (b) power spectrum in arbitrary units.

5 Driven Oscillations

Dust grains levitating in a plasma sheath under low pressures may oscillate spontaneously, as discussed in §IV, or may oscillate under the influence of an external driver. In this section, two different driving mechanisms are analyzed and the regimes of the resulting nonlinear oscillations are explored. In the first part, a direct forcing term F_f is applied to the dust grain equations of motion (14). The neutral and plasma densities are held fixed. The new qualitative feature in the dust grain dynamics is the appearance of nonlinear resonance. Subsequently, the direct forcing term is dropped and a modulation of the electrode potential is imposed. Nonlinear resonance phenomena are again exhibited but with some differences [Sorasio *et al* 2002]. In particular, a parametric resonance is excited. The following analysis should be helpful in distinguishing which mechanism predominantly influences the grain dynamics.

5.1 Forced Oscillations

In recent forced oscillation experiments [Ivlev *et al* 2000, Zafiu *et al* 2001], dust grain oscillations were driven by an external sinusoidal voltage applied to a wire above the negatively biased electrode. The wire acts on the dust grain in two different ways which, in some cases, may coexist. The sinusoidal voltage may either act as a direct force on the charged grains driving the oscillations without modifying

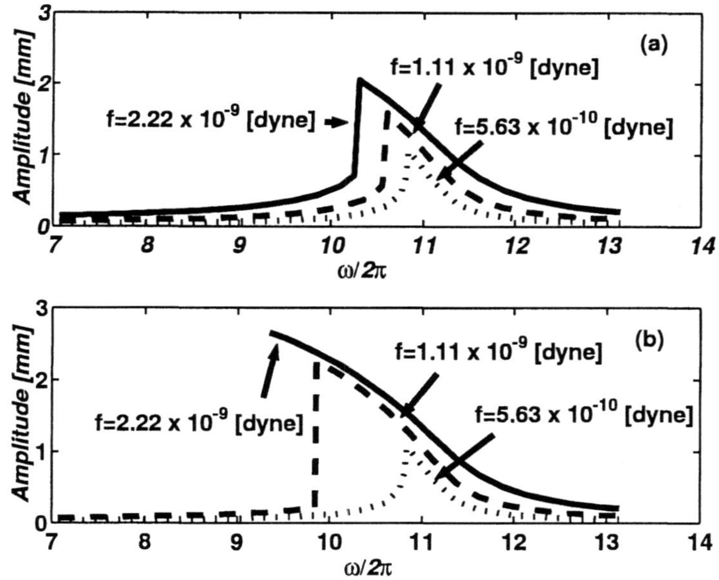

Figure 11: The oscillation amplitude at saturation as a function of the forcing frequency ω near the primary resonance for three different force amplitudes: (1) $f = 1.11 \times 10^{-10}$ dyne; (2) $f = 5.55 \times 10^{-10}$ dyne; (3) $f = 1.11 \times 10^{-9}$ dyne and (a) increasing of the excitation frequency ω, (b) decreasing of the excitation frequency ω. The parameters are: $R = 2.9$ μm, $\rho_m = 1.05$ g/cm^3, $n_0 = 1 \times 10^8$ cm^{-3}, $T_e = 1$ eV, $T_i = T_n = 0.05$ eV, and $P = 7$ mtorr, $\phi_w = 6$ V.

the sheath parameters or can influence the sheath thus acting indirectly on the grain by a modification of the trapping potential. These two effects may coexist in experimental conditions but are theoretically different giving rise to different nonlinear grain behavior. The latter exciting mechanism is not analyzed in the manuscript and the reader can find a detailed investigation in [Sorasio *et al* 2002]. In the present theoretical model, the oscillations are driven by an external force $F_f = f\cos(\omega t - \pi/2)$ that appears in the expression for the total force F_t in equation (14). Since attention is now focused on the nonlinear driven oscillations, the pressure is high enough to exclude the appearance of large amplitude self- excited oscillations. The grain radius is $R = 2.9$ μm and its mass density is $\rho = 1.05$ g/cm^3. The plasma parameters are: $n_0 = 1 \times 10^8$ cm^{-3}, $T_e = 1$ eV, $T_i = T_n = 0.05$ eV, $\phi_w = 6$ V and the pressure $P = 7$ mtorr.

Figure 11 displays the saturation amplitude of the grain oscillations near the primary resonance as a function of the excitation frequency ω for different force amplitudes. In figure 11 (a), ω increases, while in figure 11 (b) ω decreases. When the force amplitude f is small ($f = 5.63 \times 10^{-10}$ dyne), the amplitude of the oscillation increases continuously with ω up to a certain value, and then decreases. In figure 11 (a) and (b) it can be seen that the variation of the oscillation amplitude (for

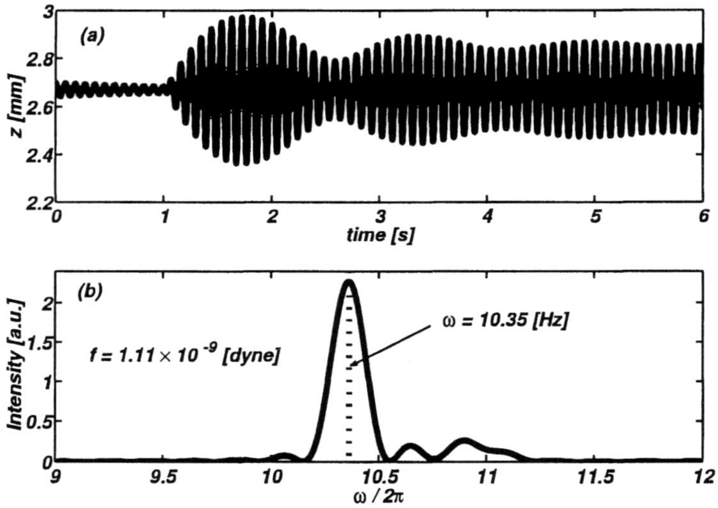

Figure 12: Dynamics of a forced dust grain oscillation with $\omega = 10.35$ Hz and $f = 1.11 \times 10^{-9}$ dyne. (a) Position versus time and (b) power spectrum. The parameters are the same as in figure 11.

$f = 5.63^{-10}$ dyne) has the same behavior in both cases. On the other hand, if f is large, the oscillation amplitude clearly shows hysteresis. The oscillation amplitude increases continuously with ω until a certain point and then jumps up to a much larger value. When ω decreases, the oscillation amplitude may show two different features: it can grow until the grain falls onto the electrode ($f = 2.22 \times 10^{-9}$ dyne) or can continuously grow until it reaches a critical value and then jumps down to a much lower oscillation amplitude ($f = 1.11 \times 10^{-9}$ dyne). The oscillation amplitude increases with the amplitude of the excitation force f. Figure 11 (b) shows that for $f = 2.22 \times 10^{-9}$ dyne, by continually decreasing the frequency ω of the excitation force, the grain oscillation amplitude continually increases and the grain eventually falls onto the electrode.

Figure 12 presents the typical dynamics of a dust grain under the influence of an external force. The forcing starts at $t = 1$ s and the dynamics reaches saturation only after about 10 s. Figure 12 shows (a) the displacement of the dust grain about the equilibrium position and (b) the power spectrum of the grain dynamics (in arbitrary units). The forcing frequency is $\omega = 10.35$ Hz and the applied force is $f = 1.11 \times 10^{-9}$ dyne. At saturation the grain oscillates at the same frequency as the driver. The frequency Ω_0 of the well, for small fluctuations, is about 11.1 Hz. For linear oscillations the grain equation of motion is [Landau and Lifshitz 1960]

$$z = a\,e^{-\lambda\,t}\cos(\Omega_0\,t + \alpha) + b\,\cos(\omega\,t), \tag{20}$$

where b is the amplitude of the oscillations at saturation and λ is the damping coefficient. It should be noted that, since our well is strongly nonlinear, the value of the eigenfrequency Ω_0 is a function of the amplitude of the oscillations. When two harmonic waves of slightly different frequency superimpose, the resulting wave amplitude displays beats [Lauterborn *et al* 1993]. This is seen in figure 12 (a). As a first approximation, it can be seen that the first envelope has a frequency of ~ 0.56 Hz, which corresponds to the superposition of the driving frequency $\omega = 10.35$ Hz with the frequency of the larger oscillations $\Omega \approx 10.9$ Hz. In higher approximation, the second envelope would have a different frequency since the amplitude of the oscillations would be lower and the eigenfrequency Ω_0 would be different. The damping time, for $P = 7$ mtorr is around 2 s; the grain motion reaches saturation after about 6 s. For longer times, the grain oscillates with the same frequency as the driving force ω. The grain dynamics thus demonstrates a nonlinear resonance in a non-parabolic potential well.

The force responsible for the observed nonlinearities is the electrostatic force since the ion drag influences the grain dynamics only marginally. In the higher pressure regimes, the grain is trapped in a potential well with a parabolic profile [Tomme *et al* 2000], because the electrostatic field increases linearly towards the electrode and the grain charge is constant [Zafiu *et al* 2001]. In the low pressure regimes, as can be seen in figure 3, there are two equilibrium points, one stable and one unstable, close to each other. In the vicinity of the stable equilibrium point, the potential well may be approximated by a parabola. When the oscillation amplitude becomes larger, the potential well no longer has a parabolic shape, but rather becomes asymmetric and changes its concavity leading to the unstable equilibrium point and hence the oscillations become strongly nonlinear.

The potential well is strongly nonlinear only on its plasma side; on the electrode side the linearity is maintained. If the particle oscillations are large enough, the grain falls onto the electrode. It should be pointed out that the grain may fall onto the electrode far before changing its polarity. The grain is pushed towards the electrode when gravity overcomes the electrostatic force and this happens when the grain is still negatively charged since the charge gradient in the sheath close to the electrode is steeper than the electric potential gradient, i.e. the charge decreases faster than the electric field increases so that the electrostatic force becomes weaker than gravity.

Figure 13 shows the dust grain charge $Q(z)$ and the electric field $E(z)$ as a function of distance from the electrode for three different pressures: $P = 0.5\ mtorr$, $P = 20\ mtorr$, $P = 60\ mtorr$. The parameters are: $T_e = 1\ eV$, $T_i = T_n = 0.05\ eV$, $\rho_d = 1.05\ g\,cm^{-3}$, $R = 2.9\ \mu m$, $\phi_w = 6\ V$, $n_0 = 1 \times 10^8\ cm^{-3}$. The charge and electric field are shown only in the z region of levitation. Under high pressure conditions the electric field varies with z much faster than the grain charge thus justifying the frequently used assumption of constant charge. Under low pressures, the electric field gradient becomes weaker and nonlinear effects introduced by the

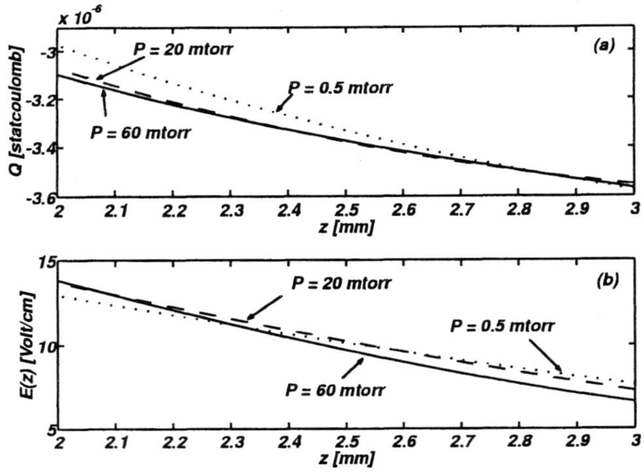

Figure 13: Dust grain charge (a) and electric potential (b) as a function of distance from the electrode, in the region of levitation, for three different pressures: $P = 0.5\ mtorr$, $P = 20\ mtorr$, $P = 60\ mtorr$. The parameters are: $T_e = 1\ eV$, $T_i = T_n = 0.05\ eV$, $\rho_d = 1.05\ g\,cm^{-3}$, $R = 2.9\ \mu m$, $\phi_w = 6\ V$, $n_0 = 1 \times 10^8\ cm^{-3}$

charge will appear [Sorasio *et al* 2002, Zafiu *et al* 2001]. As can be seen in 13 the electric field linearity observed under high pressures [Homann *et al* 1999] is maintained under low pressure conditions. The results presented in 13 are somehow different from previous investigations performed by the authors [Sorasio *et al* 2002] and present in the literature [Zafiu *et al* 2001]. In the previous investigations on nonlinear phenomena, the sheath edge was held at a fixed position $z = 0$ and the electrode potential could vary according to the sheath model. The oscillations arising from an external force were then calculated with respect to the sheath edge. It is in our understanding that, while the results obtained were theoretically correct, in order to compare theory directly with experiments, it is necessary to keep the potential on the electrode constant and to calculate the distance of the grain from the electrode. In the present model, the sheath edge oscillates with pressure or plasma density while the electrode is at a fixed position. The grain oscillations are calculated relative to the electrode. The results obtained are qualitatively similar to the previous studies [Zafiu *et al* 2001, Sorasio *et al* 2002] but quantitatively different. In particular the charge gradient, under low pressures, is highly dependent on the electrode potential.

In experiments, the external force is introduced by a low modulation of the sinusoidal voltage applied to a wire above the negatively biased electrode. The sinusoidal voltage acts directly on dust grains, as an external force, and indirectly through a induced modulation of the plasma density, as already experimentally [Schollmeyer *et al* 1999] and theoretically explained [Sorasio *et al* 2002]. If the plasma number density n_0 varies sinusoidally, the grain finds itself in an oscillating potential well and begins to oscillate. A sinusoidal variation of the plasma density not only modifies the position of the equilibrium point but also the shape of the trapping potential well. From these considerations it can be argued that the grain exhibits both nonlinear and parametric resonance regimes. In this manuscript the latter excitation mechanism, even if theoretically investigated by the authors [Sorasio *et al* 2002], is not described, while a different driving force, namely electrode potential modulation, in explored in detail in the following.

5.2 Electrode Potential Modulation

The electrode potential modulation as a way of exciting dust grain oscillation has already been experimentally explored by several groups under high [Schollmeyer *et al* 1999] and low pressure [Ohno *et al* 2002] conditions but, so far, the mechanism which drives the grain oscillation has not been theoretically identified. The main feature that differentiates the grain behavior, under these different conditions, is that no parametric resonance was detected under high pressures, while a clear parametric regime characterizes the grain dynamics under low pressures. In the present theoretical analysis we investigate theoretically the nonlinear behavior of a dust grain under the influence of a low frequency sinusoidal modulation of the voltage applied on the negatively biased electrode.

In order to compare the results of the numerical simulations with experimental data, the parameters are deduced from the experiments performed by [Ohno *et al* 2002]. The potential on the electrode oscillates around a fixed value $\phi_{wf} = 3\ V$ with the sinusoidal variation

$$\phi_w = \phi_{wf}\left(1 + \epsilon\, cos(\omega\, t - \frac{\pi}{2})\right) \tag{21}$$

where ϵ is the amplitude of the electrode modulation. The parameters are: $T_e = 0.5\ eV$, $T_i = T_n = 0.05\ eV$, $R = 2.5\ \mu_m$, $\rho_d = 1.2\ g\,cm^{-3}$, $n_0 = 2 \times 10^9\ cm^{-3}$, $P = 3\ mtorr$. In the experiment performed by [Ohno *et al* 2002], the electrode potential was $\phi_{wf} = 10\ V$ and thus much larger than the one we choose $\phi_{wf} = 3\ V$. The present choice of parameters, when the electric potential ϕ_s in the sheath reaches $3\ V$ the grain charge becomes positive. The sheath region where $\phi > 3\ V$ is thus not of interest for the analysis of the grain dynamics and has been neglected in the present study.

A variation of the electrode voltage has two effects on the confining potential. It modifies both the position z of the stable equilibrium points, and the shape of the potential itself. The primary and sub-harmonic resonance regimes are investigated. The fundamental frequency Ω_0 of the potential well is around $34\ Hz$ even if the grain oscillation frequency varies with the amplitude of the dust grain oscillations since the potential well is nonlinear [Landau and Lifshitz 1960].

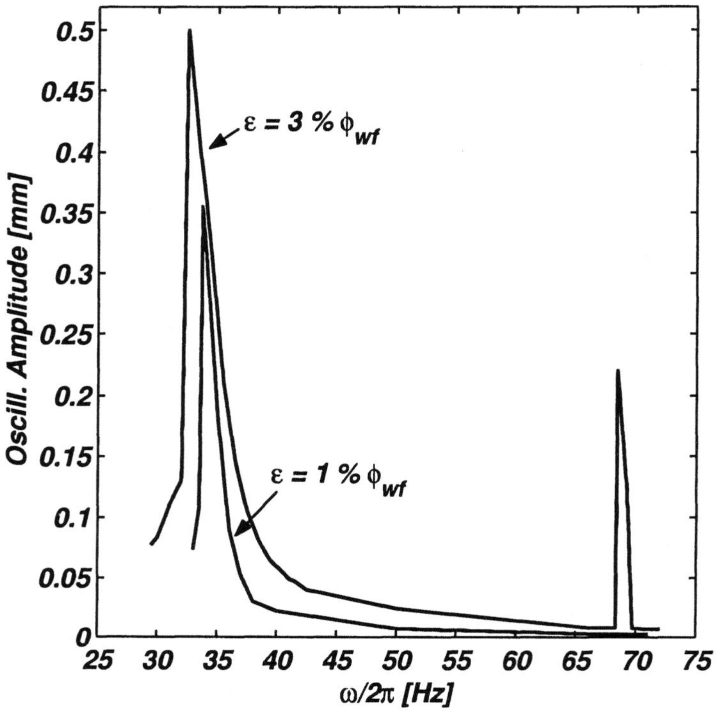

Figure 14: Saturation oscillation amplitude for two different electrode potential voltage: $\epsilon = 1\%\ \phi_{wf}$ and $\epsilon = 3\%\ \phi_{wf}$. The parameters are: $P = 3\ mtorr$, $T_e = 0.5\ eV$, $T_i = T_n = 0.05\ eV$, $\rho_d = 1.2\ g\,cm^{-3}$, $R = 2.5\ \mu m$, $\phi_{wf} = 3\ V$, $n_0 = 2 \times 10^9\ cm^{-3}$

Figure 14 shows the saturation amplitude of the induced dust grain oscillations as a function of driving frequency ω. As already pointed out by the authors under different conditions [Sorasio *et al* 2002], when the trapping potential oscillates the grain dynamics shows resonance regimes at both primary and sub-harmonic frequencies. When the driving frequency is close to the fundamental well frequency, the sinusoidal variation of the equilibrium position acts as a forcing term on the dust grain whose dynamics exhibits nonlinear features. On the other hand, at twice the natural frequency of the well, the grain dynamics is dominated by the change of the potential shape induced by the modulation of the electrode potential thus exhibiting parametric resonance.

A variation of the electrode potential induces a modification of the shape of the trapping potential as well as of its stable and unstable equilibrium points. The sinusoidal modulation of the electrode potential induces a sinusoidal modification of the shape of the trapping potential that may be understood as a time variation of the frequency of the well.

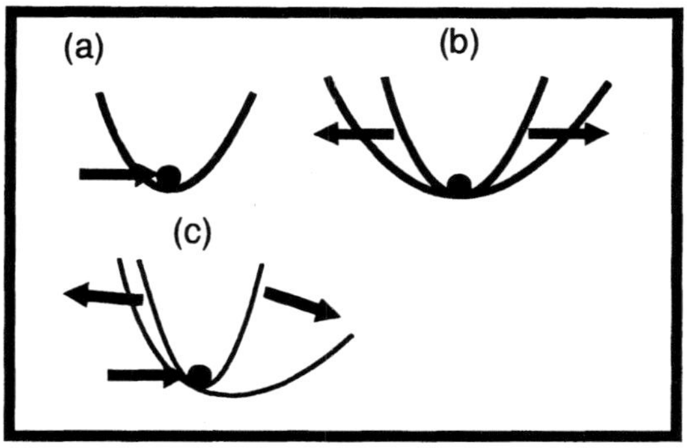

Figure 15: (a) forced (b) parametric and (c) mixed resonance

As explained by Landau ([Landau and Lifshitz 1960], page 81) in the case of "pure" parametric resonance, when the initial displacement from the equilibrium position and the velocity are zero, they remain zero unlike what happens in ordinary resonance (i.e. forced oscillations). Parametric resonance is then stronger if the driving frequency is almost twice the fundamental frequency of the well and shows up only if the parametric variation of the oscillation frequency is large enough. Figure 15 (b) presents what happens in the case of "pure" parametric variation. The potential well varies its shape with time. If the particle is at the bottom with zero velocity the particle does not move because the change in the potential well shape, no matter how intense, does not act on the grain. In contrast, in the case of forced oscillation (15 (a)), the external force induces particle motion and, if strong enough, nonlinear resonances show up. The resonance is strongest when the forcing frequency is close to the fundamental well frequency and, in order to observe resonance effects at twice the fundamental well frequency, the driving force has to be drastically (10 times) increased. The case of interest to us is presented in 15 (c) when the variation of the potential shape comes together with a displacement of the stable equilibrium point. In the case presented here, the frequency of the well varies with time (i.e. the well shape changes) and the stable equilibrium position oscillates and hence the main features of the parametric resonance described by [Landau and Lifshitz 1960] show up mixed with features of ordinary resonance.

In figure 14 it can be seen that, when $\epsilon = 3\,\%\,\phi_{wf}$ the nonlinear resonance observed at about the fundamental well frequency Ω_0 is stronger than that at twice Ω_0 as would happen in the case of pure forcing. On the other hand, with the same driver magnitude ϵ, it it possible to observe a resonance at twice Ω_0, in contrast with the case of pure forcing where the driver magnitude should be almost 10 times larger that the one needed at Ω_0. This is due to the fact that at twice Ω_0 the resonance is excited by a parametric variation of the well frequency. Further on, at twice Ω_0, no resonance is observed if the driver magnitude is lower than a threshold $\epsilon = 3\,\%\,\phi_{wf}$. The latter condition is the main feature of a parametric resonance regime [Landau

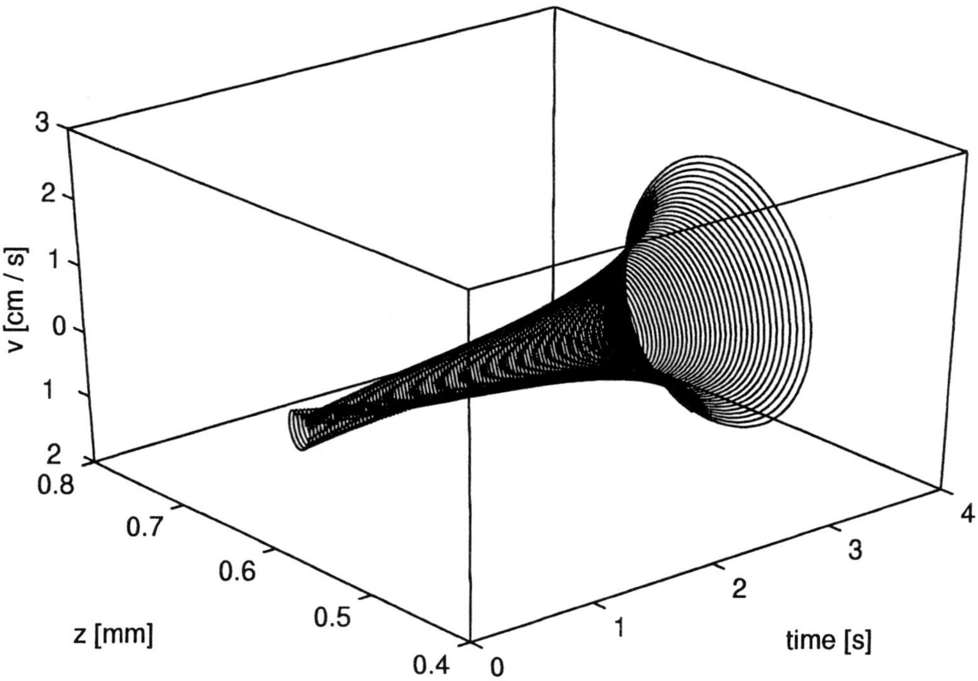

Figure 16: Dust grain dynamics under the influence of an electrode modulation of amplitude $\epsilon = 3\,\%\ \phi_{wf}$ and forcing frequency $\omega = 68.9\ Hz$. The grain path is plotted as a function of time (first 4 seconds) position z $[mm]$ and velocity v $[cm/s]$. The parameters are the same as figure 14

and Lifshitz 1960].

The dynamics of a dust grain under the influence of an electrode potential modulation at the natural well frequency Ω_0 shows almost the same features as in the case of pure forcing presented in figure 12. On the other hand, when the electrode modulation frequency is about twice Ω_0, the dust grain dynamics exhibits some peculiarities.

Figure 16 shows the grain path in the [z,v] phase space as a function of time. The forcing frequency ω is 68.9 Hz and ϵ is $3\,\%\ \phi_{wf}$. The amplitude of the oscillation increases in the first oscillation but then, as the trajectory becomes more complex, the maximum amplitude and the maximum velocity decrease. This is related to the large energy gain of the dust particle in the first oscillation. As time goes on, it becomes clear that the particle does not oscillate around a fixed point but around a line of equilibrium points since the modulation in plasma density causes the trapping potential well to oscillate. The trapping potential oscillates at double the frequency of the grain which, in the first 2 seconds, is not able to follow the well dynamics and thus creating the complex path presented in figure 16. When the grain trajectory becomes simpler, i.e. the grain reaches a dynamical equilibrium with the potential

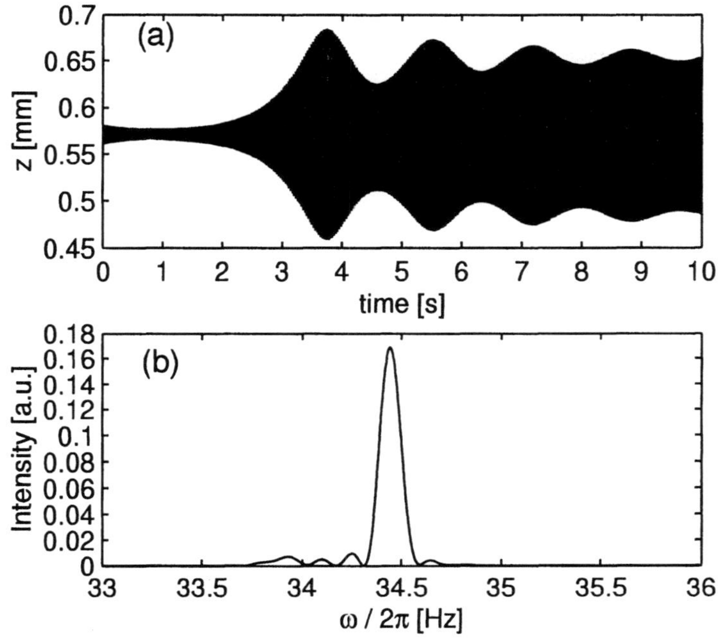

Figure 17: Dust grain (a) position and (b) oscillation spectrum under the influence of an electrode modulation of amplitude $\epsilon = 3\,\%\ \phi_{wf}$ and forcing frequency $\omega = 68.9\ Hz$. The parameters are the same as figure 14

well oscillations after about 2 seconds, the oscillations amplitude increases almost exponentially with time.

Figure 17 shows: (a) grain position as a function of time and (b) oscillation spectrum in arbitrary units for the same conditions analyzed above. After the initial 3 seconds the dust grain oscillation amplitude increases exponentially with time before reaching saturation. The frequency of the oscillation is not the electrode modulation frequency, as would be in the case of pure forcing, but is about the well fundamental frequency Ω_0; the grain dynamics exhibits parametric resonance. Figure 17 (a) shows that, after the initial growth phase, the oscillation amplitude varies for some time before reaching saturation. This effect is due to the nonlinearity of the potential well. The well eigenfrequency Ω_0 is not constant but proportional to the amplitude of the oscillations. As in the case of forced oscillations examined in the previous subsection, two harmonic waves are superimposing giving rise to beats which are damped by neutral damping whose damping rate γ is about 1.6 seconds. The first envelope has a frequency of about 0.69 Hz which corresponds to the superposition of the frequency $\omega = 33.5\ Hz$ and the main frequency $\Omega_0 = 34.14$.

6 Discussion

In this chapter a large number of experimental conditions are theoretically and numerically explored. The dynamics of dust grains oscillating in a plasma sheath under relatively low pressures is analyzed.

A new mechanism able to explain the self excited oscillations experimentally detected by several groups [Nunomura *et al* 1999, Takamura *et al* 2001, Samarian *et al* 2001] is presented. The model is able to reproduce the experimental data obtained in different situations and theoretically explained, so far, with fragmented theory based on 'ad hoc' assumptions. The proposed mechanism takes its basis on the experimentally confirmed [Quinn and Goree 2000a] existence of plasma density fluctuations in the plasma. Large emphasis has been given to the study of the plasma density fluctuation spectrum and the parameters used in the simulations have been inferred from the experiments performed by R. A. Quinn *et al.* [Quinn 2000b]. The mechanism proposed agrees with the results obtained by previous fluctuation theories [Vaulina *et al* 1999, Quinn and Goree 2000a] in the limits explained in the paper. The influence of plasma density fluctuations on grain charge have been compared with existing numerical [Cui and Goree 1994] and theoretical [Morfill *et al* 1980] studies and shows that the model considered induces charge fluctuations consistent with the ones induced by the random-like dust charging process. The new mechanism has been tested not only in comparison with the experimental data obtained by [Nunomura *et al* 1999, Takamura *et al* 2001] but also with the observations of [Samarian *et al* 2001]. The numerical results, while not presented in the manuscript, have been compared with the experimental data and it has been found that the calculated spectrum and growing rate are the same as those experimentally detected [Samarian 2002].

The nonlinear dynamics of dust grains in plasma sheaths has been explored in detail. The nonlinear oscillations detected by Ivlev it et al. [Ivlev *et al* 2000] and driven by an external sinusoidal forcing are theoretically and numerically analyzed. The results show that both electric field and dust grain equilibrium charge influence the grain dynamics within the sheath and that, under low pressures, the charge cannot be considered constant, when dealing with large amplitude oscillations. The primary and sub-harmonic resonant regimes have been numerically simulated and shows good agreement with experiments.

Large amplitude grain oscillations have been driven either by an external force (e.g. sinusoidal voltage applied to a wire above the electrode) or by a sinusoidal modulation of the voltage applied to the negatively biased electrode. Recent experiments performed by N. Ohno *et al.*[Ohno *et al* 2002] have used this driving mechanism and observed both nonlinear and parametric resonance. We have pointed out that the sinusoidal variation of the electrode potential acts on the grain through a modification of the plasma sheath and thus may induce parametric resonance. The main features of this excitation mechanism have been theoretically investigated and the numerical simulation presented shows good agreement with the experimental data [Ohno *et al* 2002].

In conclusion we presented a complete and detailed analysis of all the features characterizing the dust grain dynamics under low pressures, from the self excited oscillations through the nonlinear resonance to the parametric resonance regimes. The results contained in the manuscript cover a full range of nonlinear phenomena and describe with a general model the observations and theory previously presented.

Acknowledgements

G. Sorasio would like to thank Aldo and Margherita Sorasio for their support, G. Delzanno and A. Samarian for helpful discussions. This research was supported by the European Commission through contract No. HPRN-CT-2000-00140 for carrying out the task of the Human Potential Research Training Networks "Complex Plasmas: The Science of Laboratory Colloidal Plasmas and Mesospheric Charged Aerosols" as well as the International Space Science Institute at Bern for the international team entitled "Complex Plasmas in Space: Application to Solar System and Near Earth Space".

References

Baines M. J. *et al.* 1965, Mon. Not. R. Astron. Soc. **130**, 63-74.

Barnes M. *et al.* 1992, Phys. Rev. Lett. **68**, 313.

Cui Chunshi and Goree J. 1994, IEEE Trans. Plasma Sci. **22**, 151.

Homann A., Melzer A. and Piel A. 1999, Phys. Rev. E **59**, 3835.

Ivlev A. V. *et al.* 2000, Phys. Rev. Lett. **85**, 4060.

Landau L. D. and Lifshitz E. M. 1960, *Mechanics*, (Pergamon, Oxford), 87-91.

Landau L. D. and Lifshitz E. M. 1989, *Statistical Physics, Part 1*, (Pergamon, Oxford), 333.

Lauterborn W. , Kurtz T. and Wiesenfeldt M. , *Coherent Optics, Fundamentals and Applications*, (Springer-Verlag, Berlin).

Lieberman M. and Lichtenberg A. 1994, *Principles of Plasma Discharges and Materials Processing*, John Wiley and Sons, Inc., 303.

Misawa T. *et al.* 2000, Jap. J. Appl. Phys. **39**, L551.

Misawa T. *et al.* 2001, Phys. Rev. Lett. **86**, 1219.

Morfill G. E. *et al.* 1980, Planet. Space Sci. **28**, 1087.

Nitter T. , Aslaksen T. K. , MelandsøF. and O. Havnes 1994, IEEE Trans. Plasma Sci. **22**, 159.

Nitter T. 1996, Plasma Sources Sci. Technol. **5**, 93.

Nunomura S. *et al.* 1999, Phys. Rev. Lett. **83**, 1970.

Ohno N. , Sawai M. , Misawa T., Asano K. and Takamura S. 2002, Physica Scripta **T92**.

Quinn R. A. and Goree J. 2000, Phys. Rev. E. **61**, 3033.

Quinn R. A. , Ph.D. thesis, The University of Iowa.

Resendes D. P., Sorasio G. and Shukla P. K. 2002, Phys. Plasmas **9**, 1057.

Samarian A. A. *et al.* 2001, Phys. Rev. E **64**, 25402.

Samarian A. A. 2002, private communications.

Schollmeyer H. *et al.* 1999, Phys. Plasmas **6**, 2693.

Sheridan T. E. and Goree J. 1991, Phys. Fluids B **3**, 2796.

Shukla P.K., Mendis D. A. and V. W. Chow 1996, *The Physics of Dusty Plasmas*, edited by P. K. Shukla,World Scientific, Singapore.

Shukla P. K. and Mamun A. A. 2002, *Introduction to Dusty Plasma Physics* (Institute of Physics Publishing Lt., Bristol).

Sorasio G. ,Mendis D. A. and Rosenberg M. 2001a, Planet. Space Sci. **49**, 1257.

Sorasio G., Resendes D. P. and Shukla P. K. 2001b, JETP Lett. **74**, 77.

Sorasio G. and M. Rosenberg 2001, J. Plasma Phys. **65**,319.

Sorasio G., Resendes D. P. and Shukla P. K. 2002, Phys. Lett. A **293**, 67.

Takamura S. *et al.* 2000, in *Frontiers in Dusty Plasmas*, edited by Y. Nakamura, T. Yokota and P. K. Shukla, (Elsevier, Amsterdam), 337–346.

Takamura S. , *et al.* 2001, Phys. Plasmas **8**, 1886.

Takamura S. *et al.* 2002, Phys. Plasmas **9**, 1060.

Tomme E. B. *et al.* 2000, Phys. Rev. Lett. **85**, 2518.

Vaulina O. S. em et al. 1999, JETP **88**, 1130.

Verheest F. 2000, *Waves in Dusty Space Plasmas* (Kluwer Academic Publishers, Dordrecht).

Vladimirov S. V. and Cramer N. F. 1999, Phys. Rev E **62**, 2754–2762.

Walch B. , Horányi M. and Robertson S. 1995, Phys. Rev. Lett. **75**, 838.

Winske D. and Jones M. E. 1994, IEEE Trans. Plasma Sci. **22**, 454.

Zafiu C. *et al.* 2001, Phys. Rev. E **63**, 66403.

Motion of Nonspherical Dust Grains and Associated Wave Instability

D. D. Tskhakaya
Institute of Physics, Georgian Academy of Sciences
Tbilisi 380077, Georgia.

P. K. Shukla
Fakultät für Physik und Astronomie
Ruhr-Universität Bochum, D-44780, Germany

Abstract

In this article we discuss motions of spherical and nonspherical dust grains as well as the the dispersion properties of a dusty plasma containing irregular charged dust rods. It is shown that charged dust grains in an unmagnetized plasma execute quivering (vibrating) and bouncing motions due to the coupling of the dust dipole moment with the plasma sheath electric field. Dust periodic motions are set up by oscillating electromagnetic fields. Also examined are the dust particle motions in a partially ionized magnetoplasma, Our results reveals that the azimuthal and radial (with respect to the axial magnetic field direction) dust motions are controlled by the radial electric field, the strength of the axial magnetic field, and the collisional interactions between the neutrals and ions/dust. Furthermore, starting from an appropriate Lagrangian for elongated dust grains, we derive a kinetic equation for dust grains which are rotating. The expressions for the dust charge and dust current densities are obtained, taking into account the finite size (the dipole moment) of elongated and rotating dust grains. These charge and current densities are combined with the Maxwell-Vlasov system of equations to derive dispersion relations for the electromagnetic and electrostatic waves in a dusty plasmas. The dispersion relations are analyzed to demonstrate that the dust grain rotation introduces new classes of instabilities involving low-frequency waves. Examples of various unstable low-frequency waves include the electron whistler, the dust whistler, the dust cyclotron waves, Alfvén waves, electromagnetic ion-cyclotron waves, etc. We have found new types of instabilities when the frequencies of the normal modes are close to the dust grain rotation frequency. The results should be useful in understanding the complex dynamics of elongated and rotating dust grains as well as associated instabilities in laboratory and astrophysical environments.

1 Introduction

Studies of levitation and dynamics of charged dust grains are of significant interest in space and laboratory environments [Nitter *et al.* 1998: Chiang and Lin 1996]. Recent laboratory experiments [Chiang and Lin 1996; Law *et al.* 1998; Mohideen *et al.* 1998, 2000; Nunomura *et al.* 1998; Thomas Jr. 1999] have conclusively demonstrated the motions of charged dust clouds near negatively biased electrodes in low-temperature dusty plasma discharges. In a dusty plasma sheath, the dust grains are levitated due to the balance between the gravity and electrostatic forces, and the dust cloud is usually located [Mohideen *et al.* 1998, 2000] at a distance from the electrode several times larger than the Debye length [Shukla 1994]. The dust grains execute bouncing motions [Mohideen *et al.* 1998, 2000] which are repeatedly away and towards the electrodes. Furthermore, each grain also performs transverse quivering across the sheath electric field [Mohideen *et al.* 1998, 2000]. A number of authors [Mendis and Axford 1974; Mendis *et al.* 1983; Mendis *et al.* 1984; Grün *et al.* 1984; Horanyi 1996; Elskens *et al.* 1997; Howard *et al.* 2000, 2000] has considered the electro-gravitational dynamics of a single grain in astrophysical contexts. It is well known that large dust grains in astrophysical as well as in laboratory plasmas do not have the spherical form; the grains are typically elongated [Mohideen *et al.* 1998, 2000; Harwit 1988; Spitzer 1977]. Elongated charged grains are formed due to the coagulation of particulates in partially or fully ionized gases and clouds. Elongated charged dust grains can acquire a rotational motion due to their interaction with photons and particles of the surrounding gas, or due to the presence of an oscillating electric field in a plasma [Harwit 1988; Spitzer 1977; Tskhakaya and Shukla 2001a, 2001b]. In astrophysical plasmas, the angular frequency of the dust grain rotation can reach a rather large value $\approx 10^4 - 10^6\,\mathrm{sec}^{-1}$ for thermal dust grains, and $\approx 10^9\,\mathrm{sec}^{-1}$ for supra thermal grains [Harwit 1988; Voshchinnikov and Semenov 2000], and there is a different kind of orientation involving preferred direction (relative to the galactic disk) of the dust grain angular momentum vector.

We present a theory for the rotational, vibrational, and bouncing motions of 'elongated dust grains' in the presence of electric fields in a plasma without and with an external magnetic field [Tskhakaya *et al.* 2001]. For this purpose, we derive the Lagrangian and the corresponding equations of motions for 'elongated dust grains' in a dusty plasma. Thus, the angular momentum of the dust grains is self-consistently included. The newly derived equations of motions are analytically analyzed for three interesting cases. It is found that a nonuniform electric field in the sheath can set rotational motions of dust clouds. Furthermore, we have analytically examined the dynamics of elongated and rotating dust grains in the presence of an oscillating sheath electric field. It is shown that the governing equations of motion admits a vibrational (quivering) motion. The grain quivering frequency strongly depends on the dust dipole moment, the electric field strength, and the dust moment of inertia. Finally, we have considered the dynamics of 'elongated and rotating dust grains' in the presence of a circularly polarized electromagnetic wave. Here, the equation of motion admits analytical solutions which dictate that the dust grains rotate with the frequency of the electromagnetic fields, and at the same time they also quiver across the direction of the electric field direction. Physically, the sheath electric/electromagnetic field and the dust dipole moment provide the possibility of a potential well in which 'elongated charged dust grains' execute the complex motions, as described above. Furthermore, we also present a self-consistent theory for dust cloud motions in a partially ionized nonuniform magnetoplasma, taking into account collisional interactions between the plasma particles, pressure gradient and electromagnetic forces. It is found that dust cloud motions in the radial and azimuthal directions (with respect to the direction of a vertical dc magnetic field) occur owing to the combined effects of dust-ion and dust-neutral collisions as well as due to forces associated with a radial electric field and the gradients of the plasma and magnetic pressures.

The non-spherical charged dust grains would have, in general, a non-zero dipole moment. Hence, the dusty plasma with elongated and rotating dust grains would acquire a new characteristic frequency, which equals the angular rotation frequency of the dust grains, that can significantly affect the dispersion properties of a dusty plasma. Without specifying the origin of the rotational motion of elongated dust grains in [Mahmoodi *et al.* 2000; Tskhakaya *et al.* 2001; Shukla and Tskhakaya 2001a] the dispersion relations for unmagnetized dusty plasma are derived and the propagation of low-frequency electromagnetic waves and dust ion-acoustic waves (DIAW) are examined as well. These low-frequency waves with the frequencies close to the dust rotation frequency are found to be unstable. The specific results for the growth rates of low-frequency electromagnetic and dust acoustic waves are presented. However, cosmic and laboratory plasmas are usually embedded in an external magnetic field which can have substantial effects on the dusty plasma wave spectra when elongated and rotating dust grains are present. Accordingly, we present here the electrodynamics and dispersion properties of a dusty magnetoplasma whose constituents are electrons, ions, and finite sized elongated rotating dust grains [Tskhakaya *et al.* 2001]. We develop expressions for charge and current densities for the dust grains by including the effect of the dust dipole moment and the dust grain rotation. We calculate the forces acting on the dust grains as well as the corresponding dust kinetic equation. We derive dispersion relations for both the electromagnetic and electrostatic waves and discuss specific instability results, as well as possible applications of these results to cosmic and laboratory plasmas.

2 Equations of Motion of Dust Grains in the Plasma Sheath

In order to describe the dynamics of a single charged dust grain in electromagnetic fields, it is convenient to construct an appropriate Lagrangian. Supposing that charged dust grains constitute a discrete system of particles [Landau and Lifshitz 1971], we have for the dust particle Lagrangian

$$\mathcal{L}_d = \sum_i \frac{\Delta m_i v_i^2}{2} + \frac{1}{c}\sum_i \Delta q_i[\mathbf{v}_i \cdot \mathbf{A}(\mathbf{r}_i, t)] - \sum_i \Delta q_i \varphi(\mathbf{r}_i, t), \tag{1}$$

where the sum is carried out by all i-th part of the given grain, Δm_i and Δq_i are the mass and the charge of the i-th part of the grain, respectively, $\mathbf{r}_i$ and $\mathbf{v}_i$ are its coordinate and the velocity, $\mathbf{A}$ and φ are the vector and scalar potentials, respectively, and c is the speed of light in vacuum. Separating the motions of the center of mass and the rotation around it, we can write $\mathbf{v}_i = \mathbf{V} + \Omega \times \Delta\mathbf{r}_i$ and $\mathbf{r}_i = \mathbf{r} + \Delta\mathbf{r}_i$, where $\mathbf{V}$ and $\mathbf{r}$ are the velocity and position of the center of mass, $\Delta\mathbf{r}_i$ is the coordinate of the i-th part of the grain relative to the center of mass, and Ω is the angular velocity of the dust grain. Assuming that the scale of inhomogeneity l of the electromagnetic field is much larger than the grain size a, we can use the dipole approximation till the third order of the small parameter a/l, and expand the potentials as

$$\mathbf{A}(\mathbf{r}_i, t) = \mathbf{A}(\mathbf{r}, t) + (\Delta\mathbf{r}_i \cdot \nabla)\mathbf{A}(\mathbf{r}, t) + \frac{1}{2}(\Delta\mathbf{r}_i \cdot \nabla)^2\mathbf{A}(\mathbf{r}, t) + ..., \tag{2}$$

and

$$\varphi(\mathbf{r}_i, t) = \varphi(\mathbf{r}, t) + (\Delta\mathbf{r}_i \cdot \nabla)\varphi(\mathbf{r}, t) + \frac{1}{2}(\Delta\mathbf{r}_i \cdot \nabla)^2\varphi(\mathbf{r}, t) + ... \tag{3}$$

Accordingly, the Lagrangian (1) becomes

$$\mathcal{L}_d = \frac{m_d V^2}{2} + \frac{1}{2} I_{\alpha\beta}\Omega_\alpha\Omega_\beta + \frac{q_d}{c}\mathbf{V} \cdot \mathbf{A}(\mathbf{r}, t) - q_d\varphi(\mathbf{r}, t) + \mathbf{d} \cdot \left(\mathbf{E} + \frac{\mathbf{V}}{c} \times \mathbf{B}\right) + \mathbf{m} \cdot \mathbf{B}, \tag{4}$$

where $m_d = \sum_i \Delta m_i$ and $q_d = \sum_i \Delta q_i$ are the total mass and the charge of the grain, $I_{\alpha\beta} = \sum_i \Delta m_i \, [(\Delta\mathbf{r}_i)^2 \, \delta_{\alpha\beta} - (\Delta\mathbf{r}_i)_\alpha \cdot (\Delta\mathbf{r}_i)_\beta]$ is the tensor of the moment of inertia, $\mathbf{d} = \sum_i \Delta q_i \Delta\mathbf{r}_i$ is the dipole moment of the elongated grain, and $\mathbf{m} = (1/2c)\sum_i \Delta q_i(\Delta\mathbf{r} \times \mathbf{u}_i)$, ($\mathbf{u}_i = \Omega \times \Delta\mathbf{r}_i$ is the rotational velocity) is the magnetic moment of the grain. The electric and magnetic fields are $\mathbf{E} = -\nabla\varphi - c^{-1}\partial_t\mathbf{A}(\mathbf{r}, t)$ and $\mathbf{B} = \nabla \times \mathbf{A}(\mathbf{r}, t)$, respectively. In deriving eq. (4), we neglected the multi-dipole effect and used the relation $d\mathbf{d}/dt = \Omega \times \mathbf{d}$. In the presence of the gravity field $\mathbf{g}$, we must add the term $m_d\mathbf{g} \cdot \mathbf{r}$ in the right-hand side of eq. (4) The equation of motion for charged dust grains can be readily deduced from eq. (4), yielding

$$\frac{d\mathbf{p}}{dt} = (q_d + \mathbf{d} \cdot \nabla)\left(\mathbf{E} + \frac{1}{c}\mathbf{V} \times \mathbf{B}\right) + \frac{1}{c}(\Omega \times \mathbf{d}) \times \mathbf{B} + (\mathbf{m} \times \nabla) \times \mathbf{B} + m_d\mathbf{g}, \tag{5}$$

and

$$\begin{aligned}\frac{dM_\alpha}{dt} &= -\frac{q_d}{2m_d c} I_{\alpha\beta}\left[\frac{\partial B_\alpha}{\partial t} + (\mathbf{V}\cdot\nabla)B_\alpha\right] + \left[\mathbf{d}\times\left(\mathbf{E}+\frac{1}{c}\mathbf{V}\times\mathbf{B}\right)\right]_\alpha \\ &\quad + (\mathbf{m}\times\mathbf{B})_\alpha , \end{aligned} \tag{6}$$

where $\mathbf{p} = m_d\mathbf{V}$ is the momentum and $M_\alpha = I_{\alpha\beta}\Omega_\beta$ is the angular momentum of the grain.

In the absence of the magnetic field, the equation of motion (5) for $a/l \ll 1$ assumes the form

$$m\frac{dV_z}{dt} = -\frac{\partial U}{\partial z}, \tag{7}$$

where

$$U = q_d\varphi + d\frac{\partial\varphi}{\partial z} + m_d g z, \tag{8}$$

is the effective potential energy, and the electric field $E_z = -\partial\varphi/\partial z$ is aligned along the z axis. Since in most of the low-temperature laboratory dusty plasma discharges [Chiang and Lin 1996; Law *et al.* 1998; Mohideen *et al.* 1998. 2000; Nunomura *et al.* 1998; Thomas Jr. 1999], the charge of the dust grains is negative, we have $q_d < 0$. For negatively biased electrodes, the potential $\varphi < 0$ and tends monotonically to zero away from the electrode. In the plasma sheath, whose thickness is of the order of several Debye length λ_D, the increase of the potential φ is rather sharp. Beyond this thickness, i.e. in the pre-sheath, the effective potential φ changes smoothly. Thus, between the electrode and the unperturbed plasma, the effective potential energy U has a shape of a well with a minimum at the point ($z = z_0$) where $m_d g \approx -q_d\partial\varphi/\partial z$. Hence, the grains trapped in the potential well execute bouncing motions between the electrode and the unperturbed plasma . On the other hand, for the description of rotational motions of dust clouds [Mohideen *et al.* 1998, 2000], it is necessary to account for the transverse inhomogeneity (parallel to the electrode) of the electric field.

We now consider the dust particle motions in several interesting cases.

2.1 Quivering Motion

We consider the quivering motion of the grains in the presence of a homogeneous electric field $\mathbf{E} = \hat{\mathbf{z}}E$, where $\hat{\mathbf{z}}$ is the unit vector along the z axis. In the absence of the magnetic field, neglecting small terms that are of order a/l, we obtain for the angular momentum

$$\frac{d\mathbf{M}}{dt} = \mathbf{d}\times\mathbf{E}. \tag{9}$$

Assuming that the angular velocity $\Omega = (0, \Omega, 0)$ is directed along the y axis, we obtain

$$I\frac{d\Omega}{dt} = -dE\sin\psi, \tag{10}$$

where ψ is the angle between the dipole moment and the z axis, I is the y component of the principal moment of the dust grain inertia, and the x component of the dipole moment is $d_x = d\sin\psi$. Subsequently, we have from eq. (10)

$$\frac{d^2\psi}{dt^2} = -\Omega_0^2\sin\psi, \tag{11}$$

where $\Omega_0 = (dE/I)^{1/2}$ is a characteristic frequency of the grain quivering. Equation (11) is subjected to the initial conditions $\psi = \psi_0$ and $d\psi/dt = 0$ at $t = 0$. It describes the motion in the potential well $U = -\cos\psi$, where $-\psi_0 \leq \psi \leq \psi_0$. The solution of eq. (11) is

$$\Omega_0 t = F\left[\arcsin\left(\frac{\sin(\psi/2)}{\sin(\psi_0/2)}\right), \sin(\psi_0/2)\right] - F\left[\frac{\pi}{2}, \sin(\psi_0/2)\right], \tag{12}$$

where F is the elliptic integral of the second kind [Gradshteyn and Ryzhik 1980]. Equation (12) exhibits quivering motions across the direction of the electric field with a non-constant angular velocity

$$\Omega = \frac{d\psi}{dt} = \pm\sqrt{2(\cos\psi - \cos\psi_0)}. \tag{13}$$

It turns out that Ω varies in the range $0 \leq \Omega \leq 2\Omega_0\sin(\psi_0/2)$. The period of a single swing is

$$T = \frac{4}{\sqrt{2}\Omega_0}\int_0^{\psi_0}\frac{d\psi}{\sqrt{\cos\psi - \cos\psi_0}} \equiv \frac{4}{\Omega_0}F\left[\frac{\pi}{2}, \sin(\psi_0/2)\right]. \tag{14}$$

The averaged angular velocity is then $<\Omega>= (4/T)\psi_0$. An order of magnitude estimate for the characteristic angular velocity for the grain quivering is

$$\Omega_0 \sim 10^4\sqrt{14.4(Z_d/\rho_0)E(\mathrm{V/cm})}, \tag{15}$$

where Z_d is number of electrons residing on the dust grain surface and ρ_0 is the mass density of the dust grains. In obtaining (15) we assumed that the dust grain size is approximately 1 μ m. It follows from (15) that the grain quiver frequency can reach a value 100 kHz and more for a moderate electric field strength of a few milli-volts/cm. Thus, the present model provides a clue to the grain quivering that has been observed in [Mohideen *et al.* 1998, 2000].

2.2 Grain Motion in the Circularly Polarized Field

We consider the particulate dynamics in the presence of a magnetic field $\mathbf{B} = (0, 0, B_0)$, where B_0 is the strength of the constant magnetic field. We assume that the dust magnetic moment is along the direction of the ambient magnetic field, and we neglect the grain precision. For circularly polarized electromagnetic fields, we choose $E_x = E\cos(\omega t)$ and $E_y = E\sin(\omega t)$, where ω is the wave frequency. The x and y components of the dipole moment are $d_x = d\cos\phi$ and $d_y = d\sin\phi$, respectively. From eq. (9) we then obtain the following equation for the dust grain rotating around the z axis

$$I_z \frac{d\Omega}{dt} = -dE\sin(\phi - \omega t), \tag{16}$$

which can be put in the form

$$\frac{d^2\phi}{dt^2} = -\Omega_{0z}^2 \sin(\phi - \omega t), \tag{17}$$

where $\Omega_{0z} = (dE/I_z)^{1/2}$, and I_z is the z component of the principal moment of the dust grain inertia. The angular velocity is now defined by $\Omega = d\phi/dt$. We solve eq. (17) by introducing the initial conditions $\phi = \phi_0$ and $d\phi/dt = 0$, as used before. Accordingly, integrating eq. (17) we obtain

$$\frac{1}{2\Omega_{0z}^2}\left(\frac{\partial\Phi}{\partial t}\right)^2 - \cos\Phi = \frac{1}{2}\frac{\omega^2}{\Omega_{0z}^2} - \cos\phi_0 \equiv \varepsilon, \tag{18}$$

where

$$\Phi = \phi - \omega t. \tag{19}$$

If the effective energy ε lies between -1 and 1, then Φ satisfies $-\Phi_{max} < \Phi < \Phi_{max}$, where Φ_{max} is defined by $\sin(\Phi_{max}/2) = \sqrt{\sin^2(\phi_0/2) + \omega^2/4\Omega_{0z}^2}$. A typical solution of eq. (18) is

$$\begin{aligned}\Omega_{0z}t = F&\left[\arcsin\left(\frac{\sin(\phi_0/2)}{\sin(\Phi_{max})}\right), \sin(\Phi_{max}/2)\right] \\ -F&\left[\arcsin\left(\frac{\sin(\Phi/2)}{\sin(\Phi_{max})}\right), \sin(\Phi_{max}/2)\right].\end{aligned} \tag{20}$$

Equation (20) exhibits that in a system of reference rotating with the electric field [see, e.g. eq. (19)], the dust grain now quivers with a frequency which roughly equals Ω_{0z}. The quivering period is

$$T = \frac{4}{\Omega_{0z}} F\left[\frac{\pi}{2}, \sin(\Phi_{max}/2)\right]. \tag{21}$$

In the laboratory frame, the dust grain rotates with the frequency of the electromagnetic field, and it also quivers across the electric field direction. The angular velocity is $\Omega = \omega + d\Phi/dt$. From the latter relation, it follows that the average angular velocity of the dust grain is $<\Omega>=\omega$, where the angular bracket denotes the averaging over the quivering period, given by eq. (21). Furthermore, we note that when the effective energy ε is approximately unity, we obtain no quivering. For $t \to \infty$, the angular velocity of the dust grain exactly matches the electromagnetic field frequency, viz. $\Omega = \omega$. For the large effective energy ε , $\omega^2/\Omega_{0z}^2 \gg 1$, we have again the rotation and quivering of the dust grains. The angular velocity is

$$\Omega = \frac{\Omega_{0z}^2}{\omega}\left[\cos\phi_0 - \cos(\omega t - \phi_0)\right]. \tag{22}$$

The average angular velocity turns out to be

$$<\Omega>= \frac{\Omega_{0z}^2}{\omega}\cos\phi_0. \tag{23}$$

Our model depicts that the dust grains bounce back and forth from a negatively biased electrode and that they also quiver simultaneously across the electric field direction. A recent laboratory experiment [Mohideen *et al.* 1998, 2000] has conclusively observed such motions of elongated charged dust grains in plasmas. Physically, the present complex motions of dust clouds arise due to the combined effects of the sheath electric field, gravity and the dipole moment of the dust grain. In the presence of circularly polarized electromagnetic fields (viz. helicons/whistlers or radio-frequency waves) the particulates are set in rotation. Although our theory explains the salient features of the dust grain rotation as well as the vibrating and bouncing motions of 'elongated dust grains' in a laboratory experiment [Mohideen *et al.* 1998, 2000], the dust grain rotation and vibration that are induced by electric fields should also be visualized in astrophysical settings [Harwit 1988; Spitzer 1977] as well as in planetary rings [Grün *et al.* 1996]. It is anticipated that when the Cassini spacecraft arrives at Saturn in 2004, its cameras might possibly be able to image the vibration and the rotation of elongated charged dust grains.

3 Dust Cloud Motions in a Magnetic Field

About eight years ago, Fujiyama *et al.* [1994] experimentally studied the dynamics of silicon particles in DC and AC silane plasmas in the presence of a modulated magnetic field that is perpendicular to the discharge electric field. The experimental results demonstrated dynamical particle transport due to the balance among electrostatic force, ion drag force and force of gravity on the particles [Shukla and Mamun 2002]. The removal of dust particles from the processing plasma due to the poloidal ion drag force was also considered by Nunomura *et al.* [Nunomura *et al.* 1997]. Recently, Sato *et al.* [2000, 2001] conducted a series of experiments to study the dynamics and control of fine-particle clouds in a moderate pressure (75 mtorr-300 mtorr) partially ionized DC discharge plasma in the presence of a vertical (or axial) magnetic field whose strength varies between 400 gauss to 40 kilogauss. In the experiments of Sato *et al.* [2000, 2001] the plasma density is in the range $10^7 - 10^8$ cm^{-3} and the electron temperature is of order of a few eV. The ion temperature is $T_i \sim 0.1$ eV and the neutrals and dust grains are at room temperature. Fine dust particles used are spherical with a diameter $\sim 10\mu m$, mass of order 6×10^{-10} gms, and are strongly correlated, as typical dust charge and dust density are $q_d \sim 3 \times 10^4 e$ and $n_d \sim 10^3 - 10^4$ cm^{-3}, respectively, where e is the magnitude of the electron charge. The ion-neutral collision mean free path, of order of a fraction of a mm, is much smaller than the dust cloud size (~ 1 cm diameter). Sato *et al.* [2000, 2001] observed rotations of the cloud of negatively charged dust particles perpendicular to the magnetic field. The angular frequency of the dust cloud rotation is of the order of 0.1 rad/s, and it depends on the magnetic field strength and rf power in the plasma. On the other hand, Konopka *et al.* [2000] observed rigid and differential dust particle rotations in a nonuniform magnetic field.

Ishihara and Sato [2001] and Kaw *et al.* [2002] have presented theoretical models for understanding the motions of charged dust particles in the plasma sheath containing an external magnetic field. Ishihara and Sato Ishihara and Sato [2001] suggested that sheared ion flows in the sheath can produce self-rotation of a dust particle, while the external magnetic field causes the dust cyclotron motion. On the other hand, Kaw *et al.* [2002] suggested that the rotation of dust clouds arises spontaneously in the direction of the ion rotation; the latter, in turn, is determined by a competition between the $\mathbf{E} \times \mathbf{B}$ and diamagnetic rotations of the ions in a partially ionized magnetoplasma, where $\mathbf{E}$ is the sheath electric field, $\mathbf{B} = \hat{\mathbf{z}} B_0$ is the external magnetic field, $\hat{\mathbf{z}}$ is the unit vector along the vertical direction, and B_0 is the magnetic field strength. Although the model of Kaw *et al.* lend support to the experimental observations [Sato *et al.* 2000, 2001] involving rigid dust particle rotation, it appears that there does not exist a self-consistent theory for rigid and differential dust cloud rotations including drag forces (associated with ion-dust and dust-neutral collisions) as well as an electrostatic force caused by the radial sheath electric field and the $\mathbf{J}_d \times \mathbf{B}$ force, where $\mathbf{J}_d$ is the dust current density. Here we develop such a theory [Shukla 2002] for complex dust cloud motions in a partially ionized nonuniform dusty magnetoplasma.

We consider the dynamics of negatively charged dust particles in a nonuniform plasma sheath containing an electric field $\mathbf{E} = \hat{\mathbf{r}} E_r$ and a nonuniform magnetic field $\mathbf{B} = \hat{\mathbf{z}} B_0(r)$. The plasma also contains a nonuniform pressure gradient $\partial p/\partial r$, where $p = p_e + p_i \equiv n_e T_e + n_i T_i$ and n_j (T_j) is the density (temperature) of the particle species j (j equals e for the electrons, i for ions, d for dust grains, and n for neutrals). When the collision frequency ν_{ej} is much smaller than the electron gyrofrequency $\omega_{ce} = eB_0/m_e c$, we obtain from the inertialess electron momentum equation

$$0 = -n_e e\mathbf{E} - \nabla p_e - n_e e\frac{\mathbf{v}_e}{c} \times \mathbf{B}, \tag{24}$$

where m_e is the electron mass, c is the speed of light in vacuum, $\mathbf{v}_e$ is the electron velocity, and $|\mathbf{v}_e \cdot \nabla| \ll \omega_{ce}$. Equation (24) shows that electrons acquire an azimuthal drift speed

$$v_{e\theta} = -\frac{c}{B_0}E_r - \frac{c}{en_eB_0}\frac{\partial p_e}{\partial r} \tag{25}$$

due to the radial electric field E_r and the electron pressure gradient $\partial p_e/\partial r$. Here the subscripts θ and r stand for the azimuthal and radial components, respectively. The ions are coupled with electrons through the radial electric field. In a partially ionized dusty plasma, the ion-neutral collision frequency ν_{in} is much larger than the ion-dust collision frequency. Hence, in the steady state, we have from the ion momentum equation

$$\rho_i\nu_{in}\mathbf{v}_i = n_i e\mathbf{E} - \nabla p_i + \rho_i\nu_{id}\mathbf{v}_d + n_i e\frac{\mathbf{v}_i}{c} \times \mathbf{B}, \tag{26}$$

where $\rho_i = n_i m_i$ is the ion mass density, m_i is the ion mass, ν_{id} is the ion-dust collision frequency, $\mathbf{v}_i$ is the ion velocity, and $|\mathbf{v}_i \cdot \nabla|, (n_e m_e/n_i m_i)\nu_{ei} \ll \nu_{in}$. The neutrals are assumed to be stationary. Eliminating $\mathbf{E}$ from eqs. (24) and (26) we obtain

$$\rho_i\nu_{in}\mathbf{v}_i = Z_d n_d e\mathbf{E} - \nabla\left(p + \frac{B_0^2}{8\pi}\right) + \frac{(\mathbf{B}\cdot\nabla)\mathbf{B}}{4\pi} + \rho_i\nu_{id}\mathbf{v}_d - \frac{1}{c}\mathbf{J}_d \times \mathbf{B}, \tag{27}$$

where we have used the quasi-neutrality condition

$$n_i = n_e + Z_d n_d, \tag{28}$$

and Ampéré's law

$$\nabla \times \mathbf{B} = \frac{4\pi}{c}\mathbf{J}. \tag{29}$$

Here $\mathbf{J} = e(n_i\mathbf{v}_i - n_e\mathbf{v}_e - Z_d n_d\mathbf{v}_d) \equiv \mathbf{J}_i + \mathbf{J}_e + \mathbf{J}_d$ is sum of the ion, electron, and dust current densities, Z_d is as before the number of charges residing on the dust particle surface, and $\mathbf{v}_d$ is the dust particle velocity. It turns out that electrostatic and collisional interactions between ions, electrons and dust particles are non-trivial, as the ion velocity is now controlled by numerous forces in the right-hand side of eq. (27). The radial and azimuthal components of the ion velocity are determined from

$$v_{ir} = \frac{Z_d n_d}{n_i}\frac{eE_r}{m_i\nu_{in}} - \frac{1}{\rho_i\nu_{in}}\frac{\partial P}{\partial r} + \frac{\nu_{id}}{\nu_{dn}}v_{dr} + \frac{Z_d n_d}{n_i}\frac{\omega_{ci}}{\nu_{in}}v_{d\theta}, \tag{30}$$

and

$$v_{i\theta} = \frac{\nu_{id}}{\nu_{in}} v_{d\theta} - \frac{Z_d n_d}{n_i} \frac{\omega_{ci}}{\nu_{in}} v_{dr}, \tag{31}$$

where $P = p + B_0^2/8\pi$ is the sum of the plasma and magnetic field pressures, and $\omega_{ci} = eB_0/m_i c$ is the ion gyrofrequency. The expressions (30) and (31) show that the radial and azimuthal motions of ions are driven by forces involving the radial electric field, gradients of the plasma and magnetic pressures, the ion-dust drag, and the coupling of the dust current density with the external magnetic field. These expressions differ significantly from eqs. (3a) and (3b) of Ref. [Kaw *et al.* 2002], since the latter does not include the coupling of ions with electrons and charged dust particles in the ion momentum equation.

In a partially ionized plasma, the dust-neutral collision frequency is much larger than the dust gyrofrequency $\omega_{cd} = Z_d e B_0/m_d c$, where m_d is the dust mass. Hence, the steady state equation of motion for the dust cloud is

$$0 = -Z_d e \mathbf{E} - m_d \nu_{dn} \mathbf{v}_d - m_d \nu_{di} (\mathbf{v}_d - \mathbf{v}_i), \tag{32}$$

where the dust particle is assumed to be cold. Since $\nu_{di} \ll \nu_{dn}$, we obtain from eq. (32)

$$v_{dr} = -\frac{Z_d e}{m_d \nu_{dn}} E_r + \frac{\nu_{di}}{\nu_{dn}} v_{ir}, \tag{33}$$

and

$$v_{d\theta} = \frac{\nu_{di}}{\nu_{dn}} v_{i\theta}. \tag{34}$$

Equations (33) and (34) reveal that the radial motion of dust clouds is caused by the radial electric force and the dust-ion drag force, while the azimuthal dust rotation arises from dust-ion collisions in the presence of the azimuthal ion speed that is driven by the $\mathbf{J}_d \times \mathbf{B}$ force. Thus, the physics of the azimuthal dust rotation is quite different from what has been proposed in Ref. [Kaw *et al.* 2002].

Combining eqs. (30), (31), (33) and (34), we readily obtain

$$v_{dr} \approx \left(\frac{n_d}{n_i} \frac{\nu_{di}}{\nu_{in}} - \frac{m_i}{m_d} \right) \frac{Z_d e E_r}{m_i \nu_{dn}} - \frac{\nu_{di}}{\nu_{dn}} \frac{1}{\rho_i \nu_{in}} \frac{\partial P}{\partial r} + \frac{Z_d n_d}{n_i} \frac{\nu_{di}}{\nu_{dn}} \frac{\omega_{ci}}{\nu_{in}} v_{d\theta}, \tag{35}$$

and

$$v_{d\theta} = -\frac{Z_d \alpha}{1+\alpha^2} \left(\frac{n_d}{n_i} \frac{\nu_{di}}{\nu_{in}} - \frac{m_i}{m_d} \right) \frac{e E_r}{m_i \nu_{dn}} + \frac{\nu_{di}}{\nu_{dn}(1+\alpha^2)} \frac{\alpha}{\rho_i \nu_{in}} \frac{\partial P}{\partial r}, \tag{36}$$

where

$$\alpha = \frac{Z_d n_d}{n_i} \frac{\nu_{di}}{\nu_{dn}} \frac{\omega_{ci}}{\nu_{in}}. \tag{37}$$

Since $\nu_{di} \gg (m_i/m_d)\nu_{dn}$, the azimuthal dust speed in the absence of the pressure and magnetic field gradients takes the form

$$v_{d\theta} = \frac{\alpha}{1+\alpha^2} \frac{Z_d n_d}{n_i} \frac{\nu_{di}}{\nu_{dn}} \frac{e}{m_i \nu_{in}} \frac{\partial \phi}{\partial r}, \tag{38}$$

where $E_r = -\partial\phi/\partial r$ and ϕ is the sheath potential. We see from eq. (38) that the azimuthal dust cloud speed increases with the increase of B_0 and attains a maximum value at $\omega_{ci}/\nu_{in} \sim (1/\sqrt{2}) n_i \nu_{dn}/Z_d n_d \nu_{di}$. Furthermore, the dust rotation speed $v_{d\theta}$ is inversely proportional to the ion number density and the dust-neutral collision frequency. For rigid dust cloud rotation we can set $v_\theta \sim r\omega$, and determine the constant angular frequency ω of the dust cloud.

4 Dispersion Properties of a Plasma Containing Irregular Grains

To construct the consistent wave electrodynamics of a dusty plasma, we start with the derivation of the dust charge and dust current densities, as well as a kinetic equation for charged dust grains. The effects of the dipole moment and the principal moment of inertia of elongated and rotating dust grains are self-consistently incorporated. The newly derived dust charge and dust current densities, together with the corresponding quantities for electrons and ions, are combined with the Maxwell-Vlasov system of equations to obtain dielectric response functions for a magnetized dusty plasma [Tskhakaya *et al.* 2001]. For a cold dust gas, we have obtained explicit expressions for the permittivities associated with the dust grain rotation, as well as for those of the ambient plasmas species. The expressions for the unmagnetized plasma can be obtained by transition to the corresponding limit. The dispersion relations for transverse and longitudinal electromagnetic waves are then worked out. Our analytical results exhibit instabilities of the electron whistler, the dust whistler, the Alfvén waves, electromagnetic ion and dust cyclotron waves, as well as lower-hybrid, electrostatic ion-cyclotron, and coupled dust acoustic and dust cyclotron waves. In the unmagnetized plasma it is shown the instability of well known transversal and longitudinal waves and of dust ion-acoustic waves (DIAW) as well [Mahmoodi *et al.* 2000; Tskhakaya *et al.* 2001; Shukla and Tskhakaya 2001]. The instability arises due to the resonance interaction between waves and elongated rotating dust grains. The energy of the dust rotation can flow into low-frequency dusty plasma oscillations, driving them at non-thermal levels. However, the instability of longitudinal waves occurs only in the case when the wave vector lies in the plane of the dust grain rotation. In this case there exists a coupling between the longitudinal electric field and charge that is placed on the dust grain's surface and rotates together with the grain.

4.1 Derivation of Charge and Current Densities for the Dust Grains

We consider a multi-component dusty plasma in an external magnetic field $\hat{\mathbf{z}}B_0$, where $\hat{\mathbf{z}}$ is the unit vector along the z axis. The plasma constituents are electrons, ions, and negatively charged non-spherical, rotating dust grains. The dust sizes are much smaller than the characteristic scale sizes of the inhomogeneities (viz. wavelength of disturbances in our system). To construct the electrodynamics of charged dust grains in a magnetized dusty plasma, we have to obtain appropriate expressions for the charge and current densities of dust grains through a dust grain distribution function, taking into account the size of the dust grain. On the other hand, charge and current densities for electrons and ions assume the standard expressions.

For our purposes, we assume again that the charged dust grains are a system of discrete parts [Landau and Lifshitz 1971]. The charge micro-density of the grains is represented as

$$\rho_m = \sum_i \left[\sum_j dq_i(\mathbf{r}_j)\delta(\mathbf{r} - \mathbf{r}_j) \right], \tag{39}$$

where the first summation over i is taken over different grains, while the second over j is taken over different parts of the i-th grain. Here, $dq_i(\mathbf{r}_j)$ is the charge of the j-th part of the i-th grain, and $\delta(\mathbf{r} - \mathbf{r}_j)$ is the usual Dirac function. If there is a continuous charge distribution onto the grain, then the summation over j can be changed by an integral over the grain's volume, and introduced the density of the charge onto the grain. Hence, we have

$$\rho_m = \sum_i \int_{V_i(\mathbf{R}_i)} \bar{\rho}_i(\mathbf{r}' - \mathbf{R}_i, \mathbf{R}_i)\delta(\mathbf{r} - \mathbf{r}')d\mathbf{r}', \tag{40}$$

where $\mathbf{R}_i$ is the radius vector of the centre of mass of the grain, and integral is taken over the volume $V_i(\mathbf{R}_i)$ of the grain. In eq. (40), we introduced the density of the charge distribution onto the grain

$$dq_i(\mathbf{r}) = \frac{dq_i(\mathbf{r})}{d\mathbf{r}}d\mathbf{r} \equiv \bar{\rho}_i(\mathbf{r} - \mathbf{R}_i, \mathbf{R}_i)d\mathbf{r}. \tag{41}$$

For a point grain charge, we have

$$\bar{\rho}_i(\mathbf{r} - \mathbf{R}_i, \mathbf{R}_i) = q_i\delta(\mathbf{r} - \mathbf{R}_i), \tag{42}$$

which leads to the usual expression for the charge micro-density of the grain

$$\rho_m = \sum_i q_i\delta(\mathbf{r} - \mathbf{R}_i), \tag{43}$$

where q_i is the total charge of the i-th grain.

For the statistical description of a dust grain gas, we have to introduce the probability density D for the grain's gas state [Bogoliubov 1946; Klimontovich 1982]. If all grains are identical, we have then

$$D = D(\mathbf{R}_1, \mathbf{v}_1, \Omega_1, \theta_1, \psi_1, \varphi_1; \mathbf{R}_2, \mathbf{v}_2, \Omega_2, \theta_2, \psi_2, \varphi_2; ...., t), \tag{44}$$

where $\mathbf{v}_i$ is the velocity of the centre of mass, Ω_i is the angular velocity of the i-th grain, and θ_i, ψ_i, and φ_i (Euler angles) describe the orientation of elongated grains. For the averaged charge density of the grain, we can then write

$$\rho_d(\mathbf{r}, t) = \int d\Gamma_1, d\Gamma_2 ... d\Gamma_N D \rho_m, \tag{45}$$

where N is the total number of the grains, and $d\Gamma_i = d\mathbf{R}_i d\mathbf{v}_i d\Omega_i d\theta_i d\psi_i d\varphi_i$.

Introducing the one particle distribution function for the dust grain

$$f_d(\mathbf{R}_1, \mathbf{v}_1, \Omega_1, \theta_1, \psi_1, \varphi_1, t) = N \int d\Gamma_2 d\Gamma_3 ... d\Gamma_N D, \tag{46}$$

we can write the charge density of the grains in the following form

$$\rho_d(\mathbf{r}, t) = \int d\Gamma_1 \int_{V_1} \bar{\rho}_1(\mathbf{r}'') \delta(\mathbf{r} - \mathbf{R}_1 - \mathbf{r}'') f_d(\mathbf{R}_1, \mathbf{v}_1, \Omega_1, \theta_1, \psi_1, \varphi_1, t) d\mathbf{r}''. \tag{47}$$

In the following, we shall omit the subscript 1, and consider one dimensional grain rotation so that the angular velocity is along the external magnetic field direction, viz. $\Omega = (0, 0, \Omega)$. Equation (47) can be represented as

$$\rho_d(\mathbf{r}, t) = \int d\Gamma \hat{\rho}(\mathbf{r} - \mathbf{R}, \varphi) f_d(\mathbf{R}, \mathbf{v}, \Omega, \varphi, t), \tag{48}$$

where the integrand

$$\hat{\rho}(\mathbf{r} - \mathbf{R}, \varphi) = \int_V d\mathbf{r}' \bar{\rho}(\mathbf{r}') \delta(\mathbf{r} - \mathbf{R} - \mathbf{r}'), \tag{49}$$

describing the charge distribution onto a single grain, depends on the shape of the grain and the azimuthal orientation of the grain's elongation axis. Outside of the grain's volume, we have $\hat{\rho} = 0$. For the identical grains, we can partly determine the dependence of $\hat{\rho}$ on the azimuthal angle φ. Every given direction of the grain elongation axis, determined by the angle φ, can be considered as a final position of the turning of the axis (and simultaneously of the whole grain) from the direction with $\varphi = 0$. This allows us to write

$$\hat{\rho}(\mathbf{r} - \mathbf{R}, \varphi) = \hat{\rho}[\overleftrightarrow{\mathbf{F}}(\varphi)(\mathbf{r} - \mathbf{R}, 0)] \equiv \hat{\rho}[\overleftrightarrow{\mathbf{F}}(\varphi)(\mathbf{r} - \mathbf{R})], \tag{50}$$

where $\overleftrightarrow{\mathbf{F}}(\varphi)$ is the tensor of turning by the angle φ, viz.

$$\overleftrightarrow{\mathbf{F}} = F_{ij}(\varphi) = \begin{pmatrix} \cos\varphi & -\sin\varphi \\ \sin\varphi & \cos\varphi \end{pmatrix}. \tag{51}$$

In the dipole approximation, when the dust grain size a is much smaller than the scale length of the plasma inhomogeneity l, viz.

$$a \ll l, \tag{52}$$

we substitute eq. (50) into eq. (48) and expand the distribution function f_d around the point $\mathbf{r}$, and obtain for the grain's charge density

$$\rho_d(\mathbf{r},t) = \int (q - \mathbf{d}\cdot\nabla) f_d(\mathbf{r},\mathbf{v},\Omega,\varphi,t) d\Lambda, \tag{53}$$

where $d\Lambda = d\mathbf{v}d\Omega d\varphi$, and

$$q = \int d\mathbf{r}\hat{\rho}(\mathbf{r}), \tag{54}$$

and

$$\mathbf{d} = \overleftrightarrow{\mathbf{F}}^{-1}(\varphi) \int d\mathbf{r}\mathbf{r}\hat{\rho}(\mathbf{r}) \tag{55}$$

are the total charge and the dipole moment of the dust grain, respectively. $\overleftrightarrow{\mathbf{F}}^{-1}$ is the inverse of the tensor $\overleftrightarrow{\mathbf{F}}(\varphi)$.

Quite analogous calculations lead to the following expression for the dust current density

$$\mathbf{J}_d(\mathbf{r},t) = \int d\Lambda \left[\mathbf{v}\left(q - \mathbf{d}\cdot\nabla\right) + \Omega\times\mathbf{d}\right] f_d(\mathbf{r},\mathbf{v},\Omega,\varphi,t). \tag{56}$$

The first term on the right-hand side of eq. (56) describes the transfer of the charge [eq. (53)] and the second term describes the current arising from the dust grain rotation. Below we show that eqs. (53) and (56) are connected with the continuity equation, as it should be.

The forces acting on the dust grain are found in Section 2. Then the kinetic equation for the dust grains can be written as

$$\frac{\partial f_d}{\partial t} + \mathbf{v}\cdot\frac{\partial f_d}{\partial \mathbf{r}} + \Omega\cdot\frac{\partial f_d}{\partial \varphi} + \frac{d\mathbf{p}}{dt}\cdot\frac{\partial f_d}{\partial \mathbf{p}} + \frac{d\mathbf{M}}{dt}\cdot\frac{\partial f_d}{\partial \mathbf{M}} = 0, \tag{57}$$

where the forces $d\mathbf{p}/dt$ and $d\mathbf{M}/dt$ are defined by eq. (5) and (6). Furthermore we neglect the gravitational force in eq. (5). The kinetic equation (57) and the definitions (53) and (56) imply that the dust grain charge and current densities satisfy the continuity equation

$$\frac{\partial \rho_d}{\partial t} + \frac{\partial \mathbf{J}_d}{\partial \mathbf{r}} = 0. \tag{58}$$

Using the expressions for ρ_d and $\mathbf{J}_d$ we can construct the kinetics and electrodynamics of a dusty plasma with elongated and rotating dust grains. In what follows, we consider the wave dynamics of such a magnetized dusty plasma.

4.2 Dielectric Permittivity

We assume that the dust grain size is much smaller than the grain gyroradius and that the dust grain thermal velocity is smaller than the characteristic velocity of our problem. Under these conditions, taken together with eq. (52), the equations of motion (5) and (6) can be simplified. For simplicity, we furthermore consider the one-dimensional case of the dust grain rotation, so that $\mathbf{M} = (0, 0, M)$, where $M = I\Omega$, and I is the z component of the principal moment of inertia. The kinetic equation for the dust grain (57) then assumes the form

$$\frac{\partial f_d}{\partial t} + \mathbf{v} \cdot \frac{\partial f_d}{\partial \mathbf{r}} + \Omega \frac{\partial f_d}{\partial \varphi} + (\mathbf{d} \times \mathbf{E})_z \frac{\partial f_d}{\partial M} + q_d \left(\mathbf{E} + \frac{1}{c} \mathbf{v} \times \mathbf{B}_0 \right) \cdot \frac{\partial f_d}{\partial \mathbf{p}} = 0. \tag{59}$$

For electrons and ions, we have the well known kinetic equation

$$\frac{\partial f_\alpha}{\partial t} + \mathbf{v} \cdot \frac{\partial f_\alpha}{\partial \mathbf{r}} + e_\alpha \left[\mathbf{E} + \frac{1}{c} \mathbf{v} \times (\mathbf{B}_0 + \mathbf{B}) \right] \cdot \frac{\partial f_\alpha}{\partial \mathbf{p}} = 0, \tag{60}$$

where α equals e for electrons and i for ions, and e_α is the charge of the specie α.

Assuming that the electric and magnetic field perturbations are small, we can express the perturbed distribution function as

$$\delta f_d = f_d - f_{d0} << f_{d0} \quad \text{and} \quad \delta f_\alpha = f_\alpha - f \ll f_\alpha.$$

The equilibrium distribution functions are [Landau and Lifshitz 1989]

$$f_{d0} = \frac{n_{d0}}{2\pi(2\pi m_d T_d)^{3/2}} \frac{1}{(2\pi I T_d)^{1/2}} \exp\left[- \left(p^2/2m_d T_d \right) - \left(M - M_0 \right)^2 /2IT_d \right], \tag{61}$$

and

$$f_{\alpha 0} = \frac{n_{\alpha 0}}{(2\pi m_\alpha T_\alpha)^{3/2}} \exp - \left(p^2/2m_\alpha T_\alpha \right), \tag{62}$$

where $n_{\beta 0}$ and T_β ($\beta = e, i, d$) are the unperturbed number density and the temperature of the specie β. We assumed that the dust grains rotate with a preferred angular velocity Ω_0 so that $M_0 = I\Omega_0$.

The components of the dust dipole moment are

$$d_x = d\cos\varphi, \quad d_y = d\sin\varphi. \tag{63}$$

Thus, the perturbed dust grain distribution function is represented as

$$\delta f_d = \sum_{n=-\infty}^{\infty} \delta f_n \exp(in\varphi). \tag{64}$$

Hence eqs. (59) and (60) give [Alexandrov *et al.* 1984]

$$\begin{aligned} &\frac{\partial \delta f_n}{\partial t} + \mathbf{v}\cdot\frac{\partial \delta f_n}{\partial \mathbf{r}} + in\Omega\delta f_n - \omega_{cd}\frac{\partial \delta f_n}{\partial \psi} \\ = &-q\mathbf{E}\cdot\frac{\partial f_{d0}}{\partial \mathbf{p}}\Delta(n) - \frac{i}{2}\frac{\partial f_{d0}}{\partial M} d[(E_x - iE_y)\Delta(n-1) \\ &-(E_x + iE_y)\Delta(n+1)], \end{aligned} \tag{65}$$

and

$$\frac{\partial \delta f_\alpha}{\partial t} + \mathbf{v}\cdot\frac{\partial \delta f_\alpha}{\partial \mathbf{r}} - \omega_{c\alpha}\frac{\partial \delta f_\alpha}{\partial \psi} = -e_\alpha \mathbf{E}\cdot\frac{\partial f_{\alpha 0}}{\partial \mathbf{p}}, \tag{66}$$

where

$$\omega_{cd} = q_d B_0/m_d c \quad \text{and} \quad \omega_{c\alpha} = e_\alpha B_0/m_\alpha c$$

are the cyclotron frequencies of the dust grain and the specie α, respectively. Furthermore, $\Delta(n)$ equals 1 for $n = 0$ and 0 for $n \neq 0$. The symbol ψ is the azimuthal angle in the momentum space [Alexandrov *et al.* 1984],

$$p_x = p_\perp \cos\psi \quad \text{and} \quad p_y = p_\perp \sin\psi.$$

According to eq. (63) only $n = 0, \pm 1$ give the contribution in the summation of eq. (64).

Assuming that the perturbed quantities are proportional to $\exp(-i\omega t + i\mathbf{k}\cdot\mathbf{r})$, where ω and $\mathbf{k}$ are the frequency and the wave vector, respectively, we obtain [Alexandrov *et al.* 1984] the following solutions of eqs. (65) and (66)

$$\delta f_0 = \frac{q_d \mathbf{E}\cdot}{\omega_{cd}} \int_{\pm\infty}^{\psi} d\psi' \frac{\partial f_{d0}}{\partial \mathbf{p}} \exp\left\{-i\int_{\psi'}^{\psi} \left[\left(\omega - \mathbf{k}\cdot\mathbf{v}(\psi'')\right)/\omega_{cd}\right] d\psi''\right\}, \tag{67}$$

$$\begin{aligned} \delta f_{\pm 1} &= \pm\frac{i}{2}\frac{d(E_x \mp iE_y)}{\omega_{cd}} \int_{\pm\infty}^{\psi} d\psi' \frac{\partial f_{d0}}{\partial M}| \\ \times \exp&\left\{-i\int_{\psi'}^{\psi} \left[\left(\omega \mp \Omega - \mathbf{k}\cdot\mathbf{v}(\psi'')\right)/\omega_{cd}\right] d\psi''\right\}, \end{aligned} \tag{68}$$

$$\delta f_\alpha = \frac{e_\alpha \mathbf{E}\cdot}{\omega_{c\alpha}} \int_{\pm\infty}^{\psi} d\psi' \frac{\partial f_{\alpha 0}}{\partial \mathbf{p}} \exp\left\{-i \int_{\psi'}^{\psi} \left[\left(\omega - \mathbf{k}\cdot\mathbf{v}(\psi'')\right)/\omega_{c\alpha}\right] d\psi''\right\}. \quad (69)$$

Substituting eqs. (67) and (68) into eq. (56) as well as eq. (69) into the expression for the electron and ion current densities, viz.

$$\mathbf{J}_\alpha = e_\alpha \int d\mathbf{p}\mathbf{v} f_\alpha, \quad (70)$$

we obtain for the total current density

$$J_i = \left[\sigma_{ij}^r(\omega,\mathbf{k}) + \sum_{\beta=e,i,d} \sigma_{ij}^\beta(\omega,\mathbf{k})\right] E_j, \quad (71)$$

where the first term on the right-hand side is connected with the rotational motion of the dust grain, while the second term represents the contributions of the electrons and ions including the center of mass motion of the grains. We have denoted

$$\sigma_{ij}^r(\omega,\mathbf{k}) = \begin{pmatrix} \sigma_{xx}^r & \sigma_{xy}^r & 0 \\ -\sigma_{xy}^r & \sigma_{yy}^r & 0 \\ 0 & 0 & 0 \end{pmatrix}, \quad (72)$$

where

$$\sigma_{xx}^r = \sigma_{yy}^r = i\frac{n_{d0}d^2}{4I}\frac{1}{KV_{td}} \sum_{n=-\infty}^{\infty} \exp(-z_d) I_n(z_d) \left(\Phi_-^n + \Phi_+^n\right), \quad (73)$$

$$\sigma_{xy}^r = \frac{n_{d0}d^2}{4I}\frac{1}{KV_{td}} \sum_{n=-\infty}^{\infty} \exp(-z_d) I_n(z_d) \left(\Phi_-^n - \Phi_+^n\right), \quad (74)$$

with

$$\Phi_\pm^n = \frac{k_z^2}{K^2}\frac{KV_{td}}{\omega \pm \Omega_0 - n\omega_{cd}} J_+\left(\frac{\omega \pm \Omega_0 - n\omega_{cd}}{KV_{td}}\right)$$

$$-\left(\frac{\kappa^2}{K^2}\frac{\omega - n\omega_{cd}}{KV_{td}} \mp \frac{k_z^2}{K^2}\frac{\Omega_0}{KV_{td}}\right)\left[1 + J_+\left(\frac{\omega \pm \Omega_0 - n\omega_{cd}}{KV_{td}}\right)\right]. \quad (75)$$

Here,

$$V_{td} = (T_d/m_d)^{1/2}$$

is the grain's thermal velocity,

$$z_\beta = k_\perp^2 V_{t\beta}^2/\omega_{c\beta}^2, \quad K = (k^2+\kappa^2)^{1/2}, \quad \kappa = (m_d/I)^{1/2},$$

$I_n(z)$ is the Bessel function of an imaginary argument, and the function

$$J_+(x) = (x/\sqrt{2\pi})\int_{-\infty}^{\infty} dt \exp(-t^2/2)(x-t)^{-1}$$

has the following asymptotics

$$J_+(x) = 1 + \frac{1}{x^2} + \frac{3}{x^2} + \ldots - i\sqrt{\frac{\pi}{2}}x\exp(-x^2/2) \tag{76}$$

for $|x| \gg 1$, $|\mathrm{Re}x| \gg |\mathrm{Im}x|$, $\mathrm{Im}x < 0$; and

$$J_+(x) \approx -i\sqrt{\frac{\pi}{2}}x \tag{77}$$

for $|x| \ll 1$.

For the tensor $\sigma_{ij}^\beta(\omega, \mathbf{k})$, we have

$$\sigma_{ij}^\beta(\omega,\mathbf{k}) = \frac{e_\beta^2}{\omega_{c\beta}}\int d\mathbf{p}v_i(\psi)\int_{\pm\infty}^{\psi} d\psi' \left(\frac{\partial f_{\beta 0}}{\partial p_j}\right)_{\psi'}$$
$$\times \exp\left\{i\int_{\psi'}^{\psi}\left[\left(\omega - \mathbf{k}\cdot\mathbf{v}(\psi'')\right)/\omega_{c\beta}\right]d\psi''\right\}. \tag{78}$$

Here, e_β is q_d for $\beta = d$.

Quite straightforward calculations lead to the following expressions for the tensor of the dielectric permittivity [Alexandrov *et al.* 1984]. We have

$$\epsilon_{ij}(\omega,\mathbf{k}) = \hat{\epsilon}_{ij}(\omega,\mathbf{k}) + \epsilon_{ij}^r(\omega,\mathbf{k}), \tag{79}$$

where

$$\hat{\epsilon}_{xx} = 1 - \sum_\beta\sum_n \frac{n^2\omega_{p\beta}^2}{\omega(\omega - n\omega_{c\beta})}\frac{A_n(z_\beta)}{z_\beta}J_+(\xi_n), \tag{80}$$

$$\hat{\epsilon}_{yy} = \hat{\epsilon}_{xx} + 2\sum_\beta\sum_n \frac{\omega_{p\beta}^2 z_\beta}{\omega(\omega - n\omega_{c\beta})}A_n'(z_\beta)J_+(\xi_n), \tag{81}$$

$$\hat{\epsilon}_{xy} = -\hat{\epsilon}_{yx} = -i \sum_{\beta} \sum_{n} \frac{n^2 \omega_{p\beta}^2}{\omega(\omega - n\omega_{c\beta})} A_n'(z_\beta) J_+(\xi_n), \tag{82}$$

$$\hat{\epsilon}_{xz} = \hat{\epsilon}_{zx} = \sum_{\beta} \sum_{n} \frac{n\omega_{p\beta}^2 k_\perp}{\omega \omega_{c\beta} k_z} \frac{A_n(z_\beta)}{z_\beta} \left[1 - J_+(\xi_n)\right], \tag{83}$$

$$\hat{\epsilon}_{yz} = -\hat{\epsilon}_{zy} = -i \sum_{\beta} \sum_{n} \frac{\omega_{p\beta}^2 k_\perp}{\omega \omega_{c\beta} k_z} A_n'(z_\beta) \left[1 - J_+(\xi_n)\right], \tag{84}$$

$$\hat{\epsilon}_{zz} = 1 + \sum_{\beta} \sum_{n} \frac{\omega_{p\beta}^2 (\omega - n\omega_{c\beta})}{\omega k_z^2 V_{t\beta}^2} A_n(z_\beta) \left[1 - J_+(\xi_n)\right], \tag{85}$$

and

$$\epsilon_{ij}^r(\omega, \mathbf{k}) = \frac{4\pi i}{\omega} \sigma_{ij}^r(\omega, \mathbf{k}). \tag{86}$$

Here,

$$A_n(z) = \exp(-z) I_n(z), \qquad \xi_n = (\omega - n\omega_{c\beta})/|k_z| V_{t\beta},$$

$$\omega_{p\beta} = (4\pi e_\beta^2 n_{\beta 0}/m_\beta)^{1/2}$$

is the plasma frequency of the specie β, and $\mathbf{k} \equiv (k_\perp, 0, k_z)$

For $k_\perp^2 V_{td}^2 << \omega_{cd}^2$ and $|\omega \pm \Omega_0| >> K V_{td}$, the dust grains are assumed to be cold and the rotational part of the dielectric tensor (86) is given by

$$\epsilon_{ij}^r = \begin{pmatrix} \epsilon_\perp^r & ig^r & 0 \\ -ig^r & \epsilon_\perp^r & 0 \\ 0 & 0 & 0 \end{pmatrix}, \tag{87}$$

where

$$\epsilon_\perp^r = -\frac{\Omega_r^2}{(\omega - \Omega_0)^2} - \frac{\Omega_r^2}{(\omega + \Omega_0)^2} \tag{88}$$

and

$$g^r = \frac{\Omega_r^2}{(\omega - \Omega_0)^2} - \frac{\Omega_r^2}{(\omega + \Omega_0)^2}. \tag{89}$$

We note that this involves a new characteristic frequency

$$\Omega_r = (4\pi n_{d0} d^2/4I)^{1/2}, \tag{90}$$

for the dust grains that have a non-zero dipole moment. This frequency is of the same order as the dust plasma frequency ω_{pd}.

4.3 Dispersion Properties

The detailed analysis of the dispersion relation

$$\left| k^2 \delta_{ij} - k_i k_j - \frac{\omega^2}{c^2} \epsilon_{ij}(\omega, \mathbf{k}) \right| = 0, \tag{91}$$

for waves in a magnetized dusty plasma is rather complex, since the number of wave branches is numerous. Here, we present the dispersion properties of some most interesting modes, by pointing out the underlying approximations that are required for their existence. We first consider waves that are propagating along $\hat{\mathbf{z}}B_0$. For waves in a cold dusty plasma with

$$k_\perp V_{t\alpha} \ll \omega_{c\alpha}, \; |k_z| V_{t\alpha} \ll \omega \quad \text{and} \quad |\omega \pm n\omega_{c\alpha}| \gg |k_z| V_{t\alpha}, \tag{92}$$

we have

$$\epsilon_{xx} = \epsilon_{yy} = \epsilon_\perp = 1 - \sum_\beta \frac{\omega_{p\beta}^2}{\omega^2 - \omega_{c\beta}^2} - \frac{\omega_r^2}{(\omega - \Omega_0)^2} - \frac{\omega_r^2}{(\omega + \Omega_0)^2}, \tag{93}$$

$$\epsilon_{xy} = -\epsilon_{yx} = ig = -i \sum_\beta \frac{\omega_{p\beta}^2 \omega_{c\beta}}{\omega(\omega^2 - \omega_{c\beta}^2)} + i \frac{\Omega_r^2}{(\omega - \Omega_0)^2} - i \frac{\Omega_r^2}{(\omega + \Omega_0)^2}, \tag{94}$$

$$\epsilon_{zz} = \epsilon_\parallel = 1 - \sum_\beta \frac{\omega_{p\beta}^2}{\omega^2}, \tag{95}$$

and

$$\epsilon_{xz} = \epsilon_{zx} = \epsilon_{yz} = \epsilon_{zy} = 0. \tag{96}$$

The components of the electric field are given by a set of equations

$$\left(k_z^2 - \frac{\omega^2}{c^2} \right) E_x - i \frac{\omega^2}{c^2} g E_y - k_\perp k_z E_z = 0, \tag{97}$$

$$\frac{\omega^2}{c^2} g E_x + \left(k^2 - \frac{\omega^2}{c^2} \epsilon_\perp \right) E_y = 0, \tag{98}$$

and

$$-k_\perp k_z E_x + \left(k_\perp^2 - \frac{\omega^2}{c^2} \epsilon_\parallel \right) E_z = 0. \tag{99}$$

We note that for $k_\perp = 0$ (i.e. $\mathbf{k} = \hat{\mathbf{z}}k_z$), if $E_z \neq 0$, we have $\epsilon_\parallel = 0$, which exhibits that there is no influence of the dust grain rotation on the longitudinal waves. Obviously, the dust grain rotation can act on the waves when the electric field is situated in the plane of the rotation. The energy exchange between the dust grain rotation and such a wave is most efficient when the frequency of the rotation is close to the wave frequency.

For the circularly polarized electromagnetic waves, we have

$$\frac{k^2c^2}{\omega^2} = 1 - \sum_\beta \frac{\omega_{p\beta}^2}{\omega(\omega \mp \omega_{c\beta})} - \frac{2\Omega_r^2}{(\omega \pm \Omega_0)^2}, \tag{100}$$

where the $\pm$ in the denominators corresponds to the left/right-hand circularly polarized waves. By replacing Ω_0 by $-\Omega_0$, the dust grain rotation direction can be manipulated in order to coincide with the direction of the wave polarization.

The dispersion relation (100) can be written as

$$\frac{k^2c^2}{\omega^2} = \epsilon(\omega) - \frac{2\Omega_r^2}{(\omega - \Omega_0)^2}, \tag{101}$$

where

$$\epsilon(\omega) = 1 - \sum_\beta \frac{\omega_{p\beta}^2}{\omega(\omega + \omega_{c\beta})}. \tag{102}$$

Introducing a small frequency shift Δ around Ω_0, we let $\omega = \Omega_0 + \Delta$, where $\Delta \ll \Omega_0$, and express (101) as

$$\frac{k^2c^2}{\Omega_0^2} - \epsilon(\Omega_0) + \Delta\frac{\partial}{\partial\Omega_0}\left[\frac{k^2c^2}{\Omega_0^2} - \epsilon(\Omega_0)\right] = -\frac{2\Omega_r^2}{\Delta^2}. \tag{103}$$

If Ω_0 is far away from a characteristic frequency ω_0 of a magnetized dusty plasma, which satisfies

$$H(\omega_0) = \frac{k^2c^2}{\omega^2} - \epsilon(\omega_0) = 0, \tag{104}$$

so that the condition

$$\left|\frac{H(\Omega_0)}{\Omega_0\,(dH(\Omega_0)/d\Omega_0)}\right| \gg \frac{\Delta}{\Omega_0}, \tag{105}$$

is fulfilled (this case is referred to as a non-resonant case), we obtain

$$\Delta = \pm i\sqrt{2}\frac{\Omega_r}{kc}\Omega_0\left[1 + \frac{\Omega_0^2}{k^2c^2}\epsilon(\Omega_0)\right], \tag{106}$$

where we have assumed that $\Omega_0^2 \ll k^2c^2$. Equation (106) depicts a new type of unstable transversal waves whose frequency is close to the rotation frequency Ω_0. In the resonance case, when the inequality in eq. (105) is reversed, Ω_0 is close to some characteristic frequency of a magnetized dusty plasma, viz.

$$H(\Omega_0) = 0, \tag{107}$$

we obtain for the frequency shift

$$\Delta = \left[-\frac{2\Omega_r^2}{\Omega_0^3 \partial H(\Omega_0)/\partial\Omega_0}\right]^{1/3} \Omega_0 \begin{pmatrix} 1 \\ -\frac{1\pm i\sqrt{3}}{2} \end{pmatrix}. \tag{108}$$

Equation (108) exhibits an unstable root with a substantial growth rate, which is proportional to $\Omega_r^{2/3}$. This was expected as the dispersion relation (101) is formally similar to that dispersion relation for a two-stream instability which has been discussed in [Mikhailovskii 1974].

We now present several examples of the magnetized dusty plasma wave spectra for the resonance case. As Ω_0 is small in most of the astrophysical and terrestrial environments, we consider low-frequency regimes of the plasma oscillations.

For $|\omega_{cd}|, \omega_{ci} \ll \omega \ll |\omega_{ce}|$, we have

$$H(\omega) = \frac{k^2c^2}{\omega^2} - \frac{\omega_{pe}^2}{\omega|\omega_{ce}|}. \tag{109}$$

Setting

$$\omega = \Omega_0 + i\gamma,$$

where

$$\Omega_0 \simeq \omega_0 = k^2c^2|\omega_{ce}|/\omega_{pe}^2$$

(the electron whistler waves), we obtain for the growth rate

$$\gamma \simeq \Omega_0 \left[2\frac{\Omega_r^2}{k^2c^2}\right]^{1/3}. \tag{110}$$

In the frequency regimes, $|\omega_{cd}| \ll \omega \ll \omega_{ci}$, we have

$$H(\omega) = \frac{k^2c^2}{\omega^2} - \frac{\omega_{pd}^2}{\omega|\omega_{cd}|}. \tag{111}$$

In deriving (111) we have used the dusty plasma quasi-neutrality condition at equilibrium, viz.

$$|e|n_{e0} + |q_d|n_{d0} = e_i n_{i0}. \tag{112}$$

Setting

$$\omega = \Omega_0 + i\gamma,$$

where

$$\Omega_0 \simeq \omega_0 = k^2c^2|\omega_{cd}|/\omega_{pd}^2$$

(the dust whistler wave) [Shukla 1992; Rosenberg and Mendis 1992; Mendis and Rosenberg 1994; Shukla and Rahman 1996; Shukla 1999], we obtain for the growth rate

$$\gamma \simeq \left(2\frac{\Omega_r^2}{k^2c^2}\right)^{1/3} \Omega_0. \tag{113}$$

On the other hand, for $\omega \sim \omega_{cd}$, we have

$$H(\omega) = \frac{k^2c^2}{\omega^2} - \frac{\omega_{pd}^2}{\omega(|\omega_{cd}| - \omega)}. \tag{114}$$

For this case, setting

$$\omega = \Omega_0 + i\gamma,$$

where

$$\Omega_0 = \omega_0 = |\omega_{cd}|\left(1 - \omega_{pd}^2/k^2c^2\right)$$

(the electromagnetic dust cyclotron wave), we obtain for the growth rate

$$\gamma \simeq \left(2\frac{\Omega_r^2}{k^2c^2}\frac{\omega_{pd}^2}{k^2c^2}\right)^{1/3} |\omega_{cd}|. \tag{115}$$

For the frequency range $\omega \simeq \omega_{ci}$ (ion cyclotron waves) the growth rate is as follows

$$\gamma \simeq \left(2\frac{\Omega_r^2}{k^2c^2}\frac{\omega_{pi}^2}{k^2c^2}\right)^{1/3} \omega_{ci}. \tag{116}$$

Now we take into account thermal motion of the electrons assuming that

$$V_{td}, V_{ti} \ll \frac{\omega}{|k_z|} \ll V_{te}, \tag{117}$$

and confine ourself by the wavelengths longer than the Larmor radii

$$k_\perp^2 V_{t\alpha}^2, |k_z|^2 V_{t\alpha}^2 \ll \omega_{c\alpha}^2.$$

For the components of the dielectric permittivity tensor, we obtain from (80) to (85)

$$\epsilon_{xx} = \epsilon_{yy} = 1 + \frac{c^2}{V_A^2} - \frac{\omega_{pd}^2}{\omega^2 - \omega_{cd}^2} - \frac{\omega_r^2}{(\omega - \Omega_0)^2} - \frac{\Omega_r^2}{(\omega + \Omega_0)^2} \tag{118}$$

$$\epsilon_{xy} = -\epsilon_{yx} = -i\frac{\omega_{pd}^2}{\omega^2 - \omega_{cd}^2}\frac{\omega}{\omega_{cd}} + i\frac{\Omega_r^2}{(\omega - \Omega_0)^2} - i\frac{\Omega_r^2}{(\omega + \Omega_0)^2} \tag{119}$$

$$\epsilon_{zz} = 1 - \frac{\omega_{pi}^2}{\omega^2} - \frac{\omega_{pd}}{\omega^2} + \frac{\omega_{pe}^2}{k_z^2 V_{te}^2} \tag{120}$$

$$\epsilon_{xz} = \epsilon_{zx} = \epsilon_{yz} = \epsilon_{zy} = 0, \tag{121}$$

where we used eq. (112) and denoted

$$V_A = B_0/(\sqrt{4\pi m_i n_{i0}}).$$

We also ignored the Landau damping on electrons. The dispersion relation (91) separates into two equations

$$\epsilon_{zz}(\omega, k_z) = 0, \tag{122}$$

which is not influenced by the rotation of the grain, and

$$H(\omega) = \frac{k^2c^2}{\omega^2} - \frac{c^2}{V_A^2} + \frac{\omega_{pd}^2}{\omega_{cd}(\omega \pm \omega_{cd})} = -\frac{2\Omega_r^2}{(\omega \mp \Omega_0)^2}, \tag{123}$$

Furthermore, we assume that $\omega \ll |\omega_{cd}|$. Setting

$$\omega = \Omega_0 + i\gamma,$$

where

$$\Omega_0 = kV_A$$

(Alfven waves), and using eq. (106), we obtain for the growth rate

$$\gamma \simeq \left(\frac{\Omega_r^2}{k^2c^2}\right)^{1/3} kv_A. \tag{124}$$

Let us now consider the longitudinal waves for which the dispersion relation assumes the form

$$\frac{k_\perp^2}{k^2}\epsilon_{xx}(\omega, \mathbf{k}) + \frac{k_z^2}{k^2}\epsilon_{zz}(\omega, \mathbf{k}) = 0 \tag{125}$$

where the components ϵ_{xx} and ϵ_{zz} for the cold plasma are defined by eqs. (93) and (95). Substituting the latter into eq. (125), we obtain

$$1 - \frac{k_\perp^2}{k^2}\sum_\beta \frac{\omega_{p\beta}^2}{\omega^2 - \omega_{c\beta}^2} - \frac{k_z^2}{k^2}\sum_\beta \frac{\omega_{p\beta}^2}{\omega^2} = \frac{k_\perp^2}{k^2}\left[\frac{\Omega_r^2}{(\omega - \Omega_0)^2} + \frac{\Omega_r^2}{(\omega + \Omega_0)^2}\right]. \tag{126}$$

It follows from eq. (126) that the dust grain rotation contributes only for waves with $k_\perp \neq 0$, as in this case the electric field of the longitudinal waves has a component that lies in the plane of the dust grain rotation.

In order to obtain the growth rates for longitudinal waves, we use the same procedure as presented before for deducing eqs. (106) and (108).

We consider the lower hybrid waves with $|k_z|V_{te}, |k_z|V_{ti}, \omega_{ci} \ll \omega \ll |\omega_{ce}|$. Setting

$$\omega = \Omega_0 + i\gamma,$$

where

$$\Omega_0 = \omega_0 = \frac{\omega_{pi}\omega_{ce}}{\sqrt{\omega_{pe}^2 + \omega_{ce}^2}}, \tag{127}$$

we find the growth rate

$$\gamma = \left(\frac{1}{2}\frac{\Omega_r^2}{\omega_{pi}^2}\right)^{1/3} \Omega_0. \tag{128}$$

Next, we consider the frequency regime

$$|k_z|V_{td}, |k_z|V_{ti} \ll \omega << |k_z|V_{te}.$$

The components of the dielectric permittivity in eq. (125) now are defined by eqs. (118) and (120). Using these expressions, we obtain the dispersion relation

$$1 + \frac{1}{k^2 r_{De}^2} - \frac{k_\perp^2}{k^2}\left(\frac{\omega_{pi}^2}{\omega^2 - \omega_{ci}^2} + \frac{\omega_{pd}^2}{\omega^2 - \omega_{cd}^2}\right) - \frac{k_z^2}{k^2}\frac{\omega_{pi}^2}{\omega^2} = \frac{k_\perp^2}{k^2}\left[\frac{\Omega_r^2}{(\omega - \Omega_0)^2} + \frac{\Omega_r^2}{(\omega + \Omega_0)^2}\right]. \tag{129}$$

Equation (129) can be analyzed in two limiting cases. First, we consider the ion-cyclotron waves with $\omega_{cd} \ll \omega \simeq \Omega_{ci}$ and $k_z \ll k_\perp$. Here, setting

$$\omega \simeq \omega_0 + i\gamma,$$

we obtain the growth rate from eq. (129)

$$\gamma \simeq \left(\frac{\Omega_r^2}{2\omega_{pi}^2}\frac{k_\perp^4 c_s^4}{\Omega_0^4}\right)^{1/3} \Omega_0, \tag{130}$$

where

$$\Omega_0 = \omega_0 = (\omega_{ci}^2 + k_\perp^2 c_s^2)^{1/2} \quad \text{and} \quad c_s = (n_{i0}T_e/n_{e0}m_i)^{1/2}$$

is the ion acoustic speed.

Second, we consider the modified dust ion-acoustic waves (MDIAWs), which are characterized by $\omega_{pd}, \omega_{cd} \ll \omega \ll \omega_{ci}$, for which (129) gives

$$1 + k^2 r_{De}^2 + k_\perp^2 \rho_s^2 - \frac{k_z^2 c_{ss}^2}{\omega^2} = k_\perp^2 r_{De}^2 \left[\frac{\Omega_r^2}{(\omega - \Omega_0)^2} + \frac{\Omega_r^2}{(\omega + \Omega_0)^2} \right], \tag{131}$$

where

$$\rho_s = r_{De}\omega_{pi}/\omega_{ci} \equiv c_s/\omega_{ci}.$$

Equation (131) admits an instability of the MDIAWs with the frequency

$$\Omega_0 = \omega_0 = k_z c_s/(1 + k^2 r_{De}^2 + k_\perp^2 \rho_s^2)^{1/2},$$

and the growth rate is

$$\gamma \simeq \left(\frac{k_\perp^2}{2k^2} \frac{\Omega_r^2}{\omega_{pi}^2} \right)^{1/3} \Omega_0. \tag{132}$$

Finally, we consider coupled dust acoustic-dust cyclotron waves in a dust-electron plasma (without ions) with positive dust grains [Shukla 2000]. Here, for $kV_{td} \ll \omega \ll k_z V_{te}$, we have

$$1 + \frac{1}{k^2 r_{De}^2} - \frac{k_\perp^2}{k^2} \frac{\omega_{pd}^2}{\omega^2 - \omega_{cd}^2} - \frac{k_z^2}{k^2} \frac{\omega_{pd}^2}{\omega^2} = \frac{k_\perp^2}{k^2} \left[\frac{\Omega_r^2}{(\omega - \Omega_0)^2} + \frac{\Omega_r^2}{(\omega + \Omega_0)^2} \right]. \tag{133}$$

For $\omega \ll |\omega_{cd}|$, eq. (133) admits an instability of short wavelength DIWs when

$$\Omega_0 = \omega_0 = k r_{De} \omega_{pd} / \left(1 + k^2 r_{De}^2 + k_\perp \rho_{sd}^2\right)^{1/2},$$

where $\rho_{sd} = \lambda_{De}\omega_{pd}/\omega_{cd}$. The growth rate of this instability is

$$\gamma \simeq \left(\frac{1}{2} \frac{k_\perp^2}{k_z^2} \frac{\Omega_r^2}{\omega_{pd}^2} \right)^{1/3} \Omega_0. \tag{134}$$

On the other hand, for $\omega \sim |\omega_{cd}|$, $k_\perp \gg k_z$, and $k^2 r_{De}^2 \ll 1$, we have an instability of the dust cyclotron waves when

$$\Omega_0 = \omega_0 = (\omega_{cd}^2 + k^2 r_{De}^2 \omega_{pd}^2)^{1/2}.$$

The growth rate of the instability is

$$\gamma \simeq \left(\frac{1}{2} k_\perp^4 r_{De}^4 \frac{\Omega_r^2}{\Omega_0^2} \frac{\omega_{pd}^2}{\Omega_0^2} \right)^{1/3} \Omega_0. \tag{135}$$

It is of interest to note that a dust-electron plasma with positively charged grains can occur in the Earth's polar mesosphere [Havnes *et al.* 1996; Horanyi *et al.* 1998], where the grains are irradiated by the sun light, in which case the grains act as a source of electrons and collect ions from the ambient plasma to get positively charged. There is also a prediction [Howard *et al.* 2000] that positively charged dust grains in retrogate orbits are the most likely to be observed by the Cosmic Dust Analyzer aboard the Cassini Orbiter mission to Saturn. Furthermore, the dust electron plasma can also be created in a laboratory discharge when the dust grains are irradiated by ultra-violet (UV) radiation [Chow *et al.* 1994; Rosenberg and Mendis 1995; Rosenberg *et al.* 1999; Fortov *et al.* 1996, 1998,1999; Samarian *et al.* 2000; Sickafoose *et al.* 2000].

4.4 Instability of Waves in an Unmagnetized Dusty Plasma

Following the standard method [Alexandrov *et al.* 1984] we obtain from (80)-(85) the dielectric tensor for the dusty unmagnetized plasma [Mahmoodi *et al.* *2000*; Tskhakaya *et al.* 2001; Shukla and Tskhakaya 2001]. The method implies the transition to the limit $B_0 \to 0$ in eqs. (59),(60),(67)-(69). For the dust grains we choose again the distribution function (61). The result is

$$\varepsilon_{ij}(\omega, \mathbf{k}) = \frac{k_i k_j}{k^2}\varepsilon^l(\omega, k) + \left(\delta_{ij} - \frac{k_i k_j}{k^2}\right)\varepsilon^t(\omega, k) \tag{136}$$

$$+\left(\delta_{ij} - \frac{\Omega_{0i}\Omega_{0j}}{\Omega_0^2}\right)\varepsilon^d(\omega, k),$$

where $\varepsilon^l(\omega, k)$ and $\varepsilon^t(\omega, k)$ are the usual longitudinal and transverse dielectric permittivities, respectively. They are

$$\varepsilon^l(\omega, k) = 1 + \sum_\beta \frac{\omega_{p\beta}^2}{k^2 V_{t\beta}^2}\left[1 - J_+\left(\frac{\omega_{p\beta}}{kV_{t\beta}}\right)\right], \tag{137}$$

and

$$\varepsilon^t(\omega, k) = 1 - \sum_\beta \frac{\omega_{p\beta}^2}{\omega^2} J_+ \cdot \left(\frac{\omega}{kV_{t\beta}}\right), \tag{138}$$

where the function $J_+(x)$ and its asymptotics are defined in the previous section. The influence of the dust grain rotation is described by $\varepsilon^d(\omega, k)$ which equals

$$\varepsilon^d(\omega, k) = -\frac{\Omega_r^2}{\omega^2}\frac{k^2}{K^2}\frac{\omega}{\omega - \Omega_0} J_+ \cdot \left(\frac{\omega - \Omega_0}{KV_{t\beta}}\right) \tag{139}$$

$$+\frac{\Omega_r^2}{K^2 V_{t\beta}^2}\left(\frac{\kappa^2}{K^2}+\frac{k^2}{K^2}\frac{\Omega_0}{\omega}\right)\left[1-J_+\left(\frac{\omega-\Omega_0}{KV_{t\beta}}\right)\right],$$

where Ω_r is defined by (90), $\kappa^2 = m_d/I$ and $K = \left(k^2+\kappa^2\right)^{1/2} \approx \kappa$.

Let us now consider the frequency regimes

$$kV_{td} \ll \omega \ll kV_{ti}, kV_{te} \quad \text{and} \quad \omega \pm \Omega_0 \gg kV_{td}.$$

The participation of the charged dust gas is decisive only in the wave motion when the frequency and wave vector satisfy the latter inequalities. Without loss of generality, we may assume that the vector $\mathbf{k}$ lies in the (x, y)-plane, i.e.$\mathbf{k} = (k_\perp, 0, k_z)$. Then, from the general dispersion relation (91) and according to eqs. (136)-(139) we find for the transversal waves (which are polarized along the y-axis so that $\mathbf{E} = (0, E, 0)$) the following dispersion relation

$$\frac{k^2c^2+\omega_{pd}^2}{\omega^2} = 1 - \frac{2\Omega_r^2}{(\omega-\Omega_0)^2}. \tag{140}$$

For $\Omega_0 = 0$ the influence of the dust grain rotation disappears. However, for waves with frequencies close to Ω_0 $(\omega_p \approx \Omega_r)$, we obtain

$$\omega = \Omega_0\left[1 \pm i\frac{\sqrt{2}\Omega_r}{(k^2c^2+\omega_{pd}^2)^{1/2}}\right]. \tag{141}$$

Equation (141) reveals that the ordinary transverse waves become unstable. The growth rate is given by

$$\sqrt{2}\Omega_r\Omega_0/(k^2c^2+\omega_{pd}^2)^{1/2},$$

which strongly depends upon Ω_r (which is a function of the dipole moment as well as the moment of inertia of the dust grains) and the dust rotation frequency Ω_0. It is worth noting that in a plasma without dust grains, the low-frequency transversal oscillations decay periodically due to their collisionless absorption by electrons [Alexandrov *et al.* 1984].

Furthermore, eqs. (91) and (136) also admits low-frequency $(\omega^2 \ll k^2c^2)$ longitudinal waves. The modified dispersion relation for the dust-acoustic wave [Rao *et al.* 1990], which is deduced from eqs. (137)-(139), reads

$$1+\frac{1}{k^2r_D^2}-\frac{\omega_{pd}^2}{\omega^2}-\frac{k_\perp^2}{k}\left[\frac{\Omega_r^2}{(\omega-\Omega_0)^2}+\frac{\Omega_r^2}{(\omega+\Omega_0)^2}\right]=0, \tag{142}$$

where

$$r_D^2 = \frac{\omega_{pe}^2}{V_{te}^2}+\frac{\omega_{pi}^2}{V_{ti}^2}.$$

Equation (142) is again formally similar to that dispersion relation which has been discussed in [Mikhailovskii 1974] in connection with the two-stream instability. From (142) it follows that the dust grain rotation gives a contribution only for waves with $k_\perp \neq 0$. For $\Omega_0 = 0$ that contribution is expressed in the change of the dust-acoustic frequency

$$\omega = \left(\omega_{pd}^2 + \frac{k_\perp^2}{k^2}\Omega_r^2\right)^{1/2} \left(1 + \frac{1}{k^2 r_D^2}\right)^{-1/2}. \qquad (143)$$

However, in the presence of the dust grain rotation eq. (142) admits complex solutions for any rotating frequency Ω_0, satisfying the condition

$$\Omega_0 < \omega_{pd} \left(1 + \frac{1}{k^2 r_D^2}\right)^{-1/2} \left[1 + \left(\frac{k_\perp^2}{k^2}\frac{\Omega_r^2}{\omega_{pd}^2}\right)^{1/3}\right]^{3/2}. \qquad (144)$$

The equality of Ω_0 to the right-hand side of eq. (144) defines the boundary of the stability of the dust-acoustic wave. Setting

$$\omega \simeq \Omega_0 + i\gamma,$$

where $\gamma \ll \Omega_0$, for

$$\left(\omega_{pd}^2 + \frac{k_\perp^2}{k^2}\Omega_r^2\right) / \Omega_0^2 \approx 1 + 1/k^2 r_D^2,$$

we obtain from eq. (142) the growth rate

$$\gamma = 3^{1/2} 2^{-4/3} \left(\frac{k_\perp^2}{k^2}\frac{\Omega_r^2}{\omega_{pd}^2}\right)^{1/3} \Omega_0. \qquad (145)$$

In the similar way can we investigate the instability of dust ion-acoustic waves (DIAW). DIAWs occur in the frequency range [Shukla and Silin 1992]

$$kV_{td}, kV_{ti} \ll \omega \ll kV_{te}.$$

Hence, in the DIAWs the restoring force comes from the pressure of the inertialess electrons, while the ion mass provides the inertia. On the time scale much longer (shorter) than the ion (dust) plasma period, the dust grains remain almost immobile. From eqs. (136)-(139) we find the following dispersion relation for DIAWs in dusty plasma with rotating dust grains [Shukla and Tskhakaya 2001]

$$1 + \frac{1}{k^2 r_{De}^2} - \frac{\omega_p^2}{\omega^2} - \frac{k_\perp^2}{k}\left[\frac{\Omega_r^2}{(\omega - \Omega_0)^2} + \frac{\Omega_r^2}{(\omega + \Omega_0)^2}\right] = 0, \qquad (146)$$

where $r_{De} = \omega_{pe}/V_{te}$ and

$$\omega_p = (\omega_{pi}^2 + \omega_{pd}^2)^{1/2}.$$

Equation (146) is quite similar to eq. (142). For $\Omega_0 = 0$ the changed value of the DIAW frequency is

$$\omega_p = (\Omega_{ss}^2 + \Omega_n^2)^{1/2}, \tag{147}$$

where

$$\Omega_{ss} = \omega_{ss} \frac{(1 + \omega_{pd}^2/\omega_{pi}^2)^{1/2}}{(1+b)^{1/2}}$$

is the DIAW frequency [Shukla and Silin 1992] and

$$\omega_{ss} = k(n_{i0}/n_{e0})^{1/2} c_s, \quad b = k^2 r_{De}^2$$

$$\text{and} \quad \Omega_n = \sqrt{2}b(k_\perp/k)\Omega_r/(1+b)^{1/2}.$$

Furthermore, it can be again shown that eq. (146) admits complex solutions for any rotation frequency $\Omega_0 \neq 0$, satisfying the condition

$$\Omega_0^2 < \Omega_{ss}^2 \left[1 + \left(\frac{k_\perp^2}{k^2}\frac{\Omega_r^2}{\omega_p^2}\right)^{1/3}\right]. \tag{148}$$

The right-hand side in eq. (148) defines the boundary of the stability of the DIAWs. Setting

$$\omega \simeq \Omega_{ss} + i\gamma$$

and $\Omega_{ss} \approx \Omega_0$, where $\gamma \ll \Omega_0$, we obtain from eq. (146) the growth rate

$$\gamma = 3^{1/2} 2^{-4/3} \left(\frac{k_\perp^2}{k^2}\frac{\Omega_r^2}{\omega_p^2}\right)^{1/3} \Omega_0. \tag{149}$$

The expressions (145) and (149) exhibits that the growth rate of the instability is directly proportional to $\Omega_r^{2/3}/\Omega_0$. Thus, both the electrostatic and electromagnetic (see (141)) instabilities are associated with the rotation of elongated dust grains.

5 Discussion

In this article, we have carried out systematic studies of motions of spherical and irregular dust grains as well as associated wave instabilities associated with elongated and rotating dust grains in a collisionless plasma. First, we have analytically analyzed motions of elongated dust grains in a plasma sheath and have shown that charged dust rods execute quivering (vibrating) and bouncing motions under the combined influence of the sheath electric field and the dust dipole moment. Periodic dust motions are set up by the latter as well as by oscillating electromagnetic fields. Our results are capable of explaining the observations [Mohideen *et al.* 2000] of the dust cloud motions and its confinement by the walls of the groove. Second, we have presented a theory for the motions of charged dust clouds in a partially ionized nonuniform magnetoplasma. It is found that interactions between dust clouds, electrons and ions are non-trivial in that the dust particle motions are controlled by the combined effects of a radial electric force, the $\mathbf{J}_d \times \mathbf{B}$ and collisional drag forces. Our results show that the azimuthal dust rotation (with respect to a vertical magnetic field) is not due to the cross-field ($\mathbf{E} \times \mathbf{B}$) drift, as recognized by Sato *et al.* [Sato *et al.* 2000, 2001]. The azimuthal dust rotation speed is linearly proportional to α (see (37)) when $\alpha \ll 1$, and it scales as $(Z_d n_d/n_i)^2(\nu_{di}^2 \omega_{ci}/\nu_{dn}\nu_{in}^2)$. Thus, the rotation speed increases with the increase of B_0 and $Z_d n_d$. These features are consistent with the experimental observations [Sato *et al.* 2000, 2001]. Furthermore, the radial motion of dust clouds is caused by the radial electric field and gradients of the plasma and magnetic field pressures [Konopka *et al.* 2000]. It should be mentioned that the knowledge of differential dust motions should also help to understand the dynamics of charged dust clouds in partially ionized astrophysical environments [Kamaya and Nishi 2000, Lazarian and Yan 2002].

We have developed the electrodynamics of a magnetized dusty plasma accounting for the finite size of elongated and rotating dust grains. Starting from an appropriate Lagrangian for charged dust grains, we have derived the dust charge and dust current densities, as well as a kinetic equation for charged dust grains and the corresponding equations of motion in an external magnetic field. The effects of the dipole moment and the principal moment of inertia of the elongated and rotating dust grains are self-consistently incorporated. The newly derived charge and current densities, together with the corresponding quantities for the electrons and ions, are combined with the Vlasov-Maxwell system of equations to obtain dispersion relations for electrostatic and electromagnetic waves. The analyses of the dispersion relations exhibit instabilities of a great variety of wave modes including linearly and circularly polarized electromagnetic waves as well as dust acoustic, dust cyclotron, and lower-hybrid waves. The instability arises owing to the resonance interaction between waves and elongated rotating dust rods. The free energy stored in the dust grain rotational motion is basically coupled to both the electromagnetic and electrostatic modes, driving them at nonthermal levels. It is likely that galactic microwave emissions [Finkbeiner *et al.* 2002] could originate due to the rotational dynamics of irregular dust rods which are unevenly charged and which resonantly interact with microwaves.

Recent experiments work [Molotkov *et al.* 2000] has conclusively demonstrated the Coulomb crystallization of 300-μm highly charged ($|q| \sim 7.7 \times 10^5$) elongated cylindrical grains (with a length-to-diameter ratio of 20-40) of a mass density of 1.1 g/cm^3 and diameters of 15 and 7.5 μm in a low pressure gas discharge plasma where the electron energy ranges between 1-10 eV. Thus, a sheath electric field of order 30 V/cm can levitate the grain. In [Molotkov *et al.* 2000] is also discussed the role of induced dipole moment that can influence the grain orientation. At small pressures (0.1 torr) the oscillations with a wavelength of $\sim$ 1 mm and a frequency of 20-50 Hz are observed. The latter can be associated with the dust acoustic waves which are deduced from our Eq. (142). Furthermore, by applying an external magnetic field of 1-6 kilo Gauss and with the plasma ($n_{i0} \sim 10^9$ cm^{-3} and $n_{d0} \sim 10^3$ cm^{-3}) and dust parameters similar to those in [Molotkov *et al.* 2000], one should be able to observe the magnetization of ions as well as the electrostatic ion-cyclotron wave instability that is described by (129).

In closing, we mention that coherent or incoherent scatterings of star light and/or electromagnetic waves off nonthermal fluctuations in cosmic plasmas may yield valuable information regarding the light polarization, the dust number density and the dust charge in situ, and other plasma parameters including the ambient magnetic field strength.

Acknowledgments

This work was partially supported by the European Commission (Brussels, Belgium) through contract No. HPRN-CT-2000-00140 for carrying out the task of the Human Potential Research Training Networks "Complex Plasmas: The Science of Laboratory Colloidal Plasmas and Mesospheric Charged Aerosols", as well as by the International Space Science Institute at Bern, Switzerland.

References

Alexandrov A. F., Bogdankevich L. S. and Rukhadze A. A. *Principles of Plasma Electrodynamics* (Springer-Verlag, Berlin, 1984).

Bogoliubov N. N. *Dinamicheskie Problemy Statisticheskoj Fiziki* [Dynamic Problems of Statistical Physics (Gostekhizdat, Moscow, 1946)].

Chiang C. H. and Lin I. 1996, *Phys. Rev. Lett.* **77**, 647.

Chow V. W., Mendis D. A. and Rosenberg M. 1994, *IEEE Trans. Plasma Sci.* **22**, 179.

Elskens Y., Resendes D. P. and Mendonça J. T. 1997, *Phys. Plasmas* **4**, 4210.

Finkbeiner D. P., Schlegel D. J., Frank C. and Helies C. 2002, *Ap J.* **566**, 898.

Fortov V. E., Nefedov A. P., Petrov O. F. *et al.* 1996, *Phys. Lett.* **A 219**, 89.

Fortov V. E., Nefedov A. P. and Vaulina O. S. 1998, *Zh. Éksp. Teor. Fiz.* **114**, 2004. [1998 *JETP* **87**, 1087].

Fortov V. E., Molotkov V. I., Nefedov A. P. and Petrov O. F. 1999, *Phys. Plasmas* **6**, 1759.

Fujiyama H., Kawasaki H., Yang S. C. *et al.* 1994, *Jpn. J. Appl. Phys.* **33**, 4216.

Gradshteyn I. S. and Ryzhik I. M. *Tables of Integrals, Series, and Products*, trans. edited by Jeffrey A. (Academic Press, New York, 1980), p. 188.

Grün E., Morfill G. E. and Mendis D. A. 1984, in *Planetary Rings*, edited by Greenberg R. and Brahic A. (University of Arizona Press, Tuscon).

Grün E. *et al.* 1996, *Nature* (London) **381**, 395.

Harwit M. 1988, *Astrophysical Concepts* (Springer-Verlag, New York), p. 405.

Havnes O., Troim J., Blix T. *et al.* 1996, *J. Geophys. Res.* **101**, 10839.

Horányi M. 1996, in *The Physics of Dusty Plasmas*, edited by Shukla P. K. *et al.* (World Scientific Publishers, Singapore), p. 33.

Horányi M., Walch B., Robertson S. *et al.* 1998, *J. Geophys. Res.* **103**, 8575.

Howard J. E., Horányi M. and Stewart G. R. 1999, *Phys. Rev. Lett.* **83**, 3993.

Howard J. E., Dullin H.R. and Horányi. 2000, *Phys. Rev. Lett.* **84**, 3244.

Ishihara O. and Sato N. 2001, *IEEE Trans. Plasma Sci.* **29**, 179.

Kamaya H. and R. Nishi R. 2000, *Ap. J.* **543**, 257.

Kaw P. K., Nishikawa K. and Sato N. 2002, *Phys. Plasmas* **9**, 387.

Klimontovich Yu. L. 1982, *Kinetic Theory of Nonideal Gases and Nonideal Plasmas* (Pergamon Press, New York).

Konopka U., Samsonov D., Ivlev A. V. *et al.* 2000, *Phys. Rev. E* **61**, 1890.

Landau L. D. and Lifshitz E. M. 1971, *The Classical Theory of Fields* (Pergamon Press, Oxford).

Landau L. D. and Lifshitz E. M. 1989, *Statistical Physics* (Pergamon Press, Oxford).

Law D. A. *et al.* 1998, *Phys. Rev. Lett.* **80**, 4189.

Lazarian A. and Yan H, 2002 *ApJ* **566**, L105.

Mahmoodi J., Shukla P. K., Tsintsadze N. L. and Tskhakaya D. D. 2000, *Phys. Rev. Lett.* **84**, 2626.

Mendis D. A. and Axford W. I. 1974, *Annual Rev. of Earth and Planets* **2**, 419.

Mendis D. A. *et al* 1983, *J. Geophys. Res.* **88**, A929.

Mendis D. A. *et al.* 1984, in *Saturn,* , edited by Gehrels T. and Matthews D. (University of Arizona Press, Tuscon).

Mendis D. A. and Rosenberg M. 1994, *Ann. Rev. Astron. Astrophys.* **32**, 410.

Mikhailovskii A. B. 1974, *Theory of Plasma Instabilities: Instabilities of an Homogeneous Plasma* (Consultants Bureau, New York), Vol. 1.

Mohideen U. *et al.* 1998, *Phys. Rev. Lett.* **81**, 349.

Molotkov V. I., Nefedov A. P., Putsyl'nik M.Yu, Torchinsky V. M., Fortov V. E., Kharpak A.G. and Yoshino K. 2000, *JETP Lett.* **71**, 102.

Nitter T., Havnes O., and MelandsøF. 1998, *J. Geophys. Res.* **103**, 6605.

Nunomura S., Ohno N. and Takamura S. 1997, *Jpn. J. Appl. Phys.* **36**, 877.

Nunomura S. *et al.* 1998, *Phys. Plasmas* **5**, 3517.

Rao N. N., Shukla P. K. and Yu M. Y. 1990, *Planet. Space Sci.* **38**, 543.

Rahman H. U., Mohideen U., Smith M. A., Rosenberg M. and Mendis D. A. 2000, *Physica Scripta* **T89**, 186.

Rosenberg M. and Mendis D. A. 1992, *IEEE Trans. Plasma Sci.* **20**, 929.

Rosenberg M. and Mendis D. A. 1995, *IEEE Trans. Plasma Sci.* **23**, 177.

Rosenberg M., Mendis D. A. and Sheehan D. P. 1999, *IEEE Trans. Plasma Sci.* **27**, 239.

Samarian A. A., Vaulina O. S., Nefedov A. P., Petrov O. F. and Fortov V. E. 2000, *J. Éksp. Teor. Fiz.* **118**, 119 [2000, *JETP* **91**, 106].

Sato N., Uchida G., Ozaki R. *et al.* 2000, in *Frontiers in Dusty Plasmas*, edited by Nakamura Y., Yokota T., and Shukla P.K. (Elseveier, Amsterdam), p. 329.

Sato N., Uchida G., Kaneko T. *et al.* 2001, *Phys. Plasmas* **8**, 1786.

Shukla P. K. 1992, *Physica Scripta* **45**, 504.

Shukla P. K. and Silin V. P. 1992, *Phys. Scripta* **45**, 508.

Shukla P. K. 1994, *Phys. Plasmas* **1**, 1362.

Shukla P. K. and Rahman H. U. 1996, *Phys. Plasmas* **3**, 430.

Shukla P. K. 1999, *Phys. Lett.* **A 252**, 340.

Shukla P. K. 2000, *Phys. Rev. E* **61**, 7249.

Shukla P. K. and Tskhakaya D. D. 2001, *Phys. Plasmas* **8**, 661.

Shukla P. K. and Mamun A. A. 2002, *Introduction to Dusty plasma Physics* (Institute of Physics Publ., Bristol).

Shukla P. K. 2002, Phys. Lett. A, **299**, 258.

Sickafoose A. A., Colwell J. W., M. Horányi M. and Robertson S. 2000, *Phys. Rev. Lett.* **84**, 6034.

Spitzer Jr. L. 1977, *Physical Processes in the Interstellar Medium* (John Wiley and Sons, New York), p. 182.

Thomas Jr. E. 1999, *Phys. Plasmas* **5**, 2672.

Tskhakaya D. D., Shukla P. K., Tsintsadze N. L. and Mahmoodi J. 2001, *Physica Scripta* **T89**, 183.

Tskhakaya D. D., Shukla P. K. and Tsintsadze N. L. 2001, *JETP* **93**, 301.

Tskhakaya D. D. and Shukla P. K. 2001a, *Phys. Lett.* **A 279**, 243.

Tskhakaya D. D. and Shukla P. K. 2001b, *Phys. Lett.* **A 279**, 456.

Voshchinnikov N. V. and Semenov D. A. 2000, *Astronomy Lett.* **26**, 697.

Some Instabilities in Collisional Dusty Plasmas

M. Rosenberg
Department of Electrical and Computer Engineering,
University of California, San Diego
La Jolla, CA 92093, U. S. A.

P. K. Shukla
Theoretische Physik IV, Ruhr-Universität Bochum,
44780 Bochum, Germany

Abstract

Certain dusty plasmas in near-Earth space environments or in laboratory experiments are weakly ionized, so that neutral-charged particle collisions affect the properties of waves and instabilities. Here, we review recent selected works on several instabilities in collisional dusty plasmas, with applications to polar summer mesosphere echo regions, dusty meteor trails, and laboratory dusty plasmas. First, the effects of charged dust on the Farley-Buneman instability and the high-frequency Hall current instability are discussed, for plasma parameters representative of the low E-region at altitudes of about 90 – 100 km. Second, a low-frequency Hall current instability in the dust acoustic frequency range is discussed, with application to dusty meteor trails containing positively charged dust. Third, an ion-dust streaming instability with frequency less than the dust-neutral collision frequency is discussed. Finally, we present two examples of electrostatic instabilities that could be caused by sheared plasma flows in a nonuniform collisional dusty magnetoplasma. Applications to laboratory dusty plasma wave experiments are pointed out.

1 Introduction

Dust grains (micron to sub-micron sized solid particles) in plasma and/or radiative environments are generally electrically charged owing to processes such as plasma current collection, photoemission, or secondary electron emission. Over the last decade there has been an explosive growth of interest in how the presence of charged dust in a plasma affects the behavior of waves and instabilities, and how this applies to observed phenomena in dusty plasmas in space and laboratory environments [see e.g., Rosenberg 1996a; Verheest 1996; Verheest 2000; Shukla 2001; Shukla and Mamun 2002]. Dust occurs in many solar system plasmas including comets and planetary rings [reviews by e.g., Goertz 1989; Northrop 1992; Mendis and Rosenberg 1994; Horanyi 1996] and in near-Earth environments such as polar mesosphere summer echo regions [reviews by e.g., Cho and Kelly 1993; Cho and Rottger 1997; Havnes *et al.* 2001b] and dusty meteor trails [Kelley *et al.* 1998]. Charged dust has recently been detected in the mesosphere by rocket borne probes [e.g., Havnes 1996a, 1996b; Havnes *et al.* 2001a; Horanyi *et al.* 2000]. Dust also occurs as contaminants in various laboratory plasmas, including for example, processing plasmas used in the microelectronics industry [e.g., Selwyn 1993]. Recently there have been an increasing number of dedicated laboratory experiments to study the physics of dusty plasmas, including collective processes such as the behavior of waves and instabilities [e.g., Barkan *et al.* 1995a, 1995b, 1996; Pieper and Goree 1996; Prabhakara and Tanna 1996; Thompson *et. al* 1997, 1999; Merlino *et al.* 1998; Molotkov *et al.* 1999; Thomas and Watson 1999; Fortov *et al.* 2000; Melandsø *et al.* 2000; Nunomura *et al.* 2000; Thomas and Merlino 2001].

Certain dusty plasmas in near-Earth space environments or in laboratory experiments are weakly ionized. For example, dusty plasmas may occur in the upper mesosphere [e.g., Havnes *et al.* 2001b; Kelley *et al.* 1998], where the ratio of the ion density, n_i, to the neutral density, n_n, can be $< 10^{-9}$ [e.g., see Kelley 1989]. In laboratory dusty plasma experiments that study the properties of very low-frequency dust acoustic waves, n_i/n_n can be on the order of 10^{-6} or less [e.g., Thompson *et al.* 1997; Fortov *et al.* 2000; Thomas and Watson 1999]. In such plasmas, collisions of charged particles with neutral molecules can affect the properties of waves and instabilities.

In this article, we review some recent selected theoretical works on instabilities in collisional dusty plasmas, with application to dusty plasmas in the low E-region of the Earth's ionosphere and in laboratory environments. The paper is organized as follows. Instabilities in magnetized, collisional dusty plasmas with parameters representative of the low E-region are considered in §2. First, the effects of charged dust on the Farley-Buneman instability and the high-frequency Hall current instability are discussed, considering plasma parameters in the low E-region at altitudes of about 90 – 100 km [Rosenberg and Chow 1998; Rosenberg 2001]. Second, a low-frequency Hall current instability in the dust acoustic frequency range is discussed, with application to dusty meteor trails containing positively charged dust [Rosenberg and Shukla 2000, 2002]. Instabilities in laboratory dusty plasmas are considered in §3. First, an ion-dust streaming instability with frequency less than the dust-neutral collision frequency is discussed [Rosenberg 2002], that may have application to certain laboratory wave experiments. Then, we show that plasma sheared flows could cause instability of the Shukla-Varma mode [Shukla and Varma 1993] and drift waves in a nonuniform collisional magnetoplasma [Shukla and Rosenberg 2002]. A brief summary of these theoretical studies is presented in §4.

2 Instabilities in Dusty Plasmas in the Low E-region

2.1 General Dispersion Relation

In this section, we review recent work on the effects of dust on instabilities in the low E-region of the Earth's ionosphere, at altitudes between about $90-110$ km. At these altitudes, the plasma and neutral gas parameters are such that the electron gyrofrequency, Ω_e, is much larger than the electron-neutral collision frequency, ν_{en}, while the ion gyrofrequency, Ω_i, is much smaller than the the ion-neutral collision frequency, ν_{in} [e.g., Dimant and Sudan 1995]. In this case, the electrons can be considered to be magnetized, while the ions are non-magnetized. If there is an electric field $\mathbf{E}_0$ perpendicular to the ambient magnetic field $\mathbf{B}_0$, the electrons can acquire an $\mathbf{E}_0 \times \mathbf{B}_0$ drift relative to the ions, which can drive ion wave instabilities such as the Farley-Buneman instability [see e.g., Kelley 1989]. If charged dust is present in addition to the ions, this can affect such ion wave instabilities and also lead to the excitation of new very low frequency dust wave instabilities.

Charged dust in the low E-region is expected to be non-magnetized, since the charge-to-mass ratio of dust is generally orders of magnitude smaller than that of the ions, and the grains present a large geometric cross section for collision with neutrals. Thus the dust-neutral collision frequency is expected to be much larger than the dust gyrofrequency in the low E-region [Rosenberg and Shukla 2000].

Following Rosenberg and Shukla [2000, 2002], we consider a plasma with three charged species, namely, electrons, singly charged ions, and charged dust grains. The density of each species is n_α, with $\alpha = e, i, d$ for electrons, ions, and dust, respectively. If the dust has uniform radius a (and therefore uniform mass m_d and charge state Z_d), the equilibrium charge neutrality condition is

$$\epsilon Z_d n_d + n_i = n_e \quad , \tag{1}$$

where $\epsilon = 1$ for positively charged dust, and -1 for negatively charged dust. The relative amount of charge density carried by the dust grains can be characterized by the parameter

$$\delta = \frac{n_i}{n_e} , \tag{2}$$

where $\delta > 1$ for negatively charged dust, and $\delta < 1$ for positively charged dust.

The following equilibrium is considered [Rosenberg and Shukla 2000, 2002]. There is an external magnetic field in the z direction, $B_0\,\hat{\mathbf{z}}$, and an external electric field in the -x direction, $-E_0\,\hat{\mathbf{x}}$. Since the electrons are magnetized, they have have an $\mathbf{E}_0 \times \mathbf{B}_0$ drift in the y direction, given by $u_E\hat{\mathbf{y}} = c(E_0/B_0)\,\hat{\mathbf{y}}$. This cross-field drift of the electrons with respect to ions and dust can drive instabilities. (There are also ion and dust drifts along E_0 but these are small and can be neglected for the waves under study here). The electron distribution is assumed to be a drifting Maxwellian with temperature T_e

$$f_{0e} = n_e \left(\frac{m_e}{2\pi T_e}\right)^{3/2} exp\left[-\frac{m_e(v_x^2 + (v_y - u_E)^2 + v_z^2)}{2T_e}\right], \tag{3}$$

while the ion and dust distributions are assumed to be isotropic Maxwellians with temperatures T_i and T_d, respectively. Electrostatic waves are considered, with $\mathbf{E}_1 = -\nabla\,\phi(\mathbf{r},t)$, where $\phi(\mathbf{r},t) = \phi(x)\,exp\,[i\,(k_y y + k_z z - \omega t)]$, with $k_z \ll k_y$. In addition, the wave frequency is assumed to be $\gg$ the dust gyrofrequency, and satisfies $|\omega - k_y u_E| \ll \Omega_e$. We consider wavelengths much shorter than the ion gyroradius (and the dust gyroradius).

The linear dispersion relation is then [Rosenberg and Shukla 2000]

$$1 + \chi_e + \chi_i + \chi_d = 0\,, \tag{4}$$

where

$$\chi_e = \frac{1}{k^2\lambda_{De}^2}\,[1 + \zeta_e\,Z(\zeta_e)\Gamma_0(b_e)]\left[1 + \frac{i\nu_{en}}{\sqrt{2}k_z v_e}\Gamma_0(b_e)Z(\zeta_e)\right]^{-1}, \tag{5}$$

$$\chi_i = \frac{1}{k^2\lambda_{Di}^2}\,[1 + \zeta_i Z(\zeta_i)]\left[1 + \frac{i\nu_{in}}{\sqrt{2}k v_i}Z(\zeta_i)\right]^{-1}, \tag{6}$$

$$\chi_d = \frac{1}{k^2\lambda_{Dd}^2}\,[1 + \zeta_d Z(\zeta_d)]\left[1 + \frac{i\nu_{dn}}{\sqrt{2}k v_d}Z(\zeta_d)\right]^{-1}, \tag{7}$$

and where $\lambda_{D\alpha} = (T_\alpha/4\pi n_\alpha Z_\alpha^2 e^2)^{1/2}$ is the Debye length for each species α, with $\alpha = e, i, d$ for electrons, ions, dust, respectively, and Z is the plasma dispersion function [Fried and Conte 1961] with arguments

$$\zeta_e = \frac{\omega - k_y u_E + i\nu_{en}}{\sqrt{2}k_z v_e},$$

$$\zeta_i = \frac{\omega + i\nu_{in}}{\sqrt{2}k v_i},$$

$$\zeta_d = \frac{\omega + i\nu_{dn}}{\sqrt{2}k v_d}.$$

Here, $v_\alpha = (T_\alpha/m_\alpha)^{1/2}$ is the thermal speed of species α, $\rho_e = v_e/\Omega_e$ is the electron gyroradius, $b_e = k_y^2\rho_e^2$, and $\Gamma_0(b_e) = I_0(b_e)\,exp\,(-b_e)$, with I_0 being the modified Bessel function of zero order.

The electron-neutral and ion-neutral collision frequencies in the lower ionosphere are given by [e.g., Kelley 1989]

$$\nu_{en} = 5.4 \times 10^{-10}\,n_n T_e^{1/2}\,\mathrm{s}^{-1} \tag{8}$$

$$\nu_{in} = 2.6 \times 10^{-9}\,(n_n + n_i)\,M^{-1/2}\,\mathrm{s}^{-1}, \tag{9}$$

where the neutral and ion densities, n_n and n_i, are in units of cm^{-3}, the electron temperature T_e is in Kelvins, and M is the mean molecular weight of the neutrals and ions. The hard-sphere dust-neutral collision frequency (in the free molecular regime where the mean free path in the gas is much larger than the grain size) is given by [Baines 1965]

$$\nu_{dn} = \frac{4\pi}{3} a^2 n_n v_n \frac{m_n}{m_d} \tag{10}$$

where v_n and m_n are the neutral thermal speed and mass, respectively, and m_d is the dust mass.

2.2 Effect of Dust on Farley-Buneman Instability

The Farley-Buneman instability [Farley 1963; Buneman 1963] (also referred to as the two-stream instability [see Kelley 1989]) is an ion wave instability with frequency $\omega < \nu_{in}$ that can be excited when the electron $\mathbf{E}_0 \times \mathbf{B}_0$ drift is sufficiently larger than the ion sound speed. The instability is applicable to equatorial and auroral electrojet regions. The Farley-Buneman instability is thought to account for type-I irregularities, which are field aligned irregularities of ionization density propagating perpendicular to $\mathbf{B}_0$, that are observed using radar techniques [see Kelley 1989]. In addition, it has been suggested that the two -stream instability may play a role in certain meteor echoes [Chapin and Kudeki 1994; Pellinen-Wannberg and Wannberg 1996]. Recently, Oppenheim *et al.* [2000] and Dyrud *et al.* [2001] have reported on numerical simulation studies of two-stream or gradient-drift type instabilities in meteor trails.

The presence of charged dust can modify the mode frequency and growth rate, and the critical electron drift speed for excitation of the Farley-Buneman instability [Rosenberg and Chow 1998; Rosenberg 2001]. This is due to the lack of equilibrium charge neutrality between electrons and ions, that is, $n_i \neq n_e$, in the presence of charged dust [see eq. (1)]. The dynamics of the dust can be neglected over the time scale of this ion wave instability. This is because the heavy dust grains have a charge-to-mass ratio which is generally orders of magnitude smaller than that of the ions, and therefore have negligible response to waves with frequencies much higher than the dust plasma frequency.

Thus, neglecting the dust susceptibility, eq. (4) becomes

$$1 + \chi_e + \chi_i = 0. \tag{11}$$

Combining eq. (11) with eq. (1) for overall charge neutrality yields the general dispersion relation for the Farley-Buneman instability in a dusty plasma [Rosenberg and Chow 1998: for electron-ion plasma, see e.g., Ossakow *et al.* 1975; Schmidt and Gary 1973].

Following Rosenberg and Chow [1998] and Rosenberg [2001], we consider the limit $k_z = 0$ (so that $k = k_y$). Expanding the Z function, and in the small electron Larmor radius limit $b_e \ll 1$, where $\Gamma_0(b_e) \approx 1 - b_e$, eq. (5) for the electron susceptibility becomes

$$\chi_e \approx \frac{\omega_{pe}^2}{\Omega_e^2} \frac{\omega - ku_E + i\nu_{en}}{\omega - ku_E + i\nu_{en} b_e}. \tag{12}$$

For the ions, we consider the non-resonant regime, where $\zeta_i \gg 1$. Considering frequencies $\omega \ll \nu_{in}$, the former condition implies wavelengths much larger than the ion mean free path. For example, for parameters of the low E-region, this would correspond approximately to wavelengths $\gg 5$ cm at altitudes $h \sim 90$ km, and $\gg 50$ cm at $h \sim 100$ km. In this limit, eq. (6) for the ion susceptibility becomes

$$\chi_i \approx -\frac{\omega_{pi}^2}{A}, \tag{13}$$

where

$$A \approx \omega(\omega + i\nu_{in}) - k^2 v_i^2 \tag{14}$$

Then, inserting eqs. (12) and (13) in eq. (11), and assuming

$$A\left(1 + \frac{\omega_{pe}^2}{\Omega_e^2}\right) < \omega_{pi}^2,$$

yields the dispersion relation

$$\omega - ku_E \approx i\nu_{en}\left[\frac{\omega_{pe}^2}{\Omega_e^2}\frac{A}{\omega_{pi}^2} - b_e\right] \tag{15}$$

Equation (15) can be solved for weak growth or damping, using $\omega = \omega_r + i\gamma$, with $|\gamma| \ll \omega_r$. This yields , in the limit that $\nu_{in} \ll \omega_{pi}$ [Rosenberg 2001; Rosenberg and Chow 1998]

$$\omega_r \approx \frac{ku_E}{\left[1 + \frac{\Psi}{\delta}\right]}, \tag{16}$$

and

$$\gamma \approx \frac{\Psi}{\delta}\frac{1}{\left[1 + \frac{\Psi}{\delta}\right]}\left(\frac{\omega_r^2}{\nu_{in}} - \frac{k^2 \bar{c}_s^2}{\nu_{in}}\right). \tag{17}$$

Here the parameter Ψ is given by

$$\Psi = \frac{\nu_{en}\nu_{in}}{\Omega_e \Omega_i}, \tag{18}$$

and

$$\bar{c}_s = \left(\frac{\delta T_e + T_i}{m_i} \right)^{1/2} \tag{19}$$

is the ion sound speed modified by dust, where $\delta = n_i/n_e$ [see eq. (2)]. In this fluid limit, the critical electron drift can be obtained by setting $\gamma = 0$ which gives

$$\left(\frac{u_E}{v_i} \right)_c = \left(1 + \frac{\Psi}{\delta} \right) \left(1 + \frac{\delta T_e}{T_i} \right)^{1/2} . \tag{20}$$

It should be noted that kinetic effects, as well as larger values of (ν_{in}/ω_{pi}), modify the above estimate of the critical drift.

As discussed in Rosenberg [2001] the presence of negatively charged dust ($\delta > 1$) leads to two effects. These are (1) an increase in the mode frequency and thus the phase speed [see eq. (16)], and (2) an increase in the ion sound speed [see eq. (19)]. (Note that the presence of positively charged [$\delta < 1$] has the opposite effects.) Now, growth of the instability requires that the phase speed must be larger than the ion sound speed [see eq. (17)], but which of the above two effects dominates depends on the altitude through the parameter Ψ. It turns out that, for the parameters of the low-E region, the presence of negatively charged dust can lower the critical drift at altitudes $h < 95$ km (where $\Psi > 1$), while it increases the critical drift at altitudes $h > 95$ km (where $\Psi < 1$). Note that the opposite can occur in the presence of positively charged dust. These trends can be seen in Fig. 1 which plots the approximate critical drift speed given in eq. (20), versus δ for three values of Ψ which label the curves. Here $\Psi = 0.1$ corresponds to altitude $h \sim 100$ km, $\Psi = 1$ corresponds to $h \sim 95$ km, and $\Psi = 4$ corresponds to $h \sim 90$ km.

If the presence of negatively charged dust reduces the critical drift speed for instability at $h \sim 90$ km, this may have some relevance to the phenomena of strong radar backscatter referred to as polar summer mesosphere echoes (PMSE) [e.g., Rosenberg and Chow 1998]. It is thought that charged dust (possibly ice that can condense on meteoric particles in the cold summer mesopause) plays a role in PMSE, which occurs at altitudes between about $h \sim 80-93$ km [see e.g., reviews by Cho and Kelley 1993; Havnes *et al.* 2001b]. If Farley-Buneman modes can be excited at 90 km owing to the presence of negatively charged dust, this might result in enhanced radar backscatter from unstable waves which satisfy the Bragg condition. The Bragg condition for radar backscatter requires that the wavelength of the unstable plasma wave or irregularity is 1/2 the radar wavelength [see e.g., Kelley 1989]. For example, for PMSE at 50 MHz, the Bragg condition for backscatter requires unstable waves of wavelength $\lambda \sim 3$ m. We assume the following nominal daytime plasma parameters at $h \sim 90$ km: $n_i \sim 2 \times 10^4$ cm^{-3}, $T_e \sim T_i \sim 150$ K, and $n_n \sim 4 \times 10^{13}$ cm^{-3}. In this case the ion Debye length is $\lambda_{Di} \sim 0.7$ cm. Thus an unstable wave with $k\lambda_{Di} \sim 0.015$ would have a wavelength, $\lambda \sim 3$m. Figure 2 shows solutions of eq. (11), using eqs. (5) and (6), for the imaginary part of the frequency (normalized

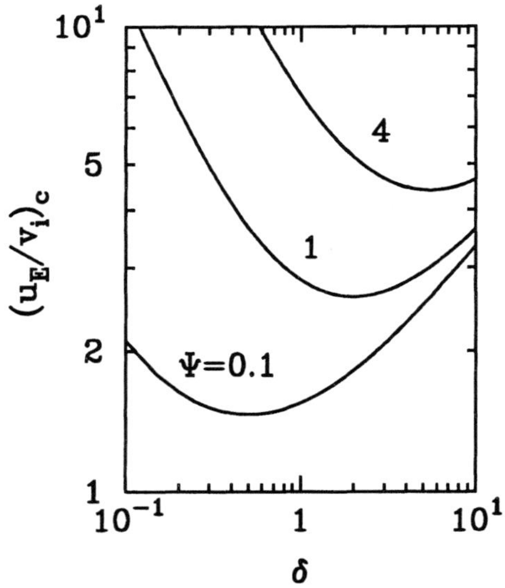

Figure 1: Critical electron drift from eq. (20) versus $\delta = n_i/n_e$, for values of Ψ which label the curves. Here $T_e = T_i$.

to ω_{pi}) versus electron drift speed u_E/v_i, at fixed $k\lambda_{Di} = 0.015$, and for various values of δ which label the curves. Nominal daytime plasma parameters at $h \sim 90$ km have been used. It can be seen that for $\delta = 2$, instability of Farley-Buneman modes with $\lambda \sim 3$ m could occur for $u_E/v_i \sim 7$. For $T_i \sim 150$ K, and $B_0 \sim 0.5$ G, this implies a perpendicular electric field strength of about $E_0 \sim 70$ mV/m. Such values of electric field strength (and even larger values) have been reported to occur sporadically in polar regions [Stauning and Olesen 1989].

2.3 Effect of Dust on High-Frequency Hall Current Instability

In addition to the Farley-Buneman instability in the frequency regime $\omega < \nu_{in}$, a shorter wavelength instability in the regime $\omega > \nu_{in}$ can be derived from eq. (11). This high-frequency Hall current instability [Lee *et al.* 1971] generally requires larger electron drift speeds for excitation than the longer wavelength Farley-Buneman instability. The presence of charged dust can also affect this instability, via its effect on equilibrium charge neutrality [Rosenberg 2001].

Following Rosenberg [2001], we again consider the flute limit $k_z = 0$ (so that $k = k_y$), assume that $\omega \gg \nu_{in}$, and that the wave phase speed $\omega/k \gg v_i$. In this case, eq. (6) for the ion susceptibility becomes

$$\chi_i \approx -\frac{\omega_{pi}^2}{\omega^2}\left[1 - \frac{i\nu_{in}}{\omega}\right], \tag{21}$$

neglecting ion Landau damping (which is small for $\omega/k \gg v_i$). For the electrons, we again use eq. (12) in the small electron Larmor radius limit.

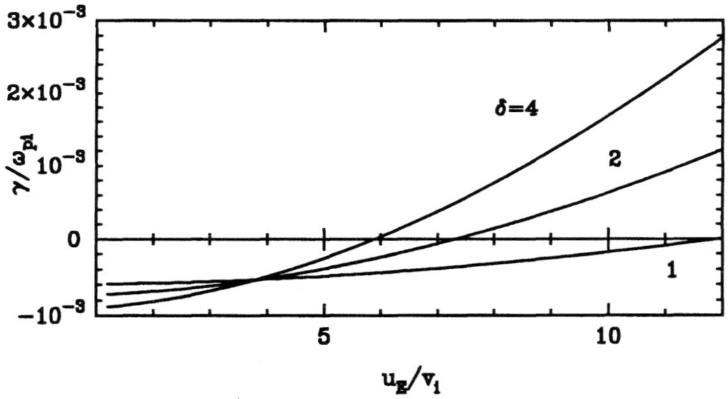

Figure 2: Solution of eq. (11) for the imaginary part of the frequency (normalized to ω_{pi}) versus u_E/v_i, for $k\lambda_{Di} = 0.015$ and various values of δ which label the curves. Other parameters are: $\omega_{pi}/\Omega_i = 210$, $m_i/m_p = 29$, $T_e = T_i$, $\nu_{en}/\omega_{pi} = 7.8$, $\nu_{in}/\omega_{pi} = 0.6$ (yielding $\Psi \approx 4$), and $k_z/k_y \approx 1.7 \times 10^{-4}$.

The dispersion relation resulting from eq. (11) with eqs. (12) and (21), in the limit $|\omega - ku_E| \gg \nu_{en} b_e^{1/2}$, is

$$1 + \frac{\omega_{pe}^2}{\Omega_e^2}\left[1 + \frac{i\nu_{en}}{\omega - ku_E}\right] - \frac{\omega_{pi}^2}{\omega^2}\left[1 - \frac{i\nu_{in}}{\omega}\right] \approx 0. \tag{22}$$

Eq. (22) an be solved for small growth or damping, with $\omega = \omega_r + i\gamma$, $|\gamma| \ll \omega_r$, to obtain [Rosenberg 2001]

$$\omega_r \approx \omega_{LH} = \frac{\omega_{pi}}{\sqrt{1+S}}, \tag{23}$$

where ω_{LH} is the ion lower hybrid frequency, with

$$S = \frac{\omega_{pe}^2}{\Omega_e^2} = \frac{\omega_{pi}^2}{\Omega_i^2}\frac{m_e}{m_i}\frac{1}{\delta} = \frac{R}{\delta}, \tag{24}$$

and

$$\frac{\gamma}{\omega_r} \approx \frac{S}{2(1+S)}\frac{\nu_{en}}{ku_E - \omega} - \frac{\nu_{in}}{2\omega_r}. \tag{25}$$

The presence of charged dust affects the instability via its effect on the the parameter S: negatively charged dust ($\delta > 1$) reduces S, while positively charged dust ($\delta < 1$) increases S [Rosenberg 2001]. From eqs. (23) and (25) it can be seen that, in the presence of negatively charged dust, the real part of the frequency, ω_r, can increase as δ increases, since S decreases as δ increases. This would be expected to result in smaller ion Landau and collisional damping. However, when δ becomes $> R$ (so that $S < 1$), the growth rate, γ, decreases, owing to a decrease in the driving term (see 1st term of eq. 25). Correspondingly, the critical drift increases. In the presence of positively charged dust, ω_r decreases as δ decreases, since S increases as δ decreases. In this case, the growth rate can also decrease, because ion Landau damping and collisional damping become relatively more important.

These trends can be seen in Figure 3 which plots the real and imaginary parts of the frequency ω_r and γ (normalized to ω_{pi}) versus $k\lambda_{Di}$, obtained by solving eq. (11) using eqs. (5) and (6), for various values of δ which label the curves. Results for some nominal plasma parameters representative of $h \sim 100$ km daytime conditions are considered, with $B \sim 0.5$ G, $m_i/m_p = 28$, $n_i \sim 10^5$ cm^{-3}, $T_e \sim T_i \sim 200$ K ($\lambda_{Di} \sim 0.5$ cm), and $n_n \sim 6 \times 10^{12}$ cm^{-3}. The electron drift speed is assumed to be large, with $u_E/v_i = 3$. Since R $(= \omega_{pe}^2/\Omega_e^2)$ is generally > 1 at altitudes about $95 - 100$ km in the low E-region, the presence of a substantial amount of negatively or positively charged dust would be required to significantly reduce growth rates [Rosenberg 2001]. Thus, it would be expected that the presence of a large amount of dust in a meteor trail, with charge density $\gtrsim$ the plasma density, would make it harder to excite this shorter wavelength instability.

2.4 Low-Frequency Hall Current Instability

The previous two subsections have focused on the effect of charged dust on ion wave instabilities driven by electron $\mathbf{E}_0 \times \mathbf{B}_0$ drifts in the low E-region. The critical electron drift for these instabilities is larger than the ion sound speed, since the unstable waves are ion waves. However, if dust is present in the lower E-region, and if it comprises sufficient charge density, there is the possibility of exciting very low frequency dust-acoustic type waves even when the electron drift speed is smaller than the ion thermal speed [Rosenberg and Shukla 2000, 2002]. Dust acoustic waves are associated with the dynamics of the charged dust grains [Rao *et al.* 1990]. The waves are sustained by the dust inertia and the pressure of the lighter plasma particles, and have phase speed $\ll v_i$.

Here we consider a very low frequency Hall current instability in the dust acoustic frequency regime, following Rosenberg and Shukla [2000, 2002]. Because the dynamics of the dust is essential for a dust acoustic mode, we retain the dust susceptibility in eq. (4) for the linear dispersion relation.

For the dust, we consider the non-resonant regime, with $\zeta_d \gg 1$, and assume weak dust-neutral collisions with $\nu_{dn} \ll \omega$. This implies that the phase speed of the dust acoustic type wave is much larger than the dust thermal speed. In this case, eq. (7) for the dust susceptibility becomes

$$\chi_d \approx -\frac{\omega_{pd}^2}{\omega^2}\left(1 - \frac{i\nu_{dn}}{\omega}\right). \tag{26}$$

For the ions, we consider the kinetic regime where $|\omega + i\nu_{in}| \ll kv_i$, (since $\omega \ll \nu_{in}$, this implies that the wavelength is $\ll$ the ion mean free path). Then eq.

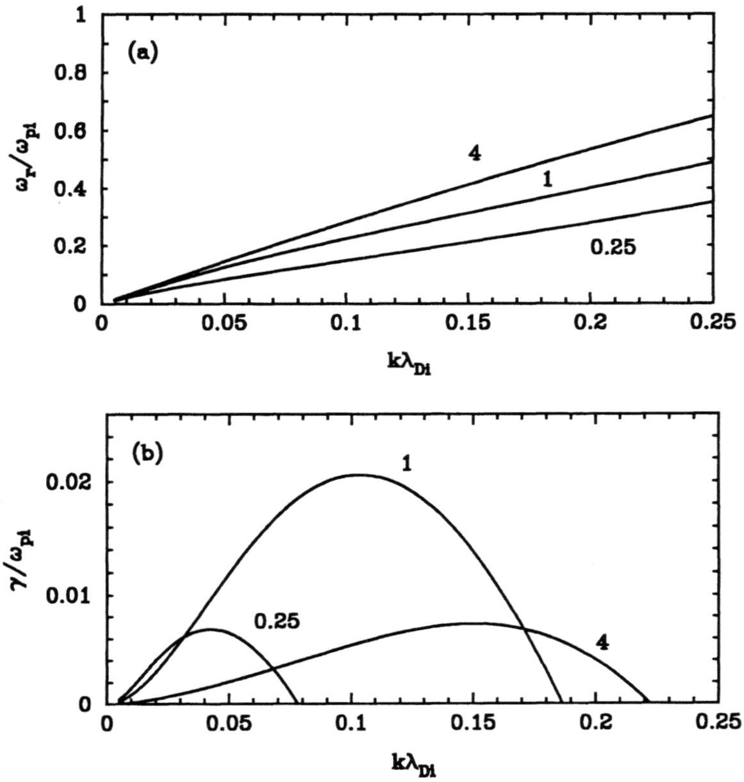

Figure 3: Solution of eq. (11) for the (a) real and (b) imaginary parts of the frequency (normalized to ω_{pi}) versus $k\lambda_{Di}$, for various values of δ which label the curves. Parameters are: $\omega_{pi}/\Omega_i = 380$, $m_i/m_p = 28$, $T_e = T_i$, $\nu_{en}/\omega_{pi} = 0.7$, $\nu_{in}/\omega_{pi} = 0.05$ (yielding $\Psi \approx 0.1$), $u_E/v_i = 3.0$, and $k_z/k_y \approx 1.7 \times 10^{-4}$.

(6) for the ion susceptibility becomes

$$\chi_i \approx \frac{1}{k^2\lambda_{Di}^2}. \tag{27}$$

In this case, the ions essentially provide Debye screening (the contribution to Landau damping being small since $\omega/k \ll v_i$).

For the electrons, we again consider the limit $\zeta_e \gg 1$, and assume $b_e \ll 1$, so that $\Gamma_0(b_e) \approx 1 - b_e$. Then, expanding the Z functions in eq. (5), we obtain for the electron susceptibility

$$\chi_e = \frac{1}{k^2\lambda_{De}^2}\left[b_e - \frac{1}{2\zeta_e^2}\right]\left[1 - \frac{i\nu_{en}[1-b_e]}{\bar{\omega} + i\nu_{en}}\left(1 + \frac{1}{2\zeta_e^2}\right)\right]^{-1}, \tag{28}$$

where $\bar{\omega} = \omega - k_y u_E$.

In the limit that $k_z = 0$, χ_e is again given by eq. (12). In this case, the dispersion relation given by eq. (4), using eqs. (12), (26) and (27), is

$$1+\frac{\omega_{pe}^2}{\Omega_e^2}\frac{\omega-ku_E+i\nu_{en}}{\omega-ku_E+i\nu_{en}b_e}+\frac{1}{k^2\lambda_{Di}^2}-\frac{\omega_{pd}^2}{\omega^2}\left(1-\frac{i\nu_{dn}}{\omega}\right)\approx 0\,. \tag{29}$$

Equation (29) can be solved for weak growth or damping, with $\omega=\omega_r+i\gamma$, $|\gamma|\ll\omega_r$. In the limit that $\nu_{en}b_e^{1/2}\ll|\bar{\omega}|$, this yields [Rosenberg and Shukla 2000]

$$\omega_r^2\approx\frac{k^2\lambda_{Di}^2\omega_{pd}^2}{1+k^2\lambda_{Di}^2\left(1+\omega_{pe}^2/\Omega_e^2\right)}, \tag{30}$$

and

$$\frac{\gamma}{\omega_r}\approx\frac{\omega_r^2}{2\omega_{pd}^2}\left(\frac{\omega_{pe}}{\Omega_e}\right)^2\frac{\nu_{en}}{(ku_E-\omega)}-\frac{\nu_{dn}}{2\omega_r}. \tag{31}$$

Instability may also occur for oblique propagation with finite $k_z\ll k_y$ [Rosenberg and Shukla 2002]. When $1/2\zeta_e^2\gg b_e$, and in the limits $\nu_{en}\gg|\bar{\omega}|\gg\nu_{en}b_e$ and $|\nu_{en}\bar{\omega}|\gg k_z^2v_e^2$, the electron susceptibility in eq. (28) becomes approximately

$$\chi_e\approx i\frac{k_z^2}{k^2}\frac{\omega_{pe}^2}{\bar{\omega}\nu_{en}}. \tag{32}$$

Then, the dispersion relation given by eq. (4), using eqs. (26), (27), and (32), becomes

$$1+i\frac{k_z^2}{k^2}\frac{\omega_{pe}^2}{\bar{\omega}\nu_{en}}+\frac{1}{k^2\lambda_{Di}^2}-\frac{\omega_{pd}^2}{\omega^2}\left(1-\frac{i\nu_{dn}}{\omega}\right)\approx 0\,. \tag{33}$$

Solving eq. (33) for weak growth or damping, the real part of the frequency is given approximately by eq. (30), while the imaginary part of the frequency is given by [Rosenberg and Shukla 2002]

$$\frac{\gamma}{\omega_r}\approx\frac{\omega_r^2}{2\omega_{pd}^2}\frac{k_z^2}{k^2}\frac{\omega_{pe}^2}{(k_yu_E-\omega)\nu_{en}}-\frac{\nu_{dn}}{2\omega_r}. \tag{34}$$

It can be seen from equations (31) and (34) for the growth rates that instability requires $ku_E>\omega_r$, i.e. that the electron drift speed must be larger than the phase speed of the dust acoustic type wave. It can also be seen by comparing these expressions that growth can maximize for wave propagation slightly oblique to $\mathbf{B}_0$, if the condition $k_z^2/k^2>(\nu_{en}/\Omega_e)^2$ is satisfied [subject also to the conditions on k_z before eqs. (30) and (32)].

There are several requirements on the dust and plasma parameters, that would need to be satisfied in order for such a dust acoustic type instability to occur [Rosenberg and Shukla 2000]. First, the dust Landau damping of the waves should be weak. Dust Landau damping is small when the phase speed of the wave is much larger than the dust thermal speed, that is, when $\omega_r/k\gg v_d$. The phase speed of the dust acoustic type wave is of the order of $\lambda_{Di}\omega_{pd}$ [from eq. (30)], which implies $\omega_r/k \sim Z_d(n_d/n_i)^{1/2}v_d$ (assuming $T_d \sim T_i$). Thus the condition $\omega_r/k \gg v_d$ generally requires that the dust charge state is large, i.e., that $Z_d \gg 1$. Second, damping due to dust-neutral collisions should be small for these waves. Collisional damping is small when the dust-neutral collision frequency is much smaller than the wave frequency, that is, when $\nu_{dn} \ll \omega_r$. Since the dust-neutral collision frequency (in the free molecular regime) scales as $\nu_{dn} \propto a^{-1}$, while ω_r is proportional to ω_{pd} which scales with grain radius a as $\omega_{pd} \propto a^{-1/2}$ (for fixed n_d), the condition $\omega_r \gg \nu_{dn}$ could be more easily satisfied for larger grain sizes. In addition, $\nu_{dn} \propto n_n$, so collisional wave damping would be smaller at lower gas pressure (i.e. higher ionospheric altitudes).

Following Rosenberg and Shukla [2000, 2002] we consider the application of this instability to dusty meteor trails in the low E-region. Dust may occur in meteor trails: it has been shown in studies by Rosinski and Snow [1961] and Hunten *et al.* [1980] that when meteors ablate, they can form dust by accretion. Recently, Kelley *et al.* [1998] have reported on the detection of a meteor contrail and meteoric dust in the Earth's upper mesosphere near about $h \sim 92$ km, presenting evidence that the ablated material in the trail may coagulate into particles of order 50 nm radius. If there is significant dust charge density in a meteor trail, that could lead to the possibility of dust wave excitation [see also Zhou and Kelley 1997].

If dust grains in the low E-region are negatively charged, Z_d would be small. For example, a dust grain of radius $a = 50$ nm that is charged negatively by plasma collection currents in a plasma with $T_e = 200$ K, would have a negative charge state on the order of $Z_d \sim 700\, a(\mu m)\, |\phi_s(V)| \sim 3$, assuming the grain surface potential $\phi_s \sim -4T_e/e$. Thus, this dust-acoustic type instability is unlikely in the case of negatively charged dust, owing to strong dust Landau damping, as discussed above. On the other hand, in the presence of solar UV radiation, dust may acquire a large positive charge, if the dust material has a low work function [e.g., Havnes *et al.* 1990; Rosenberg and Shukla 2002]. Work functions $W < 4$ eV may not be unreasonable for meteoric material, if it includes low work function metals such as Na ($W \sim 2.3$ eV), Ca ($W \sim 3.2$ eV) and Mg ($W \sim 3.7$ eV) [e.g., Rapp and Lubken 1999]. Further, the compound MgO ($W \sim 3.1$ eV) may comprise a significant fraction of the weight of a stony meteoroid [Furman 1960].

It has been shown theoretically that dust of low work function material in the upper mesosphere might be positively charged due to photoemission by solar UV radiation [see e.g., Havnes *et al.* 1990, 1996a; Rapp and Lubken 1999; Rosenberg and Shukla 2002]. However, there is some controversy about the magnitude of the positive dust charge state. Some authors [e.g., Rosenberg and Shukla 2002; Havnes *et al.* 1990] have estimated that the positive charge state Z_d could be $\gg 1$ for larger grains of size $\sim$ several tens of nm. For example, Rosenberg and Shukla [2002] estimated the equilibrium positive charge on a grain of material with work function < 4 eV, by balancing the electron and ion currents and the photoemission current to the grain together with the condition of overall charge neutrality given by eq. (1). Using parameters that may be representative of the daytime ionosphere at altitudes $h \sim 90 - 95$ km, and the solar flux of photons with energies between about $4 - 5$ eV, resulted in estimates of positive charge states $Z_d \gg 1$ (viz., $Z_d \sim 10 - 30$) for grains of size ~ 50 nm, over a range of dust densities from about 10 cm^{-3} to several thousand cm^{-3} [Rosenberg and Shukla 2002]. On the other hand, theoretical studies by Rapp and Lubken [1999] claim that the maximum positive charge on mesospheric dust is $\sim +4e$.

Here we consider a dusty meteor trail containing positively charged dust with $Z_d \gg 1$, following Rosenberg and Shukla [2000, 2002]. We use plasma parameters that may be representative of the daytime $h \sim 95$ km altitude of the low E-region, with $n_n \sim 10^{13}$ cm^{-3}, $n_i \sim 3 \times 10^4$ cm^{-3}, $B \sim 1/2$ G, and $T_i = T_e$ with $T_e \sim 200$ K [see e.g., Kelley 1989]. The ion Debye length is then $\lambda_{Di} \sim 1/2$ cm. We assume the dust grains have radius $a \sim 50$ nm, mass $m_d \sim 10^{-15}$ g, positive charge state $Z_d = 18$, and density $n_d \sim 2 \times 10^3$ cm^{-3}, [see Rosenberg and Shukla 2002], which implies $Z_d n_d / n_i \gtrsim 1$.

In Figure 4 are shown the real and imaginary parts of the frequency (normalized to ω_{pd}) versus $k\lambda_{Di}$ obtained by solving eq. (4), using eqs. (5), (6), and (7), for the following set of parameters: $m_i/m_p = 28$, $\omega_{pi}/\Omega_i = 240$, $\nu_{in}/\omega_{pi} = 0.1$, $\nu_{en}/\omega_{pi} = 1.8$, positively charged dust with $a = 0.05$ μm, $Z_d = 18$, $m_d/m_p = 4 \times 10^8$, $n_d/n_i = 0.067$, and $\nu_{dn}/\omega_{pi} = 4 \times 10^{-5}$. The electron drift speed $u_E/v_i = 1.5$. The solid lines are for $\theta = 90°$, the dashed lines are for $\theta = 89.6°$ and the dotted lines are for $\theta = 89.2°$, where $tan\,\theta = k_y/k_z$. It should be pointed out that the electron drift speed used in this figure is $\lesssim$ the the critical drift speed for the Farley-Buneman instability at these altitudes, estimated from eq. (20) [see Rosenberg and Chow 1998; Rosenberg 2001].

Figure 5 shows the real and imaginary parts of the frequency (normalized to ω_{pd}) versus u_E/v_i for the above parameters, at fixed $k\lambda_{Di} = 0.3$ and $\theta = 90°$. As can be seen, the critical drift speed can be less than the ion thermal speed [Rosenberg and Shukla 2000, 2002]. For this set of parameters, dust-acoustic waves may be excited by electron cross-field drifts u_E as low as $u_{0e}/v_i \sim 0.5$. Assuming $B \sim 0.5$ G, and $T_i = 200$ K, this implies an electric field of strength $E_0 \lesssim 10$ mV/m. While this value may be $\gtrsim$ ionospheric electric fields in the equatorial plane, high latitude ionospheric fields can be even higher, of the order of 10-50 mV/m [e.g., Mozer 1973; Kelley 1989].

The parameters used in Figs. 4 and 5 may relate to a large meteor [Rosenberg and Shukla 2002]. The meteor mass corresponding to the parameters used above is estimated as follows. Using $m_d \sim 10^{-15}$ g and $n_d \sim 2000$ cm^{-3}, this gives, for a meteor trail of length ~ 10 km and diameter ~ 40 m, a value of about 100 g for the meteoritic mass, which corresponds to a larger photographic meteor [see Ceplecha *et al.* 1998]. If the low frequency Hall current instability discussed above can occur in the dusty trail of a large meteor, this may have implications for radar scattering

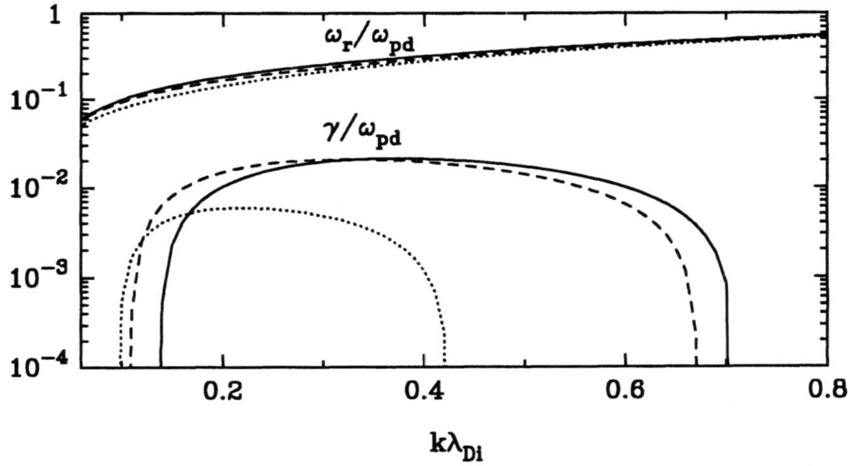

Figure 4: Solutions of eq. (4) for the real and imaginary parts of the frequency (normalized to ω_{pd}) versus $k\lambda_{Di}$, for $u_E/v_i = 1.5$ and for $\theta = 90^o$ (solid line), $\theta = 89.6^o$ (dashed line), an $\theta = 89.2^o$ (dotted line). Other parameters are: $\omega_{pi}/\Omega_i = 240$, $m_i/m_p = 28$, $T_e = T_i = T_d$, $\nu_{en}/\omega_{pi} = 1.8$, $\nu_{in}/\omega_{pi} = 0.1$, positively charged dust with $Z_d = 18$, $m_d/m_p = 4 \times 10^8$, $n_d/n_i = 0.067$, and $\nu_{dn}/\omega_{pi} = 4 \times 10^{-5}$.

[Rosenberg and Shukla 2000, 2002]. If waves with $k\lambda_{Di} \sim 0.3$ can be excited, this corresponds to unstable wavelengths λ on the order of $\lambda \sim 10$ cm. The Bragg condition for radar backscatter [see Kelley 1989] from such waves could be satisfied by radar waves with wavelength ~ 20 cm, which corresponds to a frequency $f \sim 1.5$ GHz.

3 Instabilities in Laboratory Dusty Plasma

Collisional effects also play an important role in instabilities in dusty plasmas in the laboratory. In this section, we discuss recent work by Rosenberg [2002] on a resistive ion-dust streaming instability. Then, we present two examples of electrostatic instabilities that could be caused by sheared plasma flow in a nonuniform collisional dusty plasma [Shukla and Rosenberg 2002]. Applications to laboratory dusty plasma wave experiments are pointed out.

3.1 Resistive Ion-Dust Streaming Instability

Following Rosenberg [2002], we consider a non-magnetized weakly ionized dusty plasma with three charged species: electrons, ions, and charged dust of uniform size. The condition of charge neutrality is given by eq. (1). We assume the equilibrium has an external electric field $\mathbf{E}_0$ (e.g., a discharge electric field in a glow discharge, or a sheath field in an rf discharge). The electron and ion drifts due to $\mathbf{E}_0$ are given

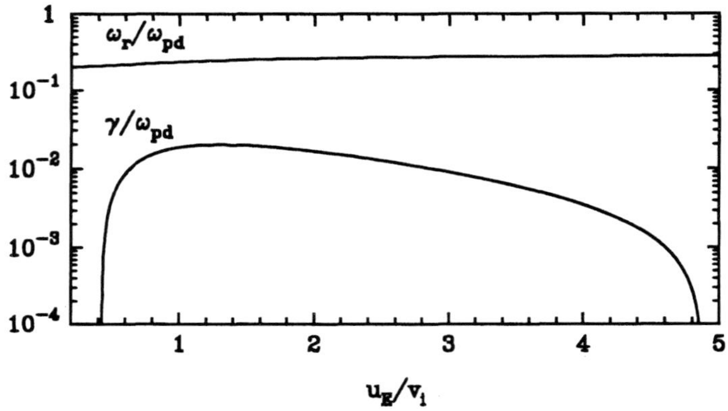

Figure 5: Solution of eq. (4) for the real and imaginary parts of the frequency (normalized to ω_{pd}) versus u_E/v_i, for $k\lambda_{Di} = 0.3$ and $\theta = 90°$. Other parameters are the same as in Fig. 4.

by [see e.g., D'Angelo and Merlino 1996]

$$\mathbf{u}_{0e} = \frac{-e\mathbf{E}_0}{m_e\nu_{en}}, \qquad (35)$$
$$\mathbf{u}_{0i} = \frac{e\mathbf{E}_0}{m_i\nu_{in}},$$

where the electron-neutral and ion-neutral collision frequencies, ν_{en} and ν_{in}, are given by

$$\nu_{en} = \sigma_{en} n_n v_e, \qquad (36)$$

$$\nu_{in} = \sigma_{in} n_n v_i, \qquad (37)$$

where σ_{en}, σ_{in} are the electron-neutral and ion-neutral collision cross sections, respectively. The dust-neutral collision frequency, ν_{dn}, in the free molecular regime is given by eq. (10).

Generally, it is expected that dust acoustic type waves could be excited primarily by ion streaming relative to the dust, since the ion drift speed can be on the order of v_i (or larger), while electron drift speeds are generally $\ll v_e$ in laboratory dusty plasmas. The instability of dust acoustic waves driven by streaming ions in collisional dusty plasmas has been investigated theoretically in various regimes by several authors [e.g., Rosenberg 1996b; D'Angelo and Merlino 1996; Kaw and Singh 1997; Mamun and Shukla 2000; Fortov *et al.* 2000]. Here, we will be concerned with a resistive instability in the frequency regime $\omega < \nu_{dn}$ [Rosenberg 2002]. This is motivated by observations of very low-frequency waves with $\omega < \nu_{dn}$ reported in several experiments [e.g., Fortov *et al.* 2000; Praburam and Goree 1996].

We neglect the small dust drift due to the electric field, and use drifting Maxwellians for electrons and ion distributions and a Maxwellian for the dust distribution. The linear dispersion relation for electrostatic waves with $\mathbf{k}$ in the $\mathbf{E}_0$ direction is given by [Rosenberg 2002; see also Miyamoto 1989]

$$1 + \sum_{\alpha} \frac{1}{k^2 \lambda_{D\alpha}^2} \frac{[1 + \zeta_\alpha Z(\zeta_\alpha)]}{\left[1 + (i\nu_{\alpha n}/\sqrt{2} k v_\alpha) Z(\zeta_\alpha)\right]} = 0\,, \tag{38}$$

where as in §2, $\alpha = e, i, d$ for electrons, ions, dust, respectively. Here $Z(\zeta_\alpha)$ is the plasma dispersion function [Fried and Conte 1961] with arguments

$$\zeta_e = \frac{\omega + k u_{0e} + i\nu_{en}}{\sqrt{2} k v_e}\,, \tag{39}$$

$$\zeta_i = \frac{\omega - k u_{0i} + i\nu_{in}}{\sqrt{2} k v_i}\,, \tag{40}$$

and

$$\zeta_d = \frac{\omega + i\nu_{dn}}{\sqrt{2} k v_d}\,. \tag{41}$$

We consider the regime $\nu_{in} \gg k v_i$ (wavelength $\gg$ the ion mean free path) and $\nu_{en} \ll k v_e$ (wavelength $\ll$ the electron mean free path). It may be possible to satisfy both these conditions in plasmas in which $\sigma_{en} \ll \sigma_{in}$, an example being certain low temperature ($T_e \sim 2$ eV) laboratory dusty plasmas using inert gases such as argon or neon [see Brown 1967]. We also consider the electron drift speed regime where $u_e/v_e \ll 1$, so that $\zeta_e \ll 1$. Further we assume non-resonant dust with $\zeta_d \gg 1$. In this case, the dispersion relation given by eq. (38) becomes

$$1 - \frac{\omega_{pi}^2}{BA_i} - \frac{\omega_{pd}^2}{BA_d} \approx 0, \tag{42}$$

where $B = 1 + (1/k^2 \lambda_{De}^2)$. Here

$$A_i \approx \Omega_i(\Omega_i + i\nu_{in}) - k^2 v_i^2, \tag{43}$$

in the limit that $\nu_{in} \gg \Omega_i$, where $\Omega_i = \omega - k u_{0i}$. In addition,

$$A_d \approx \omega(\omega + i\nu_{dn}) - k^2 v_d^2. \tag{44}$$

Now, assuming that $BA_d \ll \omega_{pd}^2$, and considering the frequency regime $\omega < \nu_{dn} \ll \nu_{in}$, eq. (42) can be solved for weak growth or damping, using $\omega = \omega_r + i\gamma$, with $|\gamma| \ll \omega_r$. This yields [Rosenberg 2002]

$$\omega_r \approx \frac{k u_{0i}}{1 + \psi}, \tag{45}$$

$$\gamma \approx \frac{\psi}{1 + \psi} \frac{1}{\nu_{dn}} \left[\omega_r^2 + \frac{\omega_{pd}^2}{\omega_{pi}^2} \left(k^2 u_{0i}^2 - k^2 v_i^2 \right) - k^2 v_d^2 \right], \tag{46}$$

where

$$\psi = \frac{\omega_{pi}^2}{\omega_{pd}^2} \frac{\nu_{dn}}{\nu_{in}}$$

To obtain eq. (46) it has also been assumed that $\Omega_i^2 \approx k^2 u_{0i}^2$, since generally the parameter $\psi \gg 1$ for laboratory dusty plasmas. It should be noted that that D'Angelo and Merlino [1996] have also considered a dust acoustic instability in higher collisionality regimes where $k < \nu_{dn}/c_d$, where c_d is the dust acoustic speed, using a fluid analysis and assuming cold dust with $T_d = 0$.

The critical ion drift speed can be obtained by setting $\gamma \approx 0$ in eq. (46). For cold dust, this yields [Rosenberg 2002]

$$\frac{u_{0i}}{v_i} \approx \left[1 + \left(\frac{\nu_{in}}{\omega_{pi}} \frac{\omega_{pd}}{\nu_{dn}}\right)^2\right]^{-1/2}, \tag{47}$$

where again $\psi \gg 1$ has been assumed. As can be seen, critical drifts $u_{0i} \gtrsim v_i$ should be sufficient for wave excitation.

It is interesting to note that this instability may persist even when the dust acoustic wave is heavily damped [Rosenberg 2002]. For example, strong dust Landau damping would occur when the phase speed of the dust acoustic wave, c_d, is on the order of the dust thermal speed, v_d [e.g., Rosenberg 1993]. For a dc glow discharge plasma in which $T_e/T_i \gg 1$, the dust acoustic speed is approximately $c_d \sim \lambda_{Di}\omega_{pd}$. Thus, $c_d \sim v_d$ implies $\lambda_{Dd} \sim \lambda_{Di}$. Now, if the ion drift speed is $u_{0i} = v_i$, it follows from eq. (46) that this instability can still occur for this value of u_{0i}, if the plasma parameters are such that roughly $(\nu_{dn}/\omega_{pd}) < (\nu_{in}/\omega_{pi})$.

Observations of fluctuations with inferred frequencies $\omega < \nu_{dn}$, in a dc glow discharge dusty plasma, were reported by Fortov *et al.* [2000]. A mechanism for wave excitation involving both ion streaming and dust charge variations was considered in the latter paper. Rosenberg [2002] pointed out that waves with $\omega < \nu_{dn}$ might be excited via the resistive instability discussed above, using a set of nominal representative parameters from Fortov *et al.* [2000]. However, it was found in Rosenberg [2002] that either the unstable frequencies or the unstable wavenumbers corresponding to this resistive instability, but not both together, may be consistent with the experimental observations reported in Fortov *et al.* [2000].

3.2 Instabilities Caused by Sheared Plasma Flows

In this section, we consider two examples of instabilities in a collisional dusty plasma, based on work that is in progress [Shukla and Rosenberg 2002]. In this case the parameters are such that both electrons and ions are magnetized, and in addition there are sheared flows. The characteristic frequency is much larger than the dust plasma frequency, so that the dust can be regarded as immobile. We assume that our nonuniform dusty magnetoplasma contains an equilibrium density gradient $(\partial n_\alpha/\partial x)$ and an equilibrium velocity gradient of the magnetic field-aligned ion flow $(\partial v_{i0}/\partial x)$ as well as the electron flow $(\partial v_{e0}/\partial x)$. The equilibrium magnetic field $B_0\hat{\mathbf{z}}$ is homogeneous.

In the presence of low-frequency ($\omega \ll \Omega_i$), long wavelength (in comparison with the ion gyroradius ρ_i) electrostatic field $\mathbf{E} = -\nabla\phi$, where ϕ is the electrostatic potential, the perpendicular components of the electron and ion fluid velocities are given by, respectively,

$$\mathbf{v}_{e\perp} \approx \frac{c}{B_0}\hat{\mathbf{z}} \times \nabla\phi - \frac{c}{eB_0 n_e}\hat{\mathbf{z}} \times \nabla p_e \,, \tag{48}$$

and

$$\begin{aligned} \mathbf{v}_{i\perp} \approx & \frac{c}{B_0}\hat{\mathbf{z}} \times \nabla\phi + \frac{c}{eB_0 n_i}\hat{\mathbf{z}} \times \nabla p_i \\ & -\frac{c}{B_0\Omega_i}\left(\frac{\partial}{\partial t} + v_{i0}\frac{\partial}{\partial z} + \mathbf{v}_{i\perp}\cdot\nabla + \nu_{in}\right)\nabla_\perp\phi \,, \end{aligned} \tag{49}$$

where $p_e = n_e T_e$ ($p_i = n_i T_i$) is the electron (ion) pressure, n_e (n_i) is the electron (ion) number density, and T_e (T_i) is the electron (ion) temperature.

Let us first derive the relevant equations for the coupled Shukla-Varma mode [Shukla and Varma 1993] and modified convective cells [Okuda and Dawson 1972] in the presence of sheared electron flows and the density gradient. Thus, substituting eqs. (48) and (49) into the equation for the modified charge density conservation

$$\frac{\partial \mathbf{E}}{\partial t} + 4\pi\nabla\cdot\mathbf{J} = 0 \,, \tag{50}$$

where the plasma current density is given by $\mathbf{J} = en_i\mathbf{v}_i - en_e\mathbf{v}_e$, we obtain the ion vorticity equation including the magnetic field-aligned electron velocity perturbation v_{ez}. We have

$$\left(\frac{\partial}{\partial t} + v_{i0}\frac{\partial}{\partial z} + V_{i*}\frac{\partial}{\partial y} + \nu_{in}\right)\nabla_\perp^2\phi + \Omega_i k_d\frac{\partial\phi}{\partial y} + \frac{\Omega_i B_0 n_{e0}}{cn_{i0}}\frac{\partial v_{ez}}{\partial z} = 0 \,, \tag{51}$$

where $V_{i*} = (cT_i/eB_0 n_i)\partial p_i/x$ is the ion diamagnetic drift speed, $k_d = n_i^{-1}\,\partial(Z_d n_d)/\partial x$, and v_{ez} is the parallel electron velocity. The parallel component of the electron momentum equation is given by

$$\left(\frac{\partial}{\partial t} + v_{e0}\frac{\partial}{\partial z} + \nu_{en}\right)v_{ez} = \frac{e}{m_e}\frac{\partial\phi}{\partial z} + \frac{c}{B_0}\frac{\partial v_{e0}}{\partial x}\frac{\partial\phi}{\partial y} \,, \tag{52}$$

where the ion plasma frequency is assumed to be much larger than the ion gyrofrequency and, for simplicity, the ion motion parallel to $\hat{\mathbf{z}}$ has been neglected. Equations (51) and (52) govern the dynamics of coupled Shukla-Varma and modified convective cells in the presence of a non-vanishing density gradient [the k_d-term in the left-hand side of eq. (51)] and the gradient of the magnetic field-aligned equilibrium electron flow velocity [cf. the second term in the right-hand side of eq. (52)].

Second, we present the governing equations for long wavelength coupled electron-drift dust-ion-acoustic waves in the presence of parallel ion flow gradients. Accordingly, we substitute eq. (49) in the ion continuity equation and make use of a Boltzmann electron response, viz.

$$\tilde{n}_e \approx n_e \frac{e\phi}{T_e}, \tag{53}$$

and assume quasineutrality $\tilde{n}_e \approx \tilde{n}_i$. Equation (53) is valid for $|\partial/\partial t| \ll \nu_{en} \ll \Omega_e$ and $\nu_{en}|\partial/\partial t| \ll v_e^2 \partial^2/\partial z^2$. To a leading order, we then obtain

$$\left(\frac{\partial}{\partial t} + v_{i0}\frac{\partial}{\partial z}\right)(\phi - \rho_s^2 \nabla_\perp^2 \phi) - \nu_{in}\rho_s^2 \nabla_\perp^2 \phi + V_{e*}\frac{\partial \phi}{\partial y} + \frac{n_i}{n_e}\frac{T_e}{e}\frac{\partial v_{iz}}{\partial z} = 0\,, \tag{54}$$

where $\rho_s = c_s/\Omega_i$ is the ion gyroradius at the electron temperature, $c_s = (n_i T_e/n_e m_i)^{1/2}$ is the dust ion-acoustic speed, $V_{e*} = \rho_s c_s/L_n$ is the electron drift speed, and $L_n = -n_i(\partial n_i/\partial x)^{-1}$ is the ion density gradient scale length.

The parallel component of the ion fluid velocity perturbation reads

$$\left(\frac{\partial}{\partial t} + v_{i0}\frac{\partial}{\partial z} + \nu_{in}\right) v_{iz} = -\frac{e}{m_i}(1+\tau)\frac{\partial \phi}{\partial z} + \frac{c}{B_0}\frac{\partial v_{i0}}{\partial x}\frac{\partial \phi}{\partial y}\,, \tag{55}$$

where $\tau = n_e T_i/n_i T_e$. Equations (54) and (55) govern the dynamics of coupled electron-drift dust ion-acoustic waves in the presence of sheared ion flows.

In order to study the linear properties of coupled Shukla-Varma and the modified convective cells as well as coupled electron-drift dust ion-acoustic waves in the presence of sheared plasma flows, we invoke the standard WKB approximation and assume that wavelengths of perturbations under consideration are much smaller than the scalelengths of the equilibrium density and plasma flow gradients, which are denoted by $n_\alpha(\partial n_\alpha/\partial x)^{-1}$ and $v_{\alpha 0}\,(\partial v_{\alpha 0}/\partial x)^{-1}$, respectively. Furthermore, we suppose that the perturbed quantities ϕ, v_{ez} and v_{iz} vary as $\exp(i\mathbf{k}\cdot\mathbf{r} - i\omega t)$. With this Ansatz, we Fourier transform eqs. (51) and (52) and combine the resultant equations to obtain

$$(\omega - k_z v_{e0} + i\nu_{en})(\omega - k_z v_{i0} - \omega_{i*} + i\nu_{in} - \omega_{sv}) = \omega_g^2 \frac{k_z^2}{k_\perp^2}\left(1 + \frac{k_y \omega_{v_e}}{k_z \Omega_e}\right), \tag{56}$$

where $\omega_{i*} = k_y V_{i*}$ is the ion drift wave frequency, $\omega_{sv} = \Omega_i k_y k_d/k_\perp^2$ is the Shukla-Varma frequency, $\omega_g = (n_e/n_i)^{1/2}(\Omega_e \Omega_i)^{1/2}$ is the modified lower-hybrid resonance frequency, and $\omega_{ve} = \partial v_{e0}/\partial x$.

It follows from eq. (56) that the modified convective cells in a collisionless dusty plasma with $\partial n_\alpha/\partial x = 0$ are stable for $\omega_{ve} > 0$. On the other hand, in the opposite limit, viz $\omega_{ve} < 0$, we have a purely growing instability if

$$\left|\frac{\partial v_{e0}}{\partial x}\right| > \frac{k_z}{k_y}\Omega_e\,. \tag{57}$$

The maximum growth rate of that instability is

$$\gamma_{max} \approx \omega_g \left(\frac{k_z k_y}{k_\perp^2}\right)^{1/2} \left|\frac{\omega_{v_e}}{\Omega_e}\right|^{1/2}. \tag{58}$$

In a collisional plasma with $|\omega - k_z v_{e0}| \ll \nu_{en}$, eq. (56) reduces to

$$\omega - k_z v_{i0} - \omega_{i*} + i\nu_{in} - \omega_{sv} = -i\frac{\omega_g^2}{\nu_{en}}\frac{k_z^2}{k_\perp^2}\left(1 + \frac{k_y\omega_{v_e}}{k_z\Omega_e}\right), \tag{59}$$

which predicts an oscillatory resistive instability if $\omega_{ve} < 0$ and eq. (57) holds. The maximum growth rate ($\gamma > \nu_{in}$) is

$$\gamma \approx \frac{\omega_g^2}{\nu_{en}}\frac{k_z k_y}{k_\perp^2}\frac{|\omega_{ve}|}{\Omega_e}. \tag{60}$$

The real part of the wave frequency is $\omega_r = k_z v_{i0} + \omega_{i*} + \omega_{sv}$. For example, consider a laboratory dusty argon plasma with the following parameters: $B_0 \sim 50$ G, $P \sim 0.1$ mtorr, $T_e \sim 1$ eV, and $T_i \sim 0.1$ eV. This gives $\Omega_e \sim 9 \times 10^8$ s^{-1}, $\Omega_i \sim 10^4$ s^{-1}, $v_e \sim 4 \times 10^7$ cm/s, $v_i \sim 5 \times 10^4$ cm/s, $\rho_e \sim 0.05$ cm, $\rho_i \sim 4$ cm, $\nu_{en} \sim 6 \times 10^4$ s^{-1}, and $\nu_{in} \sim 600$ s^{-1}. Assuming nearly perpendicular propagation, with $k_z/k_y \sim 10^{-3}$, condition (57) for instability would require an electron velocity gradient scale length of magnitude ≤ 40 cm (assuming $v_{e0} \sim v_e$). The growth rate, from eq. (60) is estimated as $\gamma_{max} \sim 10^3$ s^{-1}, using a velocity gradient scale length of ~ 5 cm.

Next, Fourier transformation of eqs. (54) and (55) and their combination yields the dispersion relation for the coupled electron-drift dust ion-acoustic waves in the presence of magnetic field-aligned ion sheared flow [see also Salimullah and Shukla 1998]

$$(\omega - k_z v_{i0})(\omega - k_z v_{i0} + i\nu_{in}) + (\omega - k_z v_{i0} + i\nu_{in})(\omega_{e*} + i\nu_{in}k_\perp^2\rho_s^2)$$
$$-\omega_s^2\left(1 + \tau - \frac{k_y\omega_{v_i}}{k_z\Omega_i}\right) = 0\,, \tag{61}$$

where $\omega_{e*} = k_y V_{e*}/(1 + k_\perp^2\rho_s^2)$ is the electron-drift wave frequency, $\omega_s = k_z c_s/(1 + k_\perp^2\rho_s^2)^{1/2}$ is the dust ion-acoustic wave frequency in a magnetized dusty plasma, and $\omega_{v_i} = \partial v_{i0}/\partial x$. The solution of eq. (61) for $|\omega - k_z v_{i0}| \gg \nu_{in}$ and $|\omega_{e*}| \gg \nu_{in}k_\perp^2\rho_s^2$ is

$$\omega - k_z v_{i0} = -\frac{\omega_{e*}}{2} \pm \frac{1}{2}\left[\omega_{e*}^2 + 4\omega_s^2\left(1 + \tau - \frac{k_y\omega_{v_i}}{k_z\Omega_i}\right)\right]^{1/2}, \tag{62}$$

which predicts an oscillatory instability if

$$\frac{\partial v_{i0}}{\partial x} > \frac{k_z}{k_y}\Omega_i \left(1 + \tau + \frac{\omega_*^2}{4\omega_s^2}\right) . \tag{63}$$

In a collisional plasma with $|\omega - k_z v_{i0}| \ll \nu_{in}$, we have from (61)

$$\omega - k_z v_{i0} + \omega_{e*} = -i\frac{\omega_s^2}{\nu_{in}}\left(1 + \tau - \frac{k_y \omega_{vi}}{k_z \Omega_i}\right) , \tag{64}$$

which exhibits a dissipative instability of the electron-drift waves if

$$\frac{\partial v_{i0}}{\partial x} > \frac{k_z}{k_y}\Omega_i (1 + \tau) . \tag{65}$$

Physically, the dissipative instabilities arise due to finite parallel fluid perturbations that cannot keep in phase with the wave potential oscillations in a plasma with velocity gradients. Thus, the free energy stored in the latter is channeled to unstable oscillations owing to collisions between the plasma particles and neutrals.

4 Summary

We have discussed recent theoretical work on several instabilities in collisional dusty plasmas, with application to dusty plasmas in the low E-region of the Earth's ionosphere and in laboratory environments. It was shown that charged dust can affect the Farley-Buneman instability in the low E-region at altitudes of about $h \sim 90 - 100$ km [Rosenberg and Chow 1998; Rosenberg 2001]. The presence of negatively charged dust can lower the critical electron drift at $h < 95$ km, while it can increase the critical drift at $h > 95$ km. This may have implications for PMSE or meteor echoes from dusty meteor trails in this low E-region [Rosenberg 2001]. Charged dust can also affect the higher frequency, shorter wavelength version of the Farley-Buneman instability, referred to as a high frequency Hall current instability. It was shown that the presence of a large amount of negatively charged or positively charged dust can reduce growth rates of this instability for daytime parameters at $h \sim 100$ km [Rosenberg 2001]. A new low-frequency Hall current instability in the dust acoustic frequency range was discussed, and applied to plasmas in the low E-region containing positively charged dust due to photoemission by solar UV radiation [Rosenberg and Shukla 2000, 2002]. Under certain conditions, including the presence of sufficient amounts of positively charged dust with charge state $Z_d \gg 1$, the critical electron drift can be smaller than the ion thermal speed. This may have application to radar scattering from dusty trails of large daytime meteors [Rosenberg and Shukla 2000, 2002]. A resistive ion-dust streaming instability in a non-magnetized dusty plasma was discussed with application to laboratory dusty plasmas. Under certain conditions, an instability with frequency less than the dust-neutral collision frequency may be excited by ions streaming relative to dust, with speed on the order of the ion thermal speed [Rosenberg 2002]. The instability may have application to observations of low frequency fluctuations in certain laboratory dc glow discharge dusty plasmas. We have also presented some preliminary results for the excitation of the Shukla-Varma mode [Shukla and Varma 1993] by sheared electron flows in a collisional dusty plasma [Shukla and Rosenberg 2002]. It is found that the combined effects of velocity shear and electron-neutral collisions may produce a Kelvin-Helmholtz like instability [see also D'Angelo 1965; D'Angelo and Song 1990]. Suggestions for observing such an instability in the laboratory have been made. Furthermore, we have also discussed the Kelvin-Helmholtz-like instability of electron-drift waves in the presence of the equilibrium ion shear flows.

Further work is planned to include effects of density gradients [e.g., Rosenberg and Krall 1994, 1996c] and boundary effects on these instabilities. For example, small scale density gradients can occur at the edges of meteor trails [e.g., Dyrud *et al.* 2001] and may also occur in dusty regions of the upper mesosphere [e.g. Havnes *et al.* 1996b]. Since dusty plasmas in the laboratory have finite extent, it is expected that boundary effects could alter the properties of ion-dust streaming instabilities in such plasmas [e.g., Rosenberg and Shukla 1998].

Acknowledgments

This work was supported in part by the following grants: DoE grant no. DE-FG03-97ER54444 (M. R.) and LANL/IGPP no. 02-1217 (M. R.). The authors acknowledge the support of the International Space Science Institute at Bern, Switzerland.

References

Baines M. F., Williams I. P. and Asebimo A. S. 1965, *Mon. Not. R. Astron. Soc.* **130**, 63.

Barkan A., Merlino R. L. and D'Angelo, N. 1995a, *Phys. Plasmas* **2**, 3563.

Barkan A., D'Angelo N. and Merlino R. L. 1995b, *Planet. Space Sci.* **43**, 905.

Barkan A., D'Angelo N. and Merlino R. L. 1996, *Planet. Space Sci.* **44**, 239.

Brown, S. C. 1967, *Basic Data of Plasma Physics* (MIT Press, Cambridge, MA).

Buneman O. 1963, *Phys. Rev. Lett.* **10**, 285.

Ceplecha Z., Borovicka J., Elford W. G., Revelle D. O., Hawkes R. L., Porubcan, V. and Simek M. 1998, *Space Sci. Rev.* **84**. 327.

Chapin E. and Kudeki E. 1994, *Geophys. Res. Lett.* **21**, 2433.

Cho J. Y. N. and Kelley M.C. 1993, *Rev. Geophys.* **31**, 243.

Cho J. Y. N. and Rottger J. 1997, *J. Geophys. Res.* **102**, 2001.

D'Angelo N. 1965, *Phys. Fluids* **8**, 1748.

D'Angelo N. and Song B. 1990, *Planet. Space Sci.* **38**, 1577.

D'Angelo N. and Merlino R. L. 1996, *Planet. Space Sci.* **44**, 1593.

Dimant Y. and Sudan R. N. 1995, *J. Geophys. Res.* **100**, 14605.

Dyrud L. P., Oppenheim M. M. and Vom Endt A. F. 2001, *Geophys. Res. Lett.* **28**, 2775.

Farley D. T. 1963, *J. Geophys. Res.* **63**, 6083.

Fortov V. E., Khrapak A. G., Khrapak S. A., Molotkov V. I., Nefedov A. P., Petrov O. F. and Torchinsky V. M. 2000, *Phys. Plasmas* **7**, 1374.

Fried, B. D. and Conte, S. D. 1961, *The Plasma Dispersion Function* (Academic Press, New York).

Furman A. M. 1960, *Soviet Astron. Astrophys. J.* **4**, 489.

Goertz C. K. 1989, *Rev. Geophys.* **27**, 271.

Havnes O., de Angelis U., Bingham R., Goertz C. K., Morfill G. E. and Tsytovich V. N. 1990, *J. Atmos. Terr. Phys.* **52**, 637.

Havnes O., Trøim J., Blix T., Mortensen W., Næsheim L. I., Thrane E. and Tønnesen T. 1996a, *J. Geophys. Res.* **101**, 10839.

Havnes O., Næsheim L. I., Hartquist T., Morfill G. E., Melandsø F., Schleicher B., Trøim J. Blix T. and Thrane E., 1996b, *Planet. Space Sci.* **44**, 1191.

Havnes O., Brattli A., Aslaksen T., Singer W., *et al.* 2001a, *Geophys. Res. Lett.* **28**, 1419.

Havnes O., Aslaksen T. and Brattli A. 2001b, *Phys. Scripta* **T89**, 133.

Horanyi M. 1996, *Annu. Rev. Astron. Astrophys.* **34**, 383.

Horanyi M., Robertson S., Smiley B., Gumbel J., *et al.* 2000, *et al. Geophys. Res. Lett.* **27**, 3825.

Hunten D. M., Turco R. P. and Toon O. B. 1980, *J. Atmos. Sci.* **37**, 1342.

Kaw P. and Singh R. 1997, *Phys. Rev. Lett.* **9**, 423.

Kelley, M. C. 1989, *The Earth's Ionosphere: Plasma Physics and Electrodynamics* (Academic Press, San Diego).

Kelley M. C., Alcala C. and Cho J. Y. N. 1998, *J. Atmos. Solar-Terr. Phys.* **60**, 359.

Lee K., Kennel C. F. and Kindel J. M. 1971, *Radio Sci.* **6**, 209.

Mamun A. A. and Shukla P. K. 2000, *Phys. Plasmas* **7**, 4412.

Melandsø F., Bjerkmo A., Morfill G., Thomas H. and Zuzic M. 2000, *Phys. Plasmas* **7**, 4368.

Mendis D. A. and Rosenberg M. 1994, *Annu. Rev. Astron. Astrophys.* **32**, 419.

Merlino R. L., Barkan A., Thompson C. and D'Angelo N. 1998, *Phys. Plasmas* **5**, 1607.

Miyamoto, K. 1989, *Plasma Physics for Nuclear Fusion* (MIT Press, Cambridge, MA)

Molotkov V. I., Nefedov A. P., Torchinskii V. M., Fortov V. E. and Khrapak A. G. 1999, *Sov. Phys. - JETP* **89**, 477.

Mozer F. S. 1973, *Rev. Geophys. Space Phys.* **11**, 755.

Northrop T. G. 1992, *Phys. Scripta* **45**, 475.

Nunomura S., Samsonov D. and Goree J. 2000, *Phys. Rev. Lett.* **84**, 5141.

Okuda H. and Dawson J. M. 1973, *Phys. Fluids* **16**, 408.

Oppenheim M. M., Vom Endt A. F. and Dyrud L. P. 2000, *Geophys. Res. Lett.* **27**, 3173.

Ossakow S. L., Papadopoulos K., Orens J. and Coffey T. 1975, *J. Geophys. Res.* **80**, 141.

Pellinen-Wannberg A. and Wannberg G. 1996, *J. Atmos. Terr. Phys.* **58**, 495.

Pieper J. B. and Goree J. 1996, *Phys. Rev. Lett.* **77**, 3137.

Prabhakara H. R. and Tanna, V. L. 1996, *Phys. Plasmas* **3**, 3176.

Praburam G. and Goree J. 1996, *Plasma Phys.* **3**, 1212.

Rao N. N., Shukla P. K. and Yu M. Y. 1990, *Planet. Space Sci.* **38**, 543.

Rapp M. and Lubken F. J. 1999, *Earth Planets Space* **51**, 799.

Rosenberg M. 1993, *Planet. Space Sci.* **41**, 229.

Rosenberg M. and Krall N. A. 1994, *Planet. Space Sci.* **42**, 889.

Rosenberg M. 1996a, in *The Physics of Dusty Plasmas*, edited by P. K. Shukla, D. A. Mendis and V. W. Chow (World Scientific, Singapore), pp. 129- 140.

Rosenberg M. 1996b, *J. Vac. Sci. Technol. A* **14**, 631.

Rosenberg M. and Krall N. A. 1996c, *Phys. Plasmas* **3**, 644.

Rosenberg M. and Chow V. W. 1998, *Planet. Space Sci.* **46**, 103.

Rosenberg M. and Shukla P. K. 1998, *Phys. Plasmas* **5**, 3786.

Rosenberg M. and Shukla P. K. 2000, *J. Geophys. Res.* **105**, 23135.

Rosenberg M. 2001, *IEEE Trans. Plasma Sci.* **29**, 261.

Rosenberg M. and Shukla P. K. 2002, *Planet. Space Sci.* **50**, 261.

Rosenberg M. 2002 "A note on ion-dust streaming instability in a collisional dusty plasma, *J. Plasma Phys.*, **67**, in press.

Rosinski J. and Snow R. H. 1961, *J. Meteorol.* **18**, 736.

Salimullah M. and Shukla P. K. 1998, *Phys. Plasmas* **5**, 4502.

Schmidt M. J. and Gary S. P. 1973, *J. Geophys. Res.* **78**, 8261.

Selwyn G. S. 1993, *Jpn. J. Appl. Phys.* **32**, 3068.

Shukla P. K. and Varma R. K. 1993, *Phys. Fluids B* **5**, 236.

Shukla P. K. and Rosenberg M. 1999, *Phys. Plasmas* **6**, 1038.

Shukla P. K. 2001, *Phys. Plasmas* **8**, 1791.

Shukla, P. K. and Mamun, A. A. 2002, *Introduction to Dusty Plasma Physics* (Institute of Physics Publishing Ltd., Bristol).

Shukla P.K. and Rosenberg M. 2002, paper in preparation.

Stauning P. and Olesen J. K. 1989, *Phys. Scripta* **40**, 325.

Thomas E. and Watson M. 1999, *Phys. Plasmas* **6**, 4111.

Thomas E. and Merlino R. L. 2001, *IEEE Trans. Plasma Sci.* **29**, 152.

Thompson C., Barkan A., D'Angelo N. and Merlino R. L. 1997, *Phys. Plasmas* **4**, 2331.

Thompson C., Barkan A., Merlino R. L. and D'Angelo N. 1999, *IEEE Trans. Plasma Sci.* **27**, 146.

Verheest F. 1996, *Space Sci. Reviews* **77**, 267.

Verheest, F. 2000, *Waves in Dusty Space Plasmas* (Kluwer Academic Publishers, Dordrecht).

Zhou Q. H. and Kelley M. C. 1997, *J. Atmos. Solar-Terr. Phys.* **59**, 739.

Self-Gravitation and Jeans Instabilities

Frank Verheest
Sterrenkundig Observatorium, Universiteit Gent,
Krijgslaan 281, B–9000 Gent, Belgium

Vladimir M. Čadež
Belgian Institute for Space Aeronomy,
Ringlaan 3, B–1180 Brussels, Belgium

Padma K. Shukla* and A. A. Mamun†
Fakultät für Physik und Astronomie,
Ruhr-Universität Bochum, D–44780 Bochum, Germany

Abstract

Self-gravitational effects are discussed in dusty plasmas, with due regard for the proper introduction of possible equilibrium configurations. An overview of physically acceptable models is given, with applications to relevant structures in space and astrophysical plasmas. For those cases where explicit solutions can be obtained, the scales over which the system is inhomogeneous are of the order of the Jeans lengths determined in the usual way by studying local perturbations of a uniform equilibrium. This has been discussed for electrostatic perturbations, where either the dust thermal velocity plays a role for neutral dust or the dust-acoustic velocity when the plasma pressure balances the dust collapse. Since the plasma pressure is much larger than that of the heavy dust grains, the system is homogeneous over much larger distances than if a neutral gas were considered. For strictly perpendicular propagation the dust-magnetosonic velocity tends to stabilize even larger distances. These results are applied to molecular dust clouds, where the relevant length scale can vary from fractions of an astronomical unit to distances much larger than the heliosphere.

1 Introduction

We aim in this chapter to discuss and combine results from two different physical domains, that of dusty plasma physics and that of self-gravitational collapse/stability

*Also at Department of Plasma Physics, Umeå University, S–90187 Umeå, Sweden
†Permanent address: Department of Physics, Jahangirnagar University, Savar, Dhaka, Bangladesh

of large scale mass distributions. This intertwining should come as no real surprise, because plasmas are so prevalent in astrophysics that Hannes Alfvén (Nobel Laureate 1970) coined the expression of a *Plasma Universe* to express what is really dominant, and it is often said that 99% of the observable universe is in the plasma state. Most of what we know about the universe has been learned by analyzing the electromagnetic radiation from these astrophysical plasmas that reaches us here on Earth.

1.1 Dusty Plasmas

One of the main other constituents of the observable universe is dust. By dust we refer to particles of dimensions that are large compared to atoms and ionized nuclei. The composition greatly varies with the astrophysical environment, and closer inspection reveals that the transition from gas to large dust particles in astrophysics is almost continuous, from macromolecules, clusters of molecules, small sub-micron sized grains, micron-sized grains, larger grains, to asteroid remnants.

The presence of dust has been known for a long time, from different types of remote observations, as for the dust around and between stars [Whittet 1992, Evans 1994, Hartquist *et al.* 1997]. There are beautiful examples of dust, like the molecular clouds seen in the Orion, Coalsack, Horsehead and Eagle nebulae. We observe these because of the attenuation and extinction of the light coming from more distant stars. There is also plenty of dust in the heliosphere, associated with planetary rings, cometary comae and tails, meteoric impacts, zodiacal light, to name only some of the more conspicuous occurrences. For the solar system, of course, we do not only rely on remote observations that still serve us well for objects outside the heliosphere, but have also collected a wealth of information from various space missions. These *in situ* measurements have been a resounding success, tremendously expanding our knowledge about space plasmas and the conditions in which they occur [Öpik 1956, Showalter 1992].

This brings us to the interplay, in space and astrophysics, between all pervading plasmas and ubiquitous dust. As dust grains are immersed in ambient plasmas and radiative environments, they become electrically charged by a variety of processes, the most simple ones being capture of plasma electrons and ions or photo-ionization. Although the idea of dust grain charging was put forward more than half a century ago by Spitzer [1941], it only came back on the agenda of space physicists when the very successful missions to planets and comets brought compelling evidence for the existence, indeed necessity of charged dust. Otherwise, if all the dust were strictly neutral, as assumed for a long time, some of the observed phenomena could not properly be explained. The term dusty plasma will refer here to the combination of ordinary plasma particles and charged dust grains. For further discussions about the many fascinating properties of dusty plasmas we can refer to reviews [Goertz 1989, Mendis and Rosenberg 1994, Horányi 1996, Verheest 1996, 1999, Shukla 2001] and books [Bliokh *et al.* 1995, Bouchoule 1999, Verheest 2000, Shukla and Mamun and books [Bliokh *et al.* 1995, Bouchoule 1999, Verheest 2000, Shukla and Mamun 2002].

This chapter is structured as follows. First of all, we emphasize in the remainder of this introduction why the interplay between self-gravitation and electromagnetic effects is important for heavier charged dust grains. Second, we recall in §2 the basic equations and discuss in §3 some of the possible stationary states, together with indications about the scale lengths on which the system varies. More specific comments about how to translate this information in the case of multispecies dusty plasmas is given in §4. Since it turns out that for those cases where explicit solutions can be obtained, the inhomogeneity length scales are of the order of the Jeans lengths, determined in the usual way by studying local perturbations of a homogeneous equilibrium, the Jeans swindle approach has been used in §5 to have an idea of the scales over which a dusty plasma is no longer truly homogeneous. This has first been done for electrostatic perturbations, where either the dust thermal velocity plays a role for neutral dust or the dust-acoustic velocity, when the plasma pressure balances the dust collapse. In the latter case, this pressure being much larger than that of the heavy dust grains, the system is homogeneous over much larger distances than if a neutral gas were considered. For strictly perpendicular propagation the dust-magnetosonic velocity comes into play, and that tends to stabilize even larger distances. As an application to get a feeling for the orders of magnitude involved, we consider in §6 molecular dust clouds, but argue that the paucity of reliable data cautions against taking the scale lengths too literally. Finally, our conclusions are recalled in §7.

1.2 Gravitational Collapse

Turning now to extended mass systems, these have been originally studied as neutral, uncharged matter, for which electromagnetic effects need not be considered. Contrary to electromagnetic forces, gravitation cannot be shielded but is always attractive, so that gravitational collapse of extended regions with distributed masses is inevitable, unless counteracted upon by thermal agitation or repulsion due to other causes. There is no way to make the gravitational potential disappear in zeroth order, as one can for the electrostatic potential or other electric field effects. This implies that a truly homogeneous equilibrium is in most cases not possible.

The resulting instability of a neutral cloud is probably the most basic instability in gravitating systems, studied in the astrophysical literature for a century, ever since Jeans obtained in 1902 the instability criterion that now bears his name, applicable to harmonic waves whose wavelength exceeds some critical value. His study was based upon the assumption that the unperturbed gaseous cloud is initially uniform [Jeans 1929]. The instability thus derived is non-oscillatory and purely growing, a macro instability corresponding to gravitational collapse of the gas or plasma. Besides the possible importance for star formation, Alfvén [1981] already expressed the idea that the solar system was thus formed out of a dusty plasma. Although this instability has now been abandoned as a possible explanation for solar system formation, it is still recognized as of fundamental importance in modern cosmology.

The main stumbling block is the determination of self-consistent but nonhomogeneous equilibrium configurations [Chandrasekhar 1961, Toomre 1964, Fridman and Polyachenko 1984, Larson 1985, Binney and Tremaine 1988, Nakano 1988, Gehman *et al.* 1996a,b, Whitworth 1998, McClements and Thyagaraja 2001]. This remains a vexing question, that is sometimes circumvented by considering local perturbations, whose wavelengths are small compared to the inhomogeneity scale lengths. In itself, there would be nothing wrong with this procedure, were it not that the locality condition can in most cases not be tested, because knowledge about the real equilibrium is lacking! This is known in the literature as the "Jeans swindle", whereby either a homogeneous equilibrium or else local perturbations are assumed, with complete disregard for how well internal consistency is obeyed. Small wonder that the limitations of such an approach were already pointed out for purely neutral gases [Spitzer 1978, Boss 1987, Čadež 1990, Vranješ and Čadež 1990].

1.3 Self-Gravitating Dusty Plasmas

The link of gravitation with dusty plasmas lies in the fact that for certain grain sizes the gravitational and electromagnetic effects can become comparable. In the heliosphere there are several instances where gravitation comes into play, e.g. the attraction between a planet and a particle of its ring system, which might easily be comparable to the electrodynamical forces between the ring particles themselves. Second, of course, there is the self-gravitational attraction between like grains, and third, there can be influences of self-gravitation on perturbations in the system, even though the Newtonian forces might be much smaller than the Coulomb forces.

To see when self-gravitational attraction is comparable to electrostatic repulsion for two identical particles with mass m and charge q that are interacting with each other at a distance r, we equate the Coulomb and Newtonian forces and get

$$\frac{q^2}{4\pi\varepsilon_0 r^2} = \frac{Gm^2}{r^2}. \tag{1}$$

Here G is the gravitational constant and ε_0 and μ_0 (to be used later) are the vacuum dielectric and magnetic permeability constants. For an ensemble of particles with mean number density N, this can be recast into the equality of plasma (ω_p) and Jeans (ω_J) frequencies through

$$\omega_p^2 \equiv \frac{Nq^2}{\varepsilon_0 m} = 4\pi GNm \equiv \omega_J^2. \tag{2}$$

For a particle with one electron charge this leads to masses of the order

$$m = \frac{e}{\sqrt{4\pi G\varepsilon_0}} = 1.8 \times 10^{-9} \text{ kg} \simeq 10^{18}\ m_p. \tag{3}$$

This very well known result shows that in ordinary electron-ion plasmas possible self-gravitational effects are so excruciatingly small that these are never even mentioned! However, for spherical charged dust grains made up of water ice with mass density ρ_{ice} the radius a corresponding to such a mass turns out to be of the order

$$a = \left(\frac{3m}{4\pi \rho_{\text{ice}}} \right)^{1/3} \simeq 80 \ \mu\text{m}. \tag{4}$$

For grains with higher charges, say a thousand electron charges or so, this gives millimeter-sized radii, rather larger than the micrometer-sized grains upon which attention has focused in solar-system plasmas. However, the existence of dust particles of this size does not seem totally unlikely, given the wide range of cosmic and astrophysical environments in which dusty plasmas are found. So the dusty plasma would be expected to be marginally stable, in the gravitational sense, to all density perturbations except those whose length scales might exceed the appropriate Jeans length, which will turn out to be typically very large — even on astrophysical scales [Verheest *et al.* 1998, Verheest 2000].

It is important to emphasize, however, that although for the stationary solutions of a self-gravitating plasma the gravitational and electrical forces can differ by orders of magnitude for smaller grains, this need not be the case for the (linear) instabilities. When more (plasma) physical detail, such as dust charge states, typical intergrain spacing versus Debye length, magnetic fields, electromagnetic effects, bulk particle flows, and plasma angular momentum is incorporated into the model, one finds that the stability of the dusty plasma (or of a plasma with dust) to collapse is far more complicated than the simple models would indicate [Mendis *et al.* 1982]. It thus becomes necessary to undertake a full investigation into all potentially stabilizing and destabilizing wave modes if we hope to understand the stability of cosmic dusty plasmas, and perhaps further our insight into the formation of protoplanets, protostars and protogalaxies. Another factor to keep in mind is rotation, as it is generally true that astrophysical bodies rotate [Toomre 1964]. Thus, in considering large astrophysical dusty plasma clouds, it is prudent to include also the effects of (differential) rotation.

2 Basic Equations

Because we will be discussing self-gravitational effects in dusty plasmas, a multi-species description is needed, even though we will also discuss neutral and simpler configurations in the one fluid model. Given the long scales involved in gravitational problems that are basically macroscopic in nature, there is no need for kinetic effects and we can use a multifluid approach [Baumjohann and Treumann 1996, Treumann and Baumjohann 1997, Choudhuri 1998, Krishan 1999].

2.1 Multifluid Description

The basic equations per species with label α [Verheest 2000] include the continuity equations

$$\frac{\partial}{\partial t}n_\alpha + \nabla \cdot (n_\alpha \mathbf{u}_\alpha) = 0, \tag{5}$$

where n_α is the number density and $\mathbf{u}_\alpha$ the fluid velocity. Similarly, the equations of motion are

$$\left(\frac{\partial}{\partial t} + \mathbf{u}_\alpha \cdot \nabla\right)\mathbf{u}_\alpha + \frac{1}{n_\alpha m_\alpha}\nabla p_\alpha = \frac{q_\alpha}{m_\alpha}(\mathbf{E} + \mathbf{u}_\alpha \times \mathbf{B}) - \nabla\psi. \tag{6}$$

Here p_α, q_α and m_α refer to the pressure, charge and mass of the respective constituents. Besides the Lorentz force involving the electric field $\mathbf{E}$ and the magnetic induction $\mathbf{B}$, possible self-gravitational effects occur through the gradient of the corresponding potential ψ. The electromagnetic fields obey Maxwell's equations

$$\nabla \times \mathbf{E} + \frac{\partial}{\partial t}\mathbf{B} = \mathbf{0}, \tag{7}$$

$$c^2\nabla \times \mathbf{B} = \frac{\partial}{\partial t}\mathbf{E} + \frac{1}{\varepsilon_0}\sum_\alpha n_\alpha q_\alpha \mathbf{u}_\alpha, \tag{8}$$

$$\nabla \cdot \mathbf{E} = \frac{1}{\varepsilon_0}\sum_\alpha n_\alpha q_\alpha, \tag{9}$$

$$\nabla \cdot \mathbf{B} = 0, \tag{10}$$

and the current and charge densities are expressed for a multispecies plasma. To close the set of basic equations, we have the gravitational Poisson's equation

$$\nabla^2\psi = 4\pi G \sum_\alpha n_\alpha m_\alpha, \tag{11}$$

and need to assume something about how the pressures behave. For simplicity we use either barotropic pressures, where the pressure of each species depends only on its density $p_\alpha(n_\alpha)$, which includes polytropic and isothermal pressures, or else assume adiabatic pressure variations. The description of neutral gases or mass systems is included in this basic framework, by simply turning of all charges and omitting electromagnetic effects.

2.2 Magnetohydrodynamic Description

For certain applications, we can look at the plasma as a whole and find it useful to go to the magnetohydrodynamic (MHD) approximation, whereby all charge separation disappears and we deal with one charged fluid. Elimination of the electric field leaves us with the reduced set of MHD equations of continuity, induction and momentum, augmented by the gravitational Poisson's equation,

$$\frac{\partial}{\partial t}\rho + \nabla \cdot (\rho \mathbf{u}) = 0, \tag{12}$$

$$\frac{\partial}{\partial t}\mathbf{B} = \nabla \times (\mathbf{u} \times \mathbf{B}), \tag{13}$$

$$\left(\frac{\partial}{\partial t} + \mathbf{u} \cdot \nabla\right)\mathbf{u} + \frac{1}{\rho}\nabla p + \nabla\psi = \frac{1}{\mu_0 \rho}(\nabla \times \mathbf{B}) \times \mathbf{B}, \tag{14}$$

$$\nabla^2\psi = 4\pi G\rho. \tag{15}$$

Now ρ, $\mathbf{u}$ and p refer to the charged fluid mass density, velocity and pressure, respectively.

3 Stationary Configurations

In this section we review three specific equilibrium configurations, among the many that might be imagined or are mathematically possible, but not all of equal physical interest. By equilibrium we will actually mean stationary states, where properties can vary from one place to the other but there is no (more) evolution in time. The states we consider are one-dimensional or slab models, or else configurations with cylindrical or spherical symmetry.

There are a couple of preliminary remarks to be made here. First of all, although one could include flows in a Cartesian one-dimensional or slab model, it is difficult to envisage a realistic astrophysical situation to which this might apply, and thus flows will be omitted here. Indeed, mass would flow in one direction, there would be depletion or a source at one end and an accumulation or sink at the other, which is hardly realistic. Similarly, spherically symmetric states are of obvious astrophysical interest, but in this configuration magnetic fields cannot be included, since these would break the spherical symmetry, in view of Gauss law. For the same reason flows are hard to imagine here, unless there is a total mass pile-up or depletion at the center of symmetry. After that, we discuss an axisymmetric state, where the only acceptable flows are azimuthal, because otherwise we would need mass sources or sinks on the axis of symmetry, or bulk motion would occur along the axis.

How these results can be extended to the dusty plasmas that we have in mind as our ultimate goal will be discussed in §4, once we understand the basic picture.

3.1 Configurations without Flows

In this subsection we will give a general discussion, with particular application to one-dimensional slab-like or spherically symmetric configurations, without flows.

3.1.1 Single fluids

In the absence of flows the equation of motion (14) reduces to

$$\nabla\left(p_0+\frac{B_0^2}{2\mu_0}\right)+\rho_0\nabla\psi_0=\frac{1}{\mu_0}(\mathbf{B}_0\cdot\nabla)\mathbf{B}_0. \tag{16}$$

Subscripts 0 refer to the stationary state. If $\mathbf{B}_0$ has a component in the direction of the one-dimensional gradient, the whole equilibrium field is a strict constant everywhere and vanishes from the description. Otherwise, the right-hand side of (16) vanishes but the gradient of B_0^2 remains on the left-hand side. Since the latter is a more general picture, we will continue with it.

To relate the information obtained so far to Poisson's gravitational law, we need to introduce some hypotheses about the pressure and about the magnetic field. We take a polytropic pressure, where $p=C\rho^\gamma$, with γ the polytropic index, and assume the magnetic field to vary in such a way that the ratio β of the plasma pressure to the magnetic pressure $B_0^2/2\mu_0$ remains constant. In this model the Alfvén speed V_A is constant, and the magnetic field and plasma pressure are stronger in regions with a higher plasma density ρ_0. This is rather realistic and may occur in highly conductive plasmas with frozen-in magnetic fields [Verheest *et al.* 2000a].

The magnetohydrostatic equilibrium balance equation reduces to

$$\frac{1+\beta}{\beta}\frac{dp_0}{dz}+\rho_0\frac{d\psi_0}{dz}=0. \tag{17}$$

Together with (15), this yields a single equation for ρ_0,

$$\rho_0\frac{d^2\rho_0}{dz^2}+(\gamma-2)\left(\frac{d\rho_0}{dz}\right)^2+\frac{4\pi G\beta}{C\gamma(1+\beta)}\rho_0^{4-\gamma}=0. \tag{18}$$

The simplest possibility is to assume isothermal pressures, in which case $\gamma=1$ and $C=c_s^2$ is the sound speed squared, and (18) has the solution

$$\rho_0(z)=\rho_{00}\mathrm{sech}^2(z/L). \tag{19}$$

The typical scale length L is defined by $L^2=(V_A^2+2c_s^2)/4\pi G\rho_{00}=(V_A^2+2c_s^2)/\omega_J^2$, where ρ_{00} is the density at the center of the cloud, with an extremum at $z=0$, because the gravitational force vanishes there. The (central) Jeans frequency ω_J is given through $\omega_J^2=4\pi G\rho_{00}$. As we will see later, the inhomogeneity scale length is very closely related to the corresponding Jeans length, obtained from linear stability analysis in a homogeneous medium.

We would like to stress that the Jeans swindle has not been used, because the present choice is fully compatible with all equilibrium constraints. It might well turn out that such a state is unstable in the lateral directions, perpendicular to

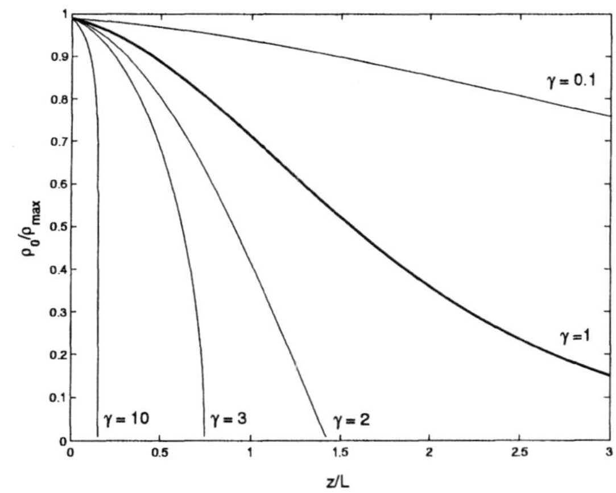

Figure 1: Density distributions vs. distance from the central location for various values of γ.

the chosen direction z, but that is quite different from starting with a completely uniform equilibrium. The magnetic field is

$$B_0(z) = B_{00}\mathrm{sech}(z/L), \tag{20}$$

where the central field strength B_{00} is given through $B_{00}^2 = \mu_0 V_A^2 \rho_{00}$, as V_A is constant.

Other possibilities have been numerically investigated, when $\gamma > 1$ and are illustrated in figure 1. Such models turn out to have well behaved pressures but to be of finite extent. This might be an indication that fragmentation could occur, but this interesting suggestion warrants further investigation. Moreover, the variation of the density near the maximum central value is then so abrupt that local perturbations would have to be on small scales that are not compatible with the usual Jeans approach. Models with $\gamma < 1$ have an unphysical pressure behaviour that increases outwardly from the center without bounds.

3.1.2 Multifluid models

The set of multifluid equations (5)–(10) indicate that for stationary states without flows the magnetic field decouples from the other physical quantities and has a vector potential that is a harmonic function. Hence, a stationary magnetic field is possible but need not be dealt with at this stage, since it does not influence the basic configuration. Furthermore, the equilibrium electric field is electrostatic and derivable from an electrostatic potential φ. The basic set of equations reduces to

$$\frac{1}{n_{\alpha 0}}\nabla p_{\alpha 0} + q_\alpha \nabla \varphi_0 + m_\alpha \nabla \psi_0 = 0,$$

$$\nabla^2 \varphi_0 + \frac{1}{\varepsilon_0} \sum_\alpha n_{\alpha 0} q_\alpha = 0,$$
$$\nabla^2 \psi_0 = 4\pi G \sum_\alpha n_{\alpha 0} m_\alpha, \tag{21}$$

and it is through the relation between pressures and densities that the set is closed. Taking again $p_\alpha = C_\alpha n_\alpha^{\gamma_\alpha}$, where γ_α is the polytropic index for species α, gives after one integration a Bernoulli type result

$$\frac{C_\alpha \gamma_\alpha}{\gamma_\alpha - 1} n_{\alpha 0}^{\gamma_\alpha - 1} + q_\alpha \varphi_0 + m_\alpha \psi_0 = D_\alpha \qquad \text{for} \quad \gamma_\alpha \neq 1,$$
$$C_\alpha \ln n_{\alpha 0} + q_\alpha \varphi_0 + m_\alpha \psi_0 = D_\alpha \qquad \text{for} \quad \gamma_\alpha = 1. \tag{22}$$

In the latter case $C_\alpha = \kappa T_\alpha$, where κ is Boltzmann's constant. The density can thus easily be expressed in terms of the potentials,

$$n_{\alpha 0} = n_{\alpha 00} \exp\left(- \frac{q_\alpha \varphi_0 + m_\alpha \psi_0}{\kappa T_\alpha} \right). \tag{23}$$

Here the constant densities $n_{\alpha 00}$ have been defined at some given reference location, preferably at the center of the cloud where e.g. the gravitational force vanishes and ψ_0 reaches an extremum. Two coupled Poisson equations remain to be solved for the potentials,

$$\nabla^2 \varphi_0 = -\sum_\alpha \frac{\omega_{p\alpha}^2 m_\alpha}{q_\alpha} \exp\left(- \frac{q_\alpha \varphi_0 + m_\alpha \psi_0}{\kappa T_\alpha} \right),$$
$$\nabla^2 \psi_0 = \sum_\alpha \omega_{J\alpha}^2 \exp\left(- \frac{q_\alpha \varphi_0 + m_\alpha \psi_0}{\kappa T_\alpha} \right). \tag{24}$$

The respective plasma ($\omega_{p\alpha}$) and Jeans ($\omega_{J\alpha}$) frequencies have been defined by using the central density values $n_{\alpha 00}$, because one has to be very careful by not assuming equilibrium densities to be constant when these quantities later crop up when investigating wave phenomena and Jeans instabilities. It is clear that analytic solutions of the coupled set (24) can only be discussed for specific cases where some simplifying assumptions help to reduce the mathematical complexity. This will be done for the usual dusty plasma model in the next section.

Another way of looking at the problem is by not immediately integrating the individual equations of motions but adding them first, so that the Poisson equations can be used in

$$\sum_\alpha \nabla p_{\alpha 0} + \left(\sum_\alpha n_\alpha q_\alpha \right) \nabla \varphi_0 + \left(\sum_\alpha n_\alpha m_\alpha \right) \nabla \psi_0 = 0 \tag{25}$$

to replace the total charge and mass densities. For strictly one-dimensional states this can be integrated to obtain a global Bernoulli type result,

$$\frac{1}{2\varepsilon_0} \left(\frac{d\varphi_0}{dz} \right)^2 - 2\pi G \left(\frac{d\psi_0}{dz} \right)^2 + \sum_\alpha p_{\alpha 0} = \mathcal{E}. \tag{26}$$

Since all $p_{\alpha 0}$ are supposed to depend on φ_0 and ψ_0 through n_α, (26) can be viewed as the energy integral for a system with an equivalent kinetic energy that is not positive definite! Unfortunately, the mechanics literature is almost silent on what happens in such situations, which again shows that the inclusion of gravitation is a much more drastic change than the simple discussion of Jeans like instabilities would lead one to believe.

3.2 Cylindrical Axisymmetry

For problems with cylindrical symmetry we go to the appropriate coordinates, assume that there is no dependence on the azimuthal angle ϑ and that all quantities, including flows, depend on the distance r from the axis of symmetry and on the coordinate z along the axis. As explained, we assume for physical plausibility that all flows are azimuthal only, to avoid mass depletion or pile up, and omit the index ϑ on the azimuthal velocity components. Some results can be taken over from the previous subsection, such as a potential electric field. The other nonzero equations give

$$\frac{1}{n_{\alpha 0}}\frac{\partial p_{\alpha 0}}{\partial r} - \frac{m_\alpha u_{\alpha 0}^2}{r} + q_\alpha \frac{\partial \varphi_0}{\partial r} - q_\alpha u_{\alpha 0} B_{z0} + m_\alpha \frac{\partial \psi_0}{\partial r} = 0, \tag{27}$$

$$\frac{1}{n_{\alpha 0}}\frac{\partial p_{\alpha 0}}{\partial z} + q_\alpha \frac{\partial \varphi_0}{\partial z} + q_\alpha u_{\alpha 0} B_{r0} + m_\alpha \frac{\partial \psi_0}{\partial z} = 0, \tag{28}$$

$$\frac{\partial B_{r0}}{\partial z} - \frac{\partial B_{z0}}{\partial r} = \mu_0 \sum_\alpha n_{\alpha 0} q_\alpha u_{\alpha 0}, \tag{29}$$

$$\frac{1}{r}\frac{\partial}{\partial r}(r B_{r0}) + \frac{\partial B_{z0}}{\partial z} = 0, \tag{30}$$

$$\frac{1}{r}\frac{\partial}{\partial r}\left(r\frac{\partial \varphi_0}{\partial r}\right) + \frac{\partial^2 \varphi_0}{\partial z^2} + \frac{1}{\varepsilon_0}\sum_\alpha n_{\alpha 0} q_\alpha = 0, \tag{31}$$

$$\frac{1}{r}\frac{\partial}{\partial r}\left(r\frac{\partial \psi_0}{\partial r}\right) + \frac{\partial^2 \psi_0}{\partial z^2} = 4\pi G \sum_\alpha n_{\alpha 0} m_\alpha. \tag{32}$$

Constant density (and pressure) models are possible, under a number of rather restrictive assumptions: there are no axial variations, no magnetic field effects, and the whole plasma is charge and current neutral. This is a very simplified physical system, and what remains is that the plasma as a whole rotates like a rigid body around the axis of symmetry, since from

$$-\frac{u_{\alpha 0}^2}{r} + \frac{d\psi_0}{dr} = 0, \tag{33}$$

$$\frac{1}{r}\frac{d}{dr}\left(r\frac{d\psi_0}{dr}\right) = 4\pi G \sum_\alpha n_{\alpha 0} m_\alpha, \tag{34}$$

there follows that $u_{\alpha 0} = r\omega_J/\sqrt{2}$, where the global Jeans frequency has been introduced through $\omega_J^2 = \sum_\alpha \omega_{J\alpha}^2 = \sum_\alpha 4\pi G n_{\alpha 0} m_\alpha$. The Jeans frequency shows up here in a different role from what it normally is supposed to govern, namely that of the collapse timescales. Moreover, it is the rotation that inhibits the inward gravitational collapse, as known already a long time from purely neutral gas considerations, because the derivations no longer refer to the charges of the constituents.

In typical dusty plasmas it is clearly the heavy dust component that will be responsible for the rotation owing to the gravitational effects, and the lighter plasma constituents are essentially tied to the dust by the charge and current neutrality. Although this case leads to stationary states that were known from neutral gases, the perturbations will yield different results because of the charged components.

For magnetized plasmas that are translationally invariant along the symmetry axis, the basic equations reduce to

$$\frac{1}{n_{\alpha 0}}\frac{dp_{\alpha 0}}{dr} - \frac{m_\alpha u_{\alpha 0}^2}{r} + q_\alpha \frac{d\varphi_0}{dr} - q_\alpha u_{\alpha 0} B_{z0} + m_\alpha \frac{d\psi_0}{dr} = 0, \tag{35}$$

$$\frac{dB_{z0}}{dr} + \mu_0 \sum_\alpha n_{\alpha 0} q_\alpha u_{\alpha 0} = 0, \tag{36}$$

$$\frac{1}{r}\frac{d}{dr}\left(r\frac{d\varphi_0}{dr}\right) + \frac{1}{\varepsilon_0}\sum_\alpha n_{\alpha 0} q_\alpha = 0, \tag{37}$$

$$\frac{1}{r}\frac{d}{dr}\left(r\frac{d\psi_0}{dr}\right) = 4\pi G \sum_\alpha n_{\alpha 0} m_\alpha. \tag{38}$$

This set of equations contains too many unknowns, so that some additional constraints need to be imposed. Without flows the magnetic field decouples from the description and the discussion rejoins that of neutral gases, and closely runs along lines that were touched upon in previous subsections. A constant magnetic field imposes current neutrality, but that requirement is in itself not enough to determine the azimuthal flow.

4 Multispecies Dusty Plasmas

In order to see the implications for dusty plasmas and at the same time keep some of the complications that always crop up in this class of problems under control, we use the standard model for low-frequency dusty plasma phenomena. The inherent simplifications on the plasma side are that the electrons and normal ions are massless, in the sense that their inertia can be neglected because the physical phenomena occur on such a slow timescale that instant balance is achieved between the various forces. We will follow here the lines of recent treatments [Tsintsadze *et al.* 2000, Rao *et al.* 2001].

To start with the stationary values, the corresponding electron and ion densities, derived in (23), reduce to the standard Boltzmann expressions as if gravitational effects were absent,

$$n_{e0} = n_{e00} \exp\left(\frac{e\varphi_0}{\kappa T_e}\right),$$

$$n_{i0} = n_{i00} \exp\left(-\frac{e\varphi_0}{\kappa T_i}\right). \tag{39}$$

On the other hand, the charged dust is treated as monodisperse, all grains having average charges and masses. The two coupled Poisson equations (24) for the equilibrium potentials are then treated in a different way. Charge neutrality is assumed to be almost perfect at all locations [Tsintsadze *et al.* 2000], so that

$$n_{i00} e \exp\left(-\frac{e\varphi_0}{\kappa T_i}\right) - n_{e00} e \exp\left(\frac{e\varphi_0}{\kappa T_e}\right) + n_{d00} q_d \exp\left(-\frac{q_d\varphi_0 + m_d\psi_0}{\kappa T_d}\right) \simeq 0. \tag{40}$$

The physical interpretation is that the electrons and ions are tied to the charged dust by electrostatic forces. It is clear that this condition means that φ_0 becomes expressible as a function of ψ_0, or vice versa. What remains of the other Poisson equation,

$$\nabla^2 \psi_0 = \omega_{Jd}^2 \exp\left(-\frac{q_d\varphi_0 + m_d\psi_0}{\kappa T_d}\right), \tag{41}$$

will reduce to an equation in ψ_0 alone when the information from (40) is inserted. Since the dust is the only constituent of the plasma mixture to feel any gravitation, that is how the mass gets distributed. The electrons and ions follow suit through their electrostatic coupling to the charged dust.

To see how acceptable this often invoked hypothesis really is, we simplify the dusty plasma model to one where all electrons are supposed to have been accreted onto the (negative) dust grains, and there are at the central location as many ions as dust charges, so that $n_{e00} \simeq 0$ and $n_{i00} \simeq Z_d n_{d00}$, with $q_d = -Z_d e$ expressed in terms of the absolute charge number. Retaining the Laplacians in both equations, but in a one-dimensional model, and using compact notations, means that we have to study solutions of

$$\begin{aligned} \lambda_{Di}^2 \frac{d^2\widetilde{\varphi}_0}{dz^2} &= \exp\left(\tau\widetilde{\varphi}_0 - \widetilde{\psi}_0\right) - \exp\left(-\widetilde{\varphi}_0\right), \\ \lambda_{Jd}^2 \frac{d^2\widetilde{\psi}_0}{dz^2} &= \exp\left(\tau\widetilde{\varphi}_0 - \widetilde{\psi}_0\right). \end{aligned} \tag{42}$$

Here λ_{Di} is the ion Debye length, given through $\lambda_{Di}^2 = \varepsilon_0\kappa T_i/n_{i00}e^2$, and $\lambda_{Jd} = v_{td}/\omega_{Jd}$ is the dust Jeans length. The normalization on the potentials is that $\widetilde{\varphi}_0 = e\varphi_0/\kappa T_i$ and $\widetilde{\psi}_0 = \psi_0/v_{td}^2$, and $\tau = Z_d T_i/T_d$ is a dimensionless temperature ratio, traditionally assumed fairly large in view of the heavy dust masses and subsequent low mobility. Following the reasoning of Tsintsadze *et al.* [2000], we use $\lambda_{Di} \ll \lambda_{Jd}$ to deduce from the first equation in (42) approximately that

$$\widetilde{\psi}_0 \simeq (1+\tau)\widetilde{\varphi}_0, \tag{43}$$

which allows a reduction of the second equation in (42) to

$$(1+\tau)\lambda_{Jd}^2 \frac{d^2\widetilde{\varphi}_0}{dz^2} = \exp\left(-\widetilde{\varphi}_0\right). \tag{44}$$

This has solutions of the sech^2 type, as will be shown later on.

Nevertheless, numerically solving the set (42) as it is, indicates that charge neutrality is not maintained throughout the dusty plasma region, contrary to what is popularly assumed [Tsintsadze *et al.* 2000]. In fact, it deviates most from charge neutrality in the inner part of the central region, showing that not all electrons can accrete onto the dust. Mathematically speaking, the electrostatic potential φ_0 and its first derivative $d\varphi_0/dz$ vanish at the central location $z = 0$. Imposing there strict charge neutrality means that from the electrostatic Poisson equation also the second derivative $d^2\varphi_0/dz^2$ vanishes at $z = 0$, and then the function $\varphi_0 = 0$ everywhere. Without gravitation that is why normal plasmas have a uniform equilibrium that is perfectly charge neutral. However, in the presence of self-gravitation $\varphi_0 = 0$ would impose that also $\psi_0 = 0$, which is not allowed by the gravitational Poisson equation, unless the Jeans swindle creeps in by the back door.

It is instructive to frame this from a different angle and go to an even more simple model, that can be solved analytically throughout and allows us to check some of the statements of the preceding paragraphs. We assume that all electrons have been accreted onto the dust grains, as already used [Mamun and Shukla 2000], that the ions are massless and the dust is so heavy that its temperature effects can be neglected. In other words, momentum balance for the ions leads to the Boltzmann expression for the ion density

$$n_{i0} = n_{i00} \exp\left(-\widetilde{\varphi}_0\right), \tag{45}$$

but dust momentum balance relates the gravitational to the electrostatic potentials,

$$\psi_0 = \frac{Z_d e}{m_d}\varphi_0 = \frac{c_{da}^2 n_{i00}}{Z_d n_{d00}}\widetilde{\varphi}_0. \tag{46}$$

In this expression the dust-acoustic velocity $c_{da} = \lambda_{Di}\omega_{pd}$ occurs, a typical velocity that will find more use later on, when different Jeans instabilities are discussed. What remains from the two Poisson equations is

$$\begin{aligned}
\lambda_{Di}^2 \frac{d^2\widetilde{\varphi}_0}{dz^2} &= \frac{Z_d n_{d0}}{n_{i00}} - \exp\left(-\widetilde{\varphi}_0\right), \\
\lambda_{da}^2 \frac{d^2\widetilde{\varphi}_0}{dz^2} &= \frac{Z_d n_{d0}}{n_{i00}},
\end{aligned} \tag{47}$$

where a modified Jeans length has been defined, $\lambda_{da} = c_{da}/\omega_{Jd}$, expressed through the dust-acoustic rather than the dust thermal velocity. Physically, this is plausible: only the electrostatic coupling between the warm ions and the cold dust can prevent total collapse of the dust, so that it is the ion pressure that counts here.

Elimination of the dust density n_{d0} between the two Poisson equations (47) gives a single equation for the electrostatic potential $\widetilde{\varphi}_0$,

$$\left(\lambda_{da}^2 - \lambda_{Di}^2\right)\frac{d^2\widetilde{\varphi}_0}{dz^2} = \exp\left(-\widetilde{\varphi}_0\right). \tag{48}$$

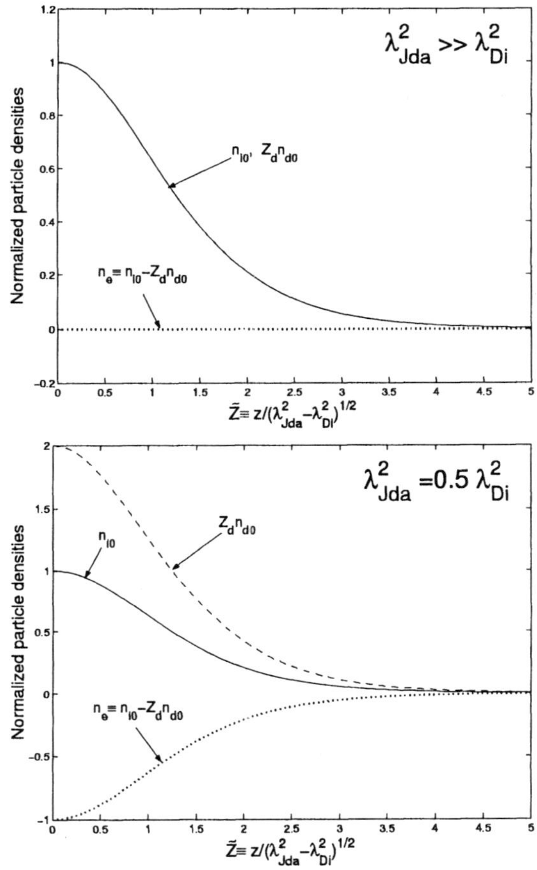

Figure 2: Densities and charge distributions vs. distance from the central location.

The solution is very simple, and leads to the ion and dust densities being expressed as

$$
\begin{aligned}
n_{i0} &= n_{i00}\mathrm{sech}^2(\tilde{z}/\sqrt{2}), \\
n_{d0} &= \frac{n_{i00}}{Z_d} \cdot \frac{\lambda_{da}^2}{\lambda_{da}^2 - \lambda_{Di}^2}\mathrm{sech}^2(\tilde{z}/\sqrt{2}).
\end{aligned} \tag{49}
$$

Here $\tilde{z}$ is a dimensionless coordinate, defined through $z^2 = \tilde{z}^2\,(\lambda_{da}^2 - \lambda_{Di}^2)$, and it has been assumed that $\lambda_{da} > \lambda_{Di}$, which is the case for almost all dusty plasmas under study. Indeed, it is straightforward to show that

$$
\frac{\lambda_{da}^2}{\lambda_{Di}^2} = \frac{\omega_{pd}^2}{\omega_{Jd}^2}. \tag{50}
$$

It thus turns out that the deviation from strict charge neutrality is

$$
n_{i0} - Z_d n_{d0} = -\frac{\lambda_{Di}^2}{\lambda_{da}^2 - \lambda_{Di}^2} n_{i00}\mathrm{sech}^2(\tilde{z}/\sqrt{2}), \tag{51}
$$

indicating that at the central location there is more negative dust than positive ions, owing to the way the self-gravitation tends to concentrate the dust, which is opposed by the coupling to the ions. Note that if the complete plasma were initially charge neutral, our results indicate that not all electrons can accrete onto the dust grains, as supposed in the mathematical description. This shows the pitfalls of simple minded approaches to self-gravitation. Nevertheless, the total charge density is small if $\lambda_{Di} \ll \lambda_{da}$. Remarkably enough, this vanishes at infinity rather than at the central location. Except for some factors, density profiles like (49) follow the same behaviour as what was obtained in the single fluid approximation (19).

The opposite cases yield discontinuities in density, showing that very heavy charged dust ($\omega_{pd} < \omega_{Jd}$) cannot be stabilized alone by electrostatic forces, imparted by massless ions, as assumed here. Again we have here a possibility of fragmentation that merits to be treated in more detail.

5 Jeans Instabilities in Dusty Plasmas

We saw for specific configurations, which we could analyze completely, that critical Jeans wavenumbers represent scales on which the stationary states are nonhomogeneous. It is interesting to put this logic on its head and derive the Jeans lengths in the naive way, in the hope to thus gain information about the inhomogeneity length scales for those systems where the exact analytical determination of stationary states cannot be pursued very far. When looking at the Jeans instabilities, we can recall results from several papers, that all assume uniform equilibria and only look at small local perturbations thereof. Two simplifying geometries are when the waves propagate along or across an external magnetic field, and these will guide us to the intermediate angles of propagation [Jacobs *et al.* 2000b, Verheest *et al.* 2000a].

5.1 Parallel Electrostatic Modes

At propagation parallel to an external magnetic field, self-gravitation only affects the longitudinal, electrostatic modes, with general dispersion law [Bliokh and Yaroshenko 1985, Bliokh *et al.* 1995, Meuris *et al.* 1997, Verheest *et al.* 1998, Verheest 1999, Verheest *et al.* 2000c, 2002]

$$\left(1-\sum_\alpha \frac{\omega_{p\alpha}^2}{(\omega-kU_\alpha)^2-k^2v_{t\alpha}^2}\right)\left(1+\sum_\alpha \frac{\omega_{J\alpha}^2}{(\omega-kU_\alpha)^2-k^2v_{t\alpha}^2}\right) + \left(\sum_\alpha \frac{\omega_{p\alpha}\omega_{J\alpha}}{(\omega-kU_\alpha)^2-k^2v_{t\alpha}^2}\right)^2 = 0, \tag{52}$$

where the sound velocities $v_{t\alpha}$ per species appear. Capital letters denote (locally constant) equilibrium values. Neutral dust species can be included by setting the appropriate charges equal to zero. Equation (52) highlights the coupling between self-gravitational effects and electrostatic plasma oscillations, and these two phenomena could be disentangled by taking the very low or very high frequency limit, respectively. The higher frequency limit of (52) is equivalent to neglecting self-gravitation altogether and leads to electrostatic plasma oscillations, including Buneman instabilities when streaming U_α between the species is present. This part of the dispersion includes the celebrated dust-acoustic mode [Rao *et al.* 1990, Barkan *et al.* 1995] as the paradigm of dusty plasma modes, where the collective behaviour of the charged dust is responsible for the inertia needed to sustain the waves.

On the other hand, the very low frequency limit of (52) is obtained when plasma effects are unimportant, and then

$$1 + \sum_\alpha \frac{\omega_{J\alpha}^2}{(\omega - kU_\alpha)^2 - k^2 v_{t\alpha}^2} \simeq 0 \tag{53}$$

describes a multispecies Jeans instability. Only sufficiently low frequencies allow the heavy dust grains to respond to the perturbations and these modes are the ones under consideration. Therefore, in the remainder of this section we shall treat the ions and *a fortiori* the electrons as being Boltzmann distributed. The implications of this were already discussed in §4, when dealing with typical multispecies dusty plasma configurations.

5.1.1 Monodisperse dust

If only one charged dust species is present and there is no streaming between species, the dispersion law (52) simplifies to

$$\omega^2 = k^2(v_{td}^2 + c_{da}^2) - \omega_{Jd}^2. \tag{54}$$

For cold dust, (54) reduces to a Jeans dust-acoustic mode [Avinash and Shukla 1994, Pandey *et al.* 1994]. If, in addition to ignoring the dust thermal speed, self-gravitational effects are excluded, the pure dust-acoustic mode is recovered [Rao *et al.* 1990]. Equation (54) is the archetypal dispersion law for "Jeans" modes. Setting $\omega^2 = 0$ determines the critical Jeans length through

$$k_{da}^2 = \frac{\omega_{Jd}^2}{v_{td}^2 + c_{da}^2}. \tag{55}$$

Since the charged dust usually has much lower temperatures than the plasma components of the dusty plasma mixture, one can safely assume that $v_{td} \ll c_{da}$ and it is essentially the dust-acoustic velocity that enters in the critical Jeans length. Compared to the purely neutral gas criterion, where

$$k_{nJ}^2 = \frac{\omega_{Jn}^2}{v_{tn}^2}, \tag{56}$$

it is obvious that neutral gases are Jeans unstable on much smaller scales than charged dust clouds. This is physically understandable because in dusty plasmas the charged dust is coupled to the plasma with a higher pressure, and in addition all like charged grains undergo an electrostatic repulsion.

Other variations on this theme result from considering streaming effects, leading to Jeans-Buneman type instabilities [Meuris *et al.* 1997, Pandey and Lakhina 1998], or by including dust charge fluctuations [Verheest *et al.* 1997b, Rao and Verheest 2000] and ion-dust collisions and drag [Verheest and Shukla 1997, Mamun and Shukla 2000, Jacobs *et al.* 2002], or from using nonlinear [Verheest and Shukla 1997, Yaroshenko and Verheest 2000] and kinetic [Mamun *et al.* 1999, Yaroshenko *et al.* 2001a] treatments.

5.1.2 Polydisperse cold dust

If the possibility of multiple charged dust species is retained, under the weak restriction that the dust species are cold, for analytical simplicity, the dispersion law turns into

$$\begin{aligned}\omega^4 &+ \left(\sum_d \omega_{Jd}^2 - k^2\lambda_D^2 \sum_d \omega_{pd}^2\right)\omega^2 \\ &- \frac{k^2\lambda_D^2}{2}\sum_d\sum_{d'}(\omega_{pd}\omega_{Jd'} - \omega_{pd'}\omega_{Jd})^2 = 0, \qquad (57)\end{aligned}$$

as discussed by Meuris *et al.* [1997]. Here λ_D is a global screening length, connected to the electron and ion Debye lengths through $\lambda_D^{-2} = \lambda_{De}^{-2} + \lambda_{Di}^{-2}$. The last term in the dispersion law clearly represents the modifications due to the presence of dust species of different characteristics. It can only vanish if all dust species have the same charge-to-mass ratio.

For two charged dust species, equation (52) reduces to [Pillay *et al.* 2000, Verheest *et al.* 2000c]:

$$\begin{aligned}&\left[\omega^2 - k^2(v_{td1}^2 + c_{da1}^2) + \omega_{Jd1}^2\right]\left[\omega^2 - k^2(v_{td2}^2 + c_{da2}^2) + \omega_{Jd2}^2\right] \\ &= (\omega_{Jd1}\omega_{Jd2} - k^2 c_{da1}c_{da2})^2, \qquad (58)\end{aligned}$$

clearly showing the coupling between the two Jeans dust modes. This is a quadratic dispersion law in ω^2 with positive discriminant, implying that both roots in ω^2 are real. Unstable modes require $\omega^2 < 0$, but this can only happen for one of the two roots, the other then being positive. A necessary and sufficient condition for instability is thus a negative constant term in (58), synonymous with a negative product of both real roots. This condition is always fulfilled for sufficiently small wavenumbers, and thus the following expression for the Jeans wave number is obtained,

$$k_J^2 = \frac{\omega_{Jd1}^2 v_{td2}^2 + \omega_{Jd2}^2 v_{td1}^2 + (\omega_{Jd1}c_{da2} - \omega_{Jd2}c_{da1})^2}{v_{td1}^2 c_{da2}^2 + v_{td2}^2 c_{da1}^2 + v_{td1}^2 v_{td2}^2}$$

$$\simeq \frac{(\omega_{Jd1}c_{da2} - \omega_{Jd2}c_{da1})^2}{v_{td1}^2 c_{da2}^2 + v_{td2}^2 c_{da1}^2}, \tag{59}$$

assuming that the pure plasma pressures are much larger than the dust pressures, so that for the dust species $v_{td\alpha} \ll c_{da\alpha}$ [Yaroshenko *et al.* 2001b].

Finally, for one charged (with subscript d) and one neutral dust or gas component (with subscript n), equation (59) simplifies to

$$k_J^2 = \frac{\omega_{Jcd}^2}{v_{tcd}^2 + c_{da}^2} + \frac{\omega_{Jnd}^2}{v_{tnd}^2}. \tag{60}$$

If the mass densities and temperatures of the charged and the neutral components are comparable, in the sense that $\omega_{Jcd} \sim \omega_{Jnd}$ and $v_{tcd} \sim v_{tnd} \ll c_{da}$, one sees that the Jeans wavenumber is essentially determined by the neutral gas. This is not surprising, because in the absence of friction between the two constituents, there is no mechanism to halt the collapse of the neutral gas but its own pressure. The charged component is coupled to the plasma, but that does not help to determine the collapse scales.

5.2 Magnetosonic Modes

Now we turn to propagation strictly perpendicular to the external magnetic field. There are two modes, the ordinary mode which is unaffected by pressure or self-gravitational effects, and the extraordinary mode, which contains all pressure and self-gravitational influences, besides the electromagnetic effects. It is the latter mode that is considered in the papers dealing with Alfvén-Jeans waves, where the low-frequency magnetosonic branch of the extraordinary mode is the one affected by self-gravitation, as recalled below [Verheest *et al.* 1997a, Mace *et al.* 1998, Verheest *et al.* 1999].

5.2.1 Average dust dispersion

If the charged dust is represented by a single species with some average properties and there is no neutral gas at all, the dispersion law for perpendicular magnetosonic or Alfvén-Jeans modes is [Verheest *et al.* 1997a, 1999]

$$\omega^2 = k^2(V_{ms}^2 + v_{td}^2) - \omega_{Jd}^2. \tag{61}$$

Here the ions and electrons are considered as quasi-inertialess, and this condition is the electromagnetic analogue of what is done for parallel electrostatic modes when electrons and ions are treated as Boltzmann distributed. The magnetosonic velocity V_{ms} is defined in general, for more than one dust species as needed further on, through

$$V_{ms}^2 = \frac{B_0^2/\mu_0 + P_e + P_i}{\sum_d N_d m_d}, \tag{62}$$

and the electron and ion pressures are given as $P_e = N_e m_e v_{te}^2$ and $P_i = N_i m_i v_{ti}^2$. It is reasonable to suppose that the effective dust pressure is smaller than the electron and ion pressures, allowing the neglect of $v_{td} \ll V_{ms}$. The critical Jeans wavenumber obtained from (61) becomes

$$k_{ms} \simeq \frac{\omega_{Jd}}{V_{ms}}. \tag{63}$$

Thus the magnetosonic velocity comes into play, expressing the dominant influence of the magnetic field, combined with the plasma pressures, on the critical Jeans lengths for perpendicular propagation. On the other hand, in (54) only plasma thermal effects, represented by c_{da}, determine k_{da}.

5.2.2 Several charged dust species

The simplest is to first consider two charged dust species, to see what modifications occur to the Jeans wavenumber. In that case, the dispersion law transforms into a biquadratic equation [Jacobs *et al.* 2000a],

$$A\omega^4 + B\omega^2 + C = 0, \tag{64}$$

with coefficients

$$\begin{aligned}
A &= \left(\Omega_1\omega_{J1}^2 + \Omega_2\omega_{J2}^2\right)^2, \\
B &= \left(\Omega_1\omega_{J1}^2 + \Omega_2\omega_{J2}^2\right)^2 (\omega_{J,total}^2 - k^2 v_{T1}^2 - k^2 v_{T2}^2) - \Omega_1^2\Omega_2^2\omega_{J,total}^4 \\
&\quad - k^2 V_{ms}^2 \omega_{J,total}^2 (\Omega_1^2\omega_{J1}^2 + \Omega_2^2\omega_{J2}^2), \\
C &= k^2 V_{ms}^2 \omega_{J,total}^2 \left[\Omega_1^2\Omega_2^2\omega_{J,total}^2(1-\mathcal{K}) + \Omega_1^2\omega_{J1}^2 k^2 v_{T2}^2 + \Omega_2^2\omega_{J2}^2 k^2 v_{T1}^2\right] \\
&\quad + k^2\left(\Omega_1\omega_{J1}^2 + \Omega_2\omega_{J2}^2\right)^2 (k^2 v_{T1}^2 v_{T2}^2 - \omega_{J1}^2 v_{T2}^2 - \omega_{J2}^2 v_{T1}^2) \\
&\quad + \Omega_1^2\Omega_2^2\omega_{J,total}^2 k^2 (\omega_{J1}^2 v_{T1}^2 + \omega_{J2}^2 v_{T2}^2) - \Omega_1^2\Omega_2^2\omega_{J,total}^6,
\end{aligned} \tag{65}$$

and having a positive discriminant. The total Jeans frequency $\omega_{J,total}$ is defined through

$$\omega_{J,total}^2 = \omega_{J1}^2 + \omega_{J2}^2. \tag{66}$$

When dust thermal speeds are neglected, the constant term can be simplified to

$$C \simeq \Omega_1^2\Omega_2^2\omega_{J,total}^4 \left[k^2 V_{ms}^2 (1-\mathcal{K}) - \omega_{J,total}^2\right]. \tag{67}$$

This is a valid approximation provided $\mathcal{K} \leq \mathcal{O}(1/3)$, where

$$\mathcal{K} = \frac{\omega_{J1}^2\omega_{J2}^2}{\omega_{J,total}^2}\left(\frac{1}{\Omega_1} - \frac{1}{\Omega_2}\right)^2. \tag{68}$$

If $C > 0$, the biquadratic dispersion law has two positive roots for ω^2 whereas $C < 0$ gives one positive and one negative root for ω^2. The latter root clearly corresponds to the Jeans unstable case, and the transition between stable and unstable modes occurs at $C = 0$, or equivalently at $k = k_{ms}$, given through

$$k_{ms}^2 = \frac{\omega_{J,total}^2}{(1-\mathcal{K})V_{ms}^2}. \tag{69}$$

Rather than discuss this for two charged dust species, we will immediately deal with the general case of different dust species.

In order to determine the critical Jeans length for general polydisperse dust, one has to consider the dispersion law as a polynomial in ω^2 with parameter k^2. Below the threshold value k_{ms} at least one of the polynomial roots leaves the positive real axis. As noted from the previous paragraphs, the Jeans criterion involves passing from $\omega^2 > 0$ to $\omega^2 < 0$ through the origin. It follows that a Taylor expansion up to first order in ω^2 will contain all the useful information, and as a consequence

$$k_{ms}^2 = \frac{\omega_{J,total}^2}{V_{ms}^2(1-\mathcal{K})} \qquad \text{at} \qquad \omega^2 = 0. \tag{70}$$

Here the definitions of $\omega_{J,total}$ and $\mathcal{K}$ have been extended to

$$\omega_{J,total}^2 = \sum_d \omega_{Jd}^2,$$
$$\mathcal{K} = \frac{1}{2\omega_{J,total}^2} \sum_d \sum_{d'} \omega_{Jd}^2 \omega_{Jd'}^2 \left(\frac{1}{\Omega_d} - \frac{1}{\Omega_{d'}}\right)^2. \tag{71}$$

The different dust species are indicated by taking d and d' as dummy indices in the summations. Under the assumptions $\omega^2 \ll c^2k^2$ and $k\lambda_D \ll 1$, the dispersion law is already linear in k^2 and up to first order the Taylor expansion simplifies to [Jacobs *et al.* 2000a]

$$\omega^2 = \frac{1-\mathcal{K}}{1+\mathcal{L}} \left[(1-\mathcal{K})k^2V_{ms}^2 - \omega_{J,total}^2\right]. \tag{72}$$

$\mathcal{L}$ is a small positive parameter, its precise form does not furnish us with valuable physical insight and is not mentioned here. If the Jeans frequencies are in some sense small compared to the gyrofrequencies, as is typically the case in astrophysical plasmas, then $0 \leq \mathcal{L} \ll \mathcal{K} \ll 1$.

5.2.3 Presence of neutral gas

If neutrals are included, the derivation becomes even more complicated. Therefore we focus on the determination of the critical Jeans lengths, instead of dealing with the formidable algebraic expressions for the dispersion laws. Previous sections indicated a shortcut for obtaining the critical Jeans lengths, whereby putting $\omega = 0$ in the elements of the dispersion tensor conserves all the necessary information required to determine the critical wave number. Following this strategy in the equivalent of

(64) and (65) indicates that the critical wave number can be expressed as [Jacobs *et al.* 2000b]

$$k_{ms}^2 = \frac{1}{1-\mathcal{K}} \left[\frac{\omega_{J,total}^2}{V_{ms}^2} + \sum_g \frac{\omega_{Jnd}^2}{v_{tnd}^2} \right], \tag{73}$$

and $\omega_{J,total}$ only refers to the charged dust components. Comparing this with (70) shows that the stabilizing effects of the charged dust components are superseded by the presence of neutral gas components. The effect of neutral gas pressure on self-gravitational instabilities in dusty plasmas is studied more in detail by Shukla and Verheest [1999] and Mamun and Shukla [2000].

Seeing that $v_{tg} \sim v_{td} \ll V_{ms}$, the neutral terms dominate (73), provided that the Jeans frequencies of charged and neutral species are comparable, or in other words, that the mass densities are of the same order.

5.3 Intermediate Angles of Propagation

For intermediate angles of propagation between parallel and strictly perpendicular, we have to revert to the full dispersion law. For simplicity, we first delve into the "classic" dusty plasma case. This model deals with only one effective dust species, and electrons and ions are treated as being inertialess ($\omega^2 \ll k^2 v_{te}^2$ and $\omega^2 \ll k^2 v_{ti}^2$). We also suppose that $\omega^2 \ll c^2 k^2$ and $k^2 \lambda_D^2 \ll 1$. All modes are coupled and incorporated in the following dispersion law [Verheest *et al.* 2000b]:

$$\begin{aligned} \omega^6 &- \left[\Omega_d^2 K^2 \left((1+K^2)\cos^2\vartheta + 1\right) + \Delta\right] \omega^4 \\ &+ \left[(\Omega_d^2 + \Delta)K^2 + 2\Delta\right] (\Omega_d^2 K^2 \cos^2\vartheta)\omega^2 - (\Omega_d^4 K^4 \cos^4\vartheta)\Delta = 0, \end{aligned} \tag{74}$$

a cubic polynomial in ω^2. Here $K = ck/\omega_{pd}$, and we note that all thermal and self-gravitational effects are contained in

$$\Delta = k^2(c_{da}^2 + v_{td}^2) - \omega_{Jd}^2. \tag{75}$$

For $\Delta > 0$ but small, all three roots of (74) are real and positive. When Δ shifts, for small absolute values, towards negative values, one of the roots ω^2 goes simultaneously with Δ through zero. The only negative root, obtained when Δ turns negative but remains small, corresponds to the purely growing Jeans instability. The latter is clearly activated through a slight change of Δ from positive to negative values. At Δ very small, the smallest root becomes $\omega^2 \simeq \Delta\cos^2\vartheta$ and of course $\Delta = 0$ gives the familiar result

$$k_{cr}^2 = \frac{\omega_{Jd}^2}{c_{da}^2 + v_{td}^2}. \tag{76}$$

Similar results are obtained when the modes are supposed to be electrostatic to begin with, but careful consideration has to be given to the transition to strictly perpendicular propagation [Salimullah and Shukla 1999, Verheest and Yaroshenko 2002].

When more dust species are added to the dusty plasma model, the full dispersion law becomes unwieldy and would require careful disentangling. It is preferable to cut this algebraic Gordian knot and go straight to the critical Jeans lengths. Previous sections mapped out the strategy of putting $\omega = 0$ in the elements of the dispersion tensor. This strategy does not allow that ϑ approaches 90^0, after having set $\omega = 0$ in the elements of the dispersion tensor. This is imputable to the fact that the limits $\omega \to 0$ and $\vartheta \to 90^\circ$ are not interchangeable, since in the model where electrons and ions are treated as massless, the limit $\omega \to 0$ and $\vartheta \to 90^0$ does not exist mathematically. Taking $k_{cr}\lambda_D \ll 1$, and using in addition the following reasonable assumptions, $Z_d m_e \ll Z_d m_i \ll m_d$, $v_{td} \ll v_{ti}, v_{te}$ and $\omega_{Je} \ll \omega_{Ji} \ll \omega_{Jd}$, gives for the critical Jeans length that

$$k_{cr}^2 \left(1 + \lambda_D^2 \sum_d \frac{\omega_{pd}^2}{v_{td}^2}\right) = \sum_d \frac{\omega_{Jd}^2}{v_{td}^2} + \frac{\lambda_D^2}{2} \sum_d \sum_{d'} \frac{(\omega_{pd}\omega_{Jd'} - \omega_{pd'}\omega_{Jd})^2}{v_{td}^2 v_{td'}^2}. \tag{77}$$

This result is independent of the angle of wave propagation ϑ, showing that the longitudinal effects determine the Jeans instability through gradient forces. For one dust species (77) matches (76) and a combination of charged and neutral dust gives

$$k_{cr}^2 = \frac{\omega_{Jcd}^2}{c_{da}^2 + v_{tcd}^2} + \frac{\omega_{Jnd}^2}{v_{tnd}^2}. \tag{78}$$

Again, because of $v_{tn} \sim v_{td} \ll c_{da}$, such critical wavenumbers are close to what is obtained for neutral dust alone [Bliokh *et al.* 1995]. Finally, we emphasize again that it is imperative to use a separate treatment for $\vartheta = 90^\circ$, as done in the previous subsection. The appropriate treatment of the extraordinary [Bliokh *et al.* 1995, Verheest *et al.* 1997a] or of (oblique) electrostatic [Salimullah and Shukla 1999, Verheest and Yaroshenko 2002] modes at strictly perpendicular propagation shows the very abrupt transition in the critical lengths.

6 Space and Astrophysical Applications

In §5 we derived a number of length scales being the criteria for the threshold for Jeans collapse of a dust gas, a dusty plasma and a dusty magnetoplasma. It should be emphasized that, owing to the wide variety of waves and instabilities that a dusty plasma can support, these are not the only possible collapse length scales. Further studies are necessary to fully understand the Jeans stability of dusty plasmas.

In order to get a feeling for the sizes of the different length scales involved, we will give some simple numerical estimates for the different cases involving (56), (55) and (63), based on numbers for interstellar dust in molecular dust clouds [Whittet 1992, Evans 1994]. These are dust formation regions where a plasma is present and hence all or part of the dust can be taken as charged. We assume that the interstellar dust has a density of $N_d = 10^{-1}$ m^{-3}, an average mass for water-ice, micron-sized grains of $m_d = 4 \times 10^{-15}$ kg and carries $Z_d = 10^4$ electron charges per grain. The interstellar magnetic field is of the order of $B_0 = 3 \times 10^{-10}$ T, and the dust temperature is $T_d = 30$ K, which has to be compared to an electron temperature of $T_e = 10^4$ K. This results in a dust Jeans frequency of $\omega_{Jd} = 5.8 \times 10^{-13}$ s^{-1} and a dust thermal velocity of $v_{td} = 3.2 \times 10^{-4}$ m/s. If the dust were neutral, this would give a critical wavenumber $k_{nJ} = 1.9 \times 10^{-9}$ m^{-1}, using (56). The corresponding Jeans length is $\lambda_{nJ} = 2\pi/k_{nJ} = 3.3 \times 10^6$ km, a small fraction of an astronomical unit.

On the other hand, if the dust is charged as indicated, the dust-acoustic velocity $c_{da} = 0.2$ m/s, clearly showing the preponderance of the plasma pressure in rendering the collapse of the heavy dust more difficult. From (55) the critical wavenumber is indeed found to be much smaller now, $k_{da} = 2.8 \times 10^{-12}$ m^{-1}, and the corresponding Jeans length is $\lambda_{da} = 2\pi/k_{da} = 2.2 \times 10^9$ km, a sizeable fraction of the solar heliosphere in the upstream direction, towards the heliopause.

Finally, for transverse instabilities, the dust Alfvén velocity is $V_{Ad} = 13.4$ m/s and the dust-magnetosonic velocity is very similar, $V_{ms} \simeq 13.4$ m/s. This clear preponderance of the magnetic effects in lowering yet again the critical wavenumber, given by (63), leads to $k_{ms} = 4.5 \times 10^{-14}$ m^{-1}. The corresponding length scales hence are $\lambda_{ms} = 1.4 \times 10^{11}$ km, some thousand astronomical units, and vastly larger than the dimensions of the solar system.

The conclusions of this simple exercise clearly point to the charged dust and the transverse magnetic field as important antagonists of gravitational collapse. Of course, the rather large unreliability of the present interstellar data about (charged) dust cautions against taking the above numbers too literally, but nevertheless the message is pointing in the right direction, as far as our theoretical developments are concerned. At the same time, we would like to recall that rather than take the Jeans lengths as indicative of collapse scales, we interpret them as inhomogeneity scale lengths, over which the system can no longer be considered as uniform.

7 Summary

To conclude, we have discussed several aspects of self-gravitation in dusty space plasmas. First of all, we emphasized why the interplay between self-gravitation and electromagnetic effects could become important for heavier charged dust grains. Second, we determined some of the possible stationary states and obtained information for the scales on which the system varies. It turns out that for the cases where explicit solutions can be obtained, these are of the order of the Jeans lengths determined in the usual way by studying local perturbations, as if the stationary state were uniform, in other words, by applying what is called the Jeans swindle. Nevertheless, we can turn the argument around and use the Jeans swindle or local perturbation approach to find the scales over which the system itself is no longer homogeneous.

This has first been discussed for electrostatic perturbations, where either the dust thermal velocity plays a role for neutral dust or the dust-acoustic velocity when the plasma pressure balances the dust collapse. Because the plasma pressure is much larger, the system is homogeneous over much greater distances than if a neutral gas were considered. Electrostatic perturbations occur for propagation parallel or oblique propagation to an external magnetic field. It is only for strictly perpendicular propagation that the dust-magnetosonic velocity enters and tends to stabilize even larger distances. To get a feeling for possible orders of magnitude, these findings were applied to molecular dust clouds, where, however, the paucity of reliable data cautions against taking the scale lengths thus obtained too literally.

Acknowledgments

F. Verheest, P.K. Shukla and A.A. Mamun thank the International Space Science Institute (Switzerland) for its kind hospitality during visits as members of the International Team on "Dust Plasma Interaction in Space", when part of this work was carried out. F. Verheest also acknowledges research grants from the Bijzonder Onderzoeksfonds (Universiteit Gent) and the Fonds voor Wetenschappelijk Onderzoek (Vlaanderen). V.M. Čadež acknowledges support by the Belgian Institute for Space Aeronomy and by the Belgian Federal Office for Scientific, Technical and Cultural Affairs. A.A. Mamun gratefully acknowledges financial support of the Alexander von Humboldt-Stiftung (Germany). It is a pleasure to thank M.A. Hellberg, G. Jacobs, R.L. Mace, S.R. Pillay, N.N. Rao and V.V. Yaroshenko for enlightening discussions.

References

Alfvén H. 1981, *Cosmic Plasma.* Reidel, Dordrecht, pp. 118–121.

Avinash K. and Shukla P.K. 1994, *Phys. Lett. A* **189**, 470.

Barkan A., Merlino R.L. and D'Angelo N. 1995, *Phys. Plasmas* **2**, 3563.

Baumjohann W. and Treumann R.A. 1996, *Basic Space Plasma Physics.* Imperial College Press, London.

Binney J. and Tremaine S. 1988, *Galactic Dynamics.* Princeton University Press, Princeton.

Bliokh P.V. and Yaroshenko V.V. 1985, *Sov. Astron.* **29**, 330.

Bliokh P., Sinitsin V. and Yaroshenko V. 1995, *Dusty and Self-gravitational Plasmas in Space.* Kluwer, Dordrecht.

Boss A.P. 1987, *Astrophys. J.* **319**, 149.

Bouchoule A. 1999, *Dusty Plasmas. Physics, Chemistry and Technological Impacts in Plasma Processing.* Wiley, Chichester.

Čadež V.M. 1990, *Astron. Astrophys.* **235**, 242.

Chandrasekhar S. 1961, *Hydrodynamic and Hydromagnetic Stability.* Clarendon Press, Oxford.

Choudhuri A.R. 1998, *The Physics of Fluids and Plasmas. An Introduction for Astrophysicists.* Cambridge University Press, Cambridge (UK).

Evans A. 1994, *The Dusty Universe.* Wiley, Chichester.

Fridman A.M. and Polyachenko V.L. 1984, *Physics of Gravitating Systems I.* Springer-Verlag, New York.

Gehman C.S., Adams F.C., Fatuzzo M. and Watkins R. 1996a, *Astrophys. J.* **457**, 718.

Gehman C.S., Adams F.C. and Watkins R. 1996b, *Astrophys. J.* **472**, 673.

Goertz C.K. 1989, *Rev. Geophys.* **27**, 271.

Hartquist T.W., Pilipp W. and Havnes O. 1997, *Astrophys. Space Sci.* **246**, 243.

Horányi M. 1996, *Annu. Rev. Astron. Astrophys.* **34**, 383.

Jacobs G., Verheest F., Hellberg M.A. and Pillay S.R. 2000a, *Phys. Plasmas* **7**, 4390.

Jacobs G., Verheest F., Hellberg M.A., Pillay S.R. and Yaroshenko V.V. 2000b, in: *Waves in Dusty, Solar and Space Plasmas.* Eds. F. Verheest, M. Goossens, M.A. Hellberg and R. Bharuthram, AIP Press, CP 537, Woodbury, NY, p. 53.

Jacobs G., Yaroshenko V.V. and Verheest F. 2002, *Phys. Rev. E* (in press).

Jeans J.H. 1929, *Astronomy and Cosmogony.* Cambridge University Press.

Krishan V. 1999, *Astrophysical Plasmas and Fluids.* Kluwer, Dordrecht.

Larson R.B. 1985, *Mon. Not. R. Astron. Soc.* **214**, 379.

Mace R.L., Verheest F. and Hellberg M.A. 1998, *Phys. Lett. A* **237**, 146.

Mamun A.A. and Shukla P.K. 2000, *Phys. Plasmas* **7**, 3762.

Mamun A.A., Salahuddin M. and Salimullah M. 1999, *Planet. Space Sci.* **47**, 79.

McClements K.G. and Thyagaraja A. 2001, *Mon. Not. R. Astron. Soc.* **323**, 733.

Mendis D.A. and Rosenberg M. 1994, *Annu. Rev. Astron. Astrophys.* **32**, 419.

Mendis D.A., Houpis H.L.F. and Hill J.R. 1982, *J. Geophys. Res.* **87**, 3449.

Meuris P., Verheest F. and Lakhina G.S. 1997, *Planet. Space Sci.* **45**, 449.

Nakano T. 1988, *Publ. Astron. Soc. Japan* **40**, 593.

Öpik E.J. 1956, *Irish Astron.* **4**, 84.

Pandey B.P. and Lakhina G.S. 1998, *Pramana* **50**, 191.

Pandey B.P., Avinash K. and Dwivedi C.B. 1994, *Phys. Rev. E* **49**, 5599.

Pillay S.R., Bharuthram R. and Verheest F. 2000, *Physica Scripta* **61**, 112.

Rao N.N. and Verheest F. 2000, *Phys. Lett. A* **268**, 390.

Rao N.N., Shukla P.K. and Yu M.Y. 1990, *Planet. Space Sci.* **38**, 543.

Rao N.N., Verheest F. and Čadež V.M. 2001, *Phys. Plasmas* **8**, 4740.

Salimullah M. and Shukla P.K. 1999, *Phys. Plasmas* **6**, 686.

Showalter M.R. 1992, in: *Origin and Evolution of Interplanetary Dust.* Eds. A.C. Levasseur-Regourd and H. Hasegawa, Kluwer, Dordrecht, p. 349.

Shukla P.K. 2001, *Phys. Plasmas* **8**, 1791.

Shukla P.K. and Mamun A.A. 2002, *Introduction to Dusty Plasma Physics.* IOP Press, London.

Shukla P.K. and Verheest F. 1999, *Astrophys. Space Sci.* **262**, 157.

Spitzer L. 1941, *Astrophys. J.* **93**, 369.

Spitzer L. 1978, *Physical Processes in Interstellar Medium.* Wiley, New York.

Toomre A. 1964, *Astrophys. J.* **139**, 1217.

Treumann R.A. and Baumjohann W. 1997, *Advanced Space Plasma Physics.* Imperial College Press, London.

Tsintsadze N.L., Mendonça J.T., Shukla P.K., Stenflo L. and Mahmoodi J. 2000, *Physica Scripta* **62**, 70.

Verheest F. 1996, *Space Sci. Rev.* **77**, 267.

Verheest F. 1999, *Plasma Phys. Control. Fusion* **41**, A445.

Verheest F. 2000, *Waves in Dusty Space Plasmas.* Kluwer, Dordrecht.

Verheest F. and Shukla P.K. 1997, *Physica Scripta* **55**, 83.

Verheest F. and Yaroshenko V.V. 2002, *Phys. Rev. E* **65**, 036415.

Verheest F., Meuris P., Mace R.L. and Hellberg M.A. 1997a, *Astrophys. Space Sci.* **254**, 253.

Verheest F., Shukla P.K., Rao N.N. and Meuris P. 1997b, *J. Plasma Phys.* **58**, 163.

Verheest F., Hellberg M.A. and Mace R.L. 1998, in: *Physics of Dusty Plasmas: Seventh Workshop.* Eds. M. Horányi, S. Robertson and B. Walch, AIP Press, CP 446, Woodbury, NY, p. 307.

Verheest F., Hellberg M.A. and Mace R.L. 1999, *Phys. Plasmas* **6**, 279.

Verheest F., Čadež V.M. and Jacobs G. 2000a, in: *Waves in Dusty, Solar and Space Plasmas.* Eds. F. Verheest, M. Goossens, M.A. Hellberg and R. Bharuthram, AIP Press, CP 537, Woodbury, NY, p. 91.

Verheest F., Jacobs G. and Hellberg M.A. 2000b, *Physica Scripta* **T84**, 171.

Verheest F., Jacobs G. and Yaroshenko V.V. 2000c, *Phys. Plasmas* **7**, 3004.

Verheest F., Yaroshenko V.V. and Hellberg M.A. 2002, *Phys. Plasmas* (in press).

Vranješ J. and Čadež V.M. 1990, *Astrophys. Space Sci.* **164**, 329.

Whittet D.C.B. 1992, *Dust in the Galactic Environment.* IOP, Bristol.

Whitworth A.P. 1998, *Mon. Not. R. Astron. Soc.* **296**, 442.

Yaroshenko V.V. and Verheest F. 2000, *J. Plasma Phys.* **64**, 359.

Yaroshenko V.V., Jacobs G. and Verheest F. 2001a, *Phys. Rev. E* **63**, 066406.

Yaroshenko V.V., Jacobs G. and Verheest F. 2001b, *Phys. Rev. E* **64**, 036401.

Magnetic Reconnection in Dusty Plasmas

D. Jovanović
Institute of Physics, P. O. Box 57, Yu–11001 Belgrade, Yugoslavia
P. K. Shukla
Institut für Theoretische Physik IV,
Ruhr–Universität Bochum, D–44780 Bochum, Germany

Abstract

Stability of a thin current sheet in an inhomogeneous dusty plasma is investigated. A new mechanism for the collisionless reconnection of the magnetic field is found. It is associated with the linear instability of a surface shear-Alfvén mode which exists at the boundary between two plasma regions with different orientations of the sheared magnetic field. This new mode is possible only in dusty plasmas in which the electron and ion densities are inhomogeneous and different from each other. The dispersion properties of the surface mode are found to be strongly dependent on the spatial profile of the current sheet. The unstable linear mode is found in the case of a current sheet with steep edges. It is identified as the bending- (or kink-) mode, whose electrostatic and vector potentials, ϕ and A_z, are symmetric and antisymmetric functions, respectively. In such a plasma configuration, the surface tearing mode (ϕ-antisymmetric, A_z-symmetric) cannot propagate as an eigenmode. It arises only as a forced mode through the resonant coupling with the bending mode and the local convective cell. The growth rate of the bending/tearing instability is estimated from the second-order contribution of the resonance.

The resistive tearing mode instabilities are also theoretically investigated in a partially ionized self-gravitating dusty magnetoplasma, taking into account the dust grain dynamics. It is found that the latter introduces new types of resistive tearing instability whose growth rate significantly differs from that of the classical tearing instability in an electron-ion plasma. Such tearing instability may be responsible for the magnetic field line reconnection in astrophysical environments.

1 Introduction

The transport and reconnection of magnetic fields play a very important role in understanding the evolution of large scale magnetic fields in many astrophysical objects such as stars, galaxies, accretion disks, molecular clouds, etc. Recently, there has been a growing interest [*Birk et al.* 1996, *Kopp et al.* 1997, *Birk and*

Wiechen 2001] in investigating the magnetic field reconnection in partially ionized plasmas whose constituents are the electrons, ions, and negatively charged dust grains. Such three-component plasmas also occur in cometary tails [*Russel et al.* 1986], in protoplanetary nebulae [*Levy and Araki* 1989, *Sonnet* 1979], as well as in young stellar objects [*Birk* 1998].

Magnetic field reconnection is one of the fundamental processes that occur in numerous current-carrying plasma environments. It consists in the magnetic field "breaking" and in the consecutive reconnection of the neighboring magnetic field lines, yielding a topologically different magnetic field structure with a lower energy content. In tokamaks, such energy release is responsible for the generation of sawteeth oscillations and for the major disruptions. The classical magnetohydrodynamic (MHD) theory describes the reconnection as a resistive process [*Furth et al.* 1963, *Rutherford* 1973] in which the violation of the topological constraints occurs due to the deviation from the frozen-in-field line regime within a narrow, viscosity dominated, region around the critical mode-rational surface. The tearing mode instability [*Furth et al.* 1963, *White* 1986] can be considered as a generic process for the annihilation of the magnetic flux in astrophysical environments [*Zweibel* 1989, 1999, *Vishniak and Lazarian* 1999].

The simple resistive-MHD theory is inadequate to describe the reconnection in high-temperature tokamaks such as Joint European Torus (JET), which occurs on a scale shorter than the electron-ion collision time [*Edwards et al.* 1986]. It is also unable to account for the similar processes in astrophysical plasmas. These are typically so rare that they are be regarded as completely collisionless. The presence of magnetic reconnection in the Earth's magnetosphere was known since the mid-1960's [*Ness* 1965, *Paschmann* 1979], and it has been explored since then by a number of satellite missions, notably the GEOTAIL, see e. g. [*Lee and Russell* 1994] and [*Matsumoto et al.* 1994]. The rapid bursts of the magnetic storms, often called the magnetic explosions in space, result from the reconnection of magnetic field lines in the vicinity of the magnetic X-points. The typical duration of such bursts is less than one hour, which is many orders of magnitude shorter than the corresponding resistive-MHD time for similar processes.

The early theory of the fast (collisionless) reconnection, based on the Landau-damping by the resonant electrons, was presented in Ref. [*Coppi et al.* 1966] and later refined in Ref. *Mahajan et al.* [1978] to allow for the tearing mode that is wider than the current channel. However, these also failed to account for the large growth rates of the magnetic storms. A purely two-fluid theory of fast (collisionless) reconnection was proposed in Refs. *Ottaviani and Porcelli* [1993] and *Calfaro et al.* [1998], with the primary goal to address the problem of the fast sawteething in the hot plasma of JET. In this two-fluid treatment, the usual role of the viscosity in resistive-MHD is played by the finite electron inertia. The detailed computer simulations [*Grasso et al.* 2001] revealed that the apparent violation of the topological constraints of the Hamiltonian system is the consequence of the existence of (near-)singularities of the electric current and the plasma vorticity. To preserve the entropy, the growth of the magnetic islands is accompanied by the simultaneous production of finer and finer structures in the current and the vorticity. Eventually, the saturation occurs due to the appropriate phase-averaging of the short scales. While the emergence of near-singularities of the current and vorticity at the same spatial location occur naturally in finite systems with conductive boundaries [*Ottaviani and Porcelli* 1993, *Calfaro et al.* 1998], this is not the case in unbounded systems, and this theory is difficult to apply to the astrophysical phenomena.

Recently, another two-fluid theory, in which the local violation of the frozen-in condition and the rapid plasma removal from the magnetic X-point is mediated by the coupling between the shear-Alfvén and whistler waves via Hall current, was put forward in Ref. *Drake et al.* [1997]. The existence of a magnetic structure consistent with that of whistlers, possessing the magnetic field components both in the plane of reconnection and perpendicular to it, was recently identified [*Deng and Matsumoto* 2001] in the magnetic data collected by the GEOTAIL satellite, and was labelled by the authors as the "smoking gun" proof of the whistler-reconnection theory.

In section 2 we propose an alternative theory for the collisionless magnetic reconnection in dusty plasmas. Due to the presence of macroscopic charged dust particulates, a new shear-Alfvén-type surface wave is enabled on a thin current sheet. The process consists predominantly in the bending (or kink) of the current sheet, which is destabilized by the linear mode-coupling with the tearing- (or the reconnection-) surface mode with negative energy. The latter cannot propagate as an eigenmode, and thus appears only as a forced mode, with smaller amplitude than the bending mode. We study in detail the conditions for the surface-wave tearing/bending instability, and calculate the growth rate. We also find that the magnetic reconnection is accompanied by an out-of-plane component of the magnetic field created by the finite $\vec{E} \times \vec{B}$ current that arises due to the asymmetry of the electrons and ions in the presence of the charged dust. The present theory may explain the magnetic reconnection in interplanetary and interstellar plasmas, particularly in the interstellar molecular clouds where the standard MHD and fast-reconnection theories are not applicable due to the absence of collisions and the singularities in the plasma vorticity. After some modifications it may also be applied to the Earth's magnetosphere.

In section 3 we present analytical studies of the resistive tearing mode instabilities in a self-gravitating dusty magnetoplasma, taking into account the dynamics of charged dust grains. For this purpose, we derive a set of equations which governs the coupling between the scalar potential, the parallel component of the vector potential, and the dust number density perturbation. The newly derived equations are then solved in the constant A_z approximation to derive a general dispersion relation for the resistive tearing mode. New tearing mode instabilities in highly collisional- and dust-inertia-dominated- regimes are found. Our results are summarized in section 4.

2 Collisionless reconnection

2.1 Governing equations

We study nonlinear electromagnetic perturbations in a very low-β^* plasma, $\beta^* \ll m_e/m_i$, where $\beta^*(= n_0 T_e/(\epsilon_0 c^2 B^2)$ is the ratio of the plasma and magnetic pressures, and m_e and m_i are the electron and ion masses, respectively. The plasma consists of the electrons, ions and dust particulates whose respective charges are $-e$, $+e$ and $-z_d e$, and whose spatially inhomogeneous unperturbed densities are given by

$$n_{0,j}(x) = n_{0,j}\left(1 - \frac{x}{L_j}\right), \quad j = e, i, d, \tag{1}$$

where the subscripts e, i and d denote, respectively, the electrons, ions and dust particles. The characteristic inhomogeneity scalelengths L_j are mutually related via the plasma quasineutrality. The plasma is immersed in a weakly sheared magnetic field $\vec{B}_0(x)$

$$\vec{B}_0(x) = \vec{e}_z B_{z,0} + \vec{e}_y B_{y,0}(x), \quad B_{z,0}\ (= \text{constant}) \gg B_{y,0}(x). \tag{2}$$

In the very low-β^* regime studied here, the electrons and ions are described by the cold hydrodynamic equations of the continuity and momentum

$$\left(\frac{\partial}{\partial t} + \vec{v}_j \cdot \nabla\right) n_j + n_j \nabla \cdot \vec{v}_j = 0, \tag{3}$$

and

$$\left(\frac{\partial}{\partial t} + \vec{v}_j \cdot \nabla\right) \vec{v}_j = \frac{q_j}{m_j}\left(\vec{E} + \vec{v}_j \times \vec{B}\right) + \nu \vec{v}_j, \tag{4}$$

where ν is the electron-ion collision frequency. The huge mass difference between the macroscopic dust grains and the microscopic plasma constituents (e, i) provides a good separation of the corresponding temporal scales. We consider a process whose time variation is slow compared to the ion gyrofrequency, but fast compared to the dust-related dynamics (viz the dust inertia and dust gyro-motion). We also assume strong convective nonlinearities and weak dependence along the magnetic field lines. Hence we focus in the regimes

$$\omega_{p,d}\Omega_d \ll \partial/\partial t \sim \vec{v}_e \cdot \nabla \sim \vec{v}_i \cdot \nabla \ll \Omega_i, \qquad \nabla_\parallel \ll \nabla_\perp, \tag{5}$$

where $\omega_{p,d}$ and Ω_d are the dust plasma and dust gyrofrequencies, respectively, and Ω_i is the ion gyrofrequency. To the leading order in the small parameters $(1/\Omega_i)(d/dt)$ and m_e/m_i, the perpendicular (to $\vec{e}_z$) component of the electron and ion fluid velocities are obtained from Eq. (4). We have

$$\vec{v_\perp}_e = \frac{1}{B_{z,0}} \vec{e}_z \times (\nabla_\perp \phi - v_{z,e} \nabla_\perp A_z), \tag{6}$$

and

$$\vec{v_{\perp i}} = \frac{1}{B_{z,0}}\vec{e}_z \times \nabla_\perp \phi - \frac{1}{B_{z,0}\Omega_i}\left[\frac{\partial}{\partial t} + \frac{1}{B_{z,0}}\left(\vec{e}_z \times \nabla_\perp \phi\right)\cdot\nabla\right]\nabla_\perp \phi. \tag{7}$$

Here the electric and the magnetic fields are expressed in terms of the electrostatic and vector potentials ϕ, $\vec{A}$ as

$$\vec{E} = -\nabla\phi - \vec{e}_z\frac{\partial A_z}{\partial t}, \quad \vec{B} = \vec{e}_z B_{z,0} - \vec{e}_z \times \nabla A_z, \tag{8}$$

and the compressional magnetic field $\nabla_\perp \times \vec{A}_\perp$ is neglected in the very low-β^* regime. We substitute Eqs. (6)-(8) into the hydrodynamic equations, assume relatively small perturbations of the plasma number density and the magnetic field, viz. $|n - n_0| \ll n_0$ and $|\vec{B} - \vec{e}_z B_{z,0}| \ll B_{z,0}$, and close the system of equations by means of the quasineutrality and the parallel component of Ampère's law, which are, respectively,

$$n_e - n_i = n_{0,e}(x) - n_{0,i}(x) = \frac{n_{0,d}\, z_d}{L_d}x, \qquad v_{z,e} = \frac{e}{m_e}d^2\nabla_\perp^2 A_z, \tag{9}$$

and which readily yield

$$\left[\frac{\partial}{\partial t} + \frac{1}{B_{z,0}}\left(\vec{e}_z \times \nabla\phi\right)\cdot\nabla\right]\left(\nabla_\perp^2\phi - \frac{B_{z,0}v_* x}{d^2}\right) +$$

$$c_A^2\left[\frac{\partial}{\partial z} - \frac{1}{B_{z,0}}\left(\vec{e}_z \times \nabla A_z\right)\cdot\nabla\right]\nabla_\perp^2 A_z = 0, \tag{10}$$

and

$$\left[\frac{\partial}{\partial t} + \frac{1}{B_{z,0}}\left(\vec{e}_z \times \nabla\phi\right)\cdot\nabla\right]\left(1 - \nabla_\perp^2\right)A_z + \frac{\partial\phi}{\partial z} = \nu\nabla_\perp^2 A_z. \tag{11}$$

Here d is the collisionless skin depth, $d \equiv c/\omega_{p,e} = (\varepsilon_0 m_e c^2/n_{0,e}e^2)^{1/2}$, $c_A = c\,(\Omega_i/\omega_{p,i})$ is the Alfvén speed, $\omega_{p,i}$ is the ion plasma frequency, and v_* is the dust-drift speed associated with the inhomogeneity of the number density of the dust grains through $v_* = (c^2/\omega_{p,e}^2)(n_{0,d}z_d/L_d)(\Omega_i/n_{0,i})$. Equations (10) and (11) constitute our basic system of equations, which we analyze below.

2.2 Conservation laws

Integrating the appropriate linear combinations of our basic equations for the entire x, y plane, we obtain the conservation laws

$$\frac{\partial W}{\partial t} + c_A\frac{\partial P}{\partial z} = -\nu R, \tag{12}$$

and

$$\frac{\partial Q}{\partial t} + c_A\frac{\partial W}{\partial z} = -\nu S, \tag{13}$$

where

$$W = \int_{-\infty}^{\infty} dx\, dy \left[(\nabla_\perp \phi)^2 + \frac{B_{z,0} v_*}{d^2} x\phi + c_A^2 (\nabla_\perp A_z)^2 + c_A^2 d^2 \left(\nabla_\perp^2 A_z\right)^2 \right], \quad (14)$$

$$P = \int_{-\infty}^{\infty} dx\, dy\, c_A \nabla_\perp A_z \cdot \nabla_\perp \phi, \quad (15)$$

$$Q = \int_{-\infty}^{\infty} dx\, dy \left(\nabla_\perp A_z \cdot \nabla_\perp \phi + \frac{B_{z,0} v_*}{d^2} x A_z + d^2 \nabla_\perp^2 A_z \nabla_\perp^2 \phi \right), \quad (16)$$

$$R = \int_{-\infty}^{\infty} dx\, dy\, c_A^2 d^2 \left(\nabla_\perp A_z\right)^2, \quad (17)$$

and

$$S = \int_{-\infty}^{\infty} dx\, dy\, c_A^2 d^2\, \nabla_\perp A_z^2\, \nabla_\perp \phi^2. \quad (18)$$

Separating the zero-order component of the vector potential and its perturbation, $A_z = -\int dx B_{y,0} + \delta A_z$, one readily verifies that the energy W associated with the perturbations ϕ and δA_z may be negative either in the presence of the magnetic shear, $(d/dx)B_{y,0} \neq 0$, or of the inhomogeneity of the unperturbed densities, $v_* \neq 0$. Thus, two different linear instabilities may be expected that are excited by the collisional dissipation of the perturbations with negative energy. These are the standard resistive magnetic reconnection (the tearing mode) and the beam-plasma instability that arises due to the presence of a zero-order drift velocity v_*. We will show below that an instability is possible also in a collisionless plasma. Such process is based on the linear wave transformation in which the negative energy mode transfers its energy to another mode with positive energy, which yields the simultaneous growth of their amplitudes.

2.3 Linearized equations for the surface wave

In the collisionless limit, $\nu = 0$, we linearize Eqs. (10) and (11) and seek the solution in the form of a plane wave, $\propto \exp[-i(\omega t - k_y y - k_z z)]$ whose amplitude is x-dependent. After some straightforward algebra, the linearized equations may be written in the form

$$\alpha\phi + \left(\frac{d^2}{dx^2} - k_y^2 - 1 \right) \psi = 0, \quad (19)$$

and

$$\left(\frac{d^2}{dx^2} - k_y^2 - V_* + \alpha\beta \right) \phi - \alpha\psi = 0, \quad (20)$$

where we have conveniently normalized the length by the skin depth, $\vec{r} \to \vec{r}/d$, and introduced the notations

$$\psi = c_A \delta A_z, \quad V_* = \frac{k_y v_*}{\omega}, \quad \alpha = \left(1 - \frac{d^2}{dx^2}\right)\beta, \quad \beta = \frac{c_A k_y}{\omega} \left[\frac{k_z}{k_y} + \frac{B_{0,y}(x)}{B_{0,z}} \right].$$

The linearized equations (19) and (20) may be decoupled in two equivalent ways

$$\left(\frac{d^2}{dx^2} - k_y^2 - 1\right)\left[\frac{1}{\alpha}\left(\frac{d^2}{dx^2} - k_y^2 - V_* + \alpha\beta\right)\phi\right] + \alpha\phi = 0, \tag{21}$$

and

$$\left(\frac{d^2}{dx^2} - k_y^2 - V_* + \alpha\beta\right)\left[\frac{1}{\alpha}\left(\frac{d^2}{dx^2} - k_y^2 - 1\right)\psi\right] + \alpha\psi = 0. \tag{22}$$

In the absence of the magnetic shear, viz. β = constant, we obtain from Eq. (21) or (22) the dispersion relation for the electromagnetic waves in a cold dusty plasma. We have

$$k_\perp^2 = -\frac{1}{2}\left\{1 + V_* - \beta^2 \pm \left[\left(1 + V_* - \beta^2\right)^2 - 4V_*\right]^{1/2}\right\}. \tag{23}$$

It is worth noting that in an inhomogeneous dusty plasma, $V_* \neq 0$, there exist two different propagating modes, which for $V_* = 0$ reduce to the standard electromagnetic modes in a cold plasma, namely the $k_{\perp,1} = 0$ quasimode and the shear-Alfvén mode, $k_{\perp,2}^2 = 1 - \beta^2$. Equations (21) and (22) are difficult to solve in general, i. e. for arbitrary profiles of the sheared magnetic field. However, here we study the case when the sheared field, $B_{y,0}(x)$, is produced by a current along the z-axis that is located in the spatial region $-a < x < a$. We further restrict our analysis to the perturbations whose wavelength is much larger than the transverse dimension of the current layer, i. e. $k_y a \ll 1$. In such a geometry, the magnetic field $B_{y,0}(x)$ is homogeneous everywhere except for $x \in (-a, a)$, where it undergoes a sharp jump. The solution is expected to be spatially localized in the vicinity of the inhomogeneous transitional region, i. e. to have the form of a surface wave. A typical surface mode consists of two evanescent plane waves, for $x < -a$ and $x > a$, which are appropriately connected through a thin transitional layer.

Setting $B_{y,0}$ = constant we readily find that the 'outside' solution, $|x| > a$, of Eqs. (19) and (20) consists of two linearly independent, evanescent modes. The first mode, in which ψ is an even function of x and ϕ is odd, corresponds to the reconnection of the magnetic field lines and the filamentation of the current sheet, and it will be labelled below as the 'tearing mode'. It is given by

$$\psi_{tear} = \Psi_1 e^{-\kappa_1(|x|-a)} + \Psi_2 e^{-\kappa_2(|x|-a)}, \tag{24}$$

and

$$\phi_{tear} = \hat{t}\psi_{tear}, \quad \hat{t} = -\frac{1}{\alpha}\left(\frac{d^2}{dx^2} - k_y^2 - 1\right), \tag{25}$$

where

$$\kappa_x^2 = k_y^2 + \frac{1}{2}\left(1 + V_* - \beta^2(\pm a) \pm \left\{\left[1 + V_* - \beta^2(\pm a)\right]^2 - 4V_*\right\}^{1/2}\right). \tag{26}$$

The second linear mode, with ψ being odd and ϕ even function of x describes the bending of the magnetic field lines, i. e. a perturbation in the form of the kink of the current sheet, and it will be labeled as the 'bending mode'

$$\phi_{bend} = \Phi_1 e^{-\kappa_1(|x|-a)} + \Phi_2 e^{-\kappa_2(|x|-a)}, \tag{27}$$

$$\psi_{bend} = \hat{b}\phi_{bend}, \quad \hat{b} = \frac{1}{\alpha}\left(\frac{d^2}{dx^2} - k_y^2 - V_* + \alpha\beta\right). \tag{28}$$

Inside the transitional layer, $x \in (-a, a)$, with any desired accuracy, the solution can be calculated approximately making use of the small parameter $\epsilon = k_y a \ll 1$. A simple iterative scheme is constructed by integrating Eqs. (19) and (20) with respect to x twice, and making some obvious linear combinations. We obtain

$$\left[G_i\left(x, \phi, \psi\right)\right]_{-a}^{x} = I_i\left(x, \phi, \psi\right), \quad i = 1, 2, 3, 4., \tag{29}$$

where the square bracket denotes the variation of a function between $-a$ and x, $[f(x)]_{-a}^{x} = f(x) - f(-a)$, and

$$G_1 = \frac{d\psi}{dx}, \quad I_1 = \int_{-a}^{x} dx' \left[\left(k_y^2 + 1\right)\psi - \alpha\phi\right], \tag{30}$$

$$G_2 = \frac{d\phi}{dx}, \quad I_2 = \int_{-a}^{x} dx' \left[\left(k_y^2 + V_* - \alpha\beta\right)\phi + \alpha\psi\right], \tag{31}$$

$$G_3 = \phi - x\frac{d\phi}{dx} \quad I_3 = -\int_{-a}^{x} dx'\, x' \left[\left(k_y^2 + V_* - \alpha\beta\right)\phi + \alpha\psi\right], \tag{32}$$

and

$$G_4 = \psi - x\frac{d\psi}{dx} \quad I_4 = -\int_{-a}^{x} dx'\, x' \left[\left(k_y^2 + 1\right)\psi - \alpha\phi\right]. \tag{33}$$

For a thin current sheet, one expects that all the above integrals are small quantities of the order $\mathcal{O}(k_y a)$. However, that is not always the case. First, one should note that the function $\beta(x)$ is proportional to the unperturbed magnetic field, and thus it is expected to be a well behaved and smooth function, whose characteristic length of inhomogeneity is of the order a. Conversely, the shape of the current profile is physically far less restricted, and it may feature much larger gradients than the magnetic field. As a consequence, the function $\alpha(x) = \beta(x) - \beta''(x)$ may contain sharp peaks related with the jumps in the current profile, that contribute significantly to the integrals defined in Eqs. (30)-(33). Thus, the nature of the 'inside' solution strongly depends on the properties of the current sheet. Before we proceed with the solution, it is necessary to determine the regimes of interest and to construct the appropriate iterative schemes for the solution of the linear problem in different regimes.

Qualitatively, the following regimes may distinguished:

(*i*) Current sheet with gentle slopes characterized by

$$\int_{-a}^{+a} dx\, x\alpha \sim a \ll 1 \quad \text{or} \quad \int_{-a}^{+a} dx\, \alpha\beta \sim a \ll 1, \tag{34}$$

where the first condition applies to Eqs. (30) and (33) and the second to Eqs. (31) and (32). In other words, a reliable iterating scheme for the 'inside' solution may be based on the expressions (30)-(33) only if all the integrals that are involved scale as $\mathcal{O}(k_y a)$, i. e. for a current sheet with gentle slopes. To emphasize this fact, in the rest of the text the quantities G_i and I_i, defined in Eqs. (30)-(33) will be denoted by G_i^G and I_i^G

(*ii*) Current sheet with steep edges characterized by

$$\int_{-a}^{+a} dx\, x\alpha \sim 1 \quad \text{or} \quad \int_{-a}^{+a} dx\, \alpha\beta \sim 1. \tag{35}$$

In this case, the terms proportional to α make contributions of the order $\mathcal{O}(1)$ to the integrals in Eqs. (30)-(33). In order to calculate these contribution, we integrate by part all the terms proportional to $\beta''(x)$, which yields the expression of the same form as Eq. (29), but with the following definitions for the functions G_i and I_i

$$G_1^S = \frac{d}{dx}(\psi + \beta\phi) - 2\frac{d\beta}{dx}\phi,$$

$$I_1^S = \int_{-a}^{x} dx' \left[\left(k_y^2 + 1\right)\psi - \beta\left(1 - \frac{d^2}{dx'^2}\right)\phi\right] \tag{36}$$

$$G_2^S = \frac{d}{dx}(\phi - \beta\psi) + 2\frac{d\beta}{dx}\psi,$$

$$I_2^S = \int_{-a}^{x} dx' \left[\left(k_y^2 + V_*\right)\phi - \beta k_y^2\psi\right], \tag{37}$$

$$G_3^S = \left(1 - x\frac{d}{dx}\right)\phi - x\frac{d\beta}{dx}\psi,$$

$$I_3^S = -\int_{-a}^{x} dx' \left\{x'\left[\left(k_y^2 + V_*\right)\phi + \beta\left(\frac{d^2}{dx'^2} - k_y^2\right)\psi\right] + \frac{d\beta}{dx'}(x'\psi)\right\}, \tag{38}$$

$$G_4^S = \left(1 - x\frac{d}{dx}\right)\psi + x\frac{d\beta}{dx}\phi,$$

$$I_4^S = -\int_{-a}^{x} dx' \left\{x'\left[\left(k_y^2 + 1\right)\psi - \beta\phi\right] - \frac{d\beta}{dx'}(x'\phi)\right\}, \tag{39}$$

where the superscript "S" refers to the steep current profile. As all the above integrals scale as $\mathcal{O}(k_y a)$, Eqs. (36)-(39) provide the basis for an iterating scheme for the 'inside' solution in the case of a current sheet with steep edges.

(*iii*) Current sheet with very sharp edges characterized by

$$\int_{-a}^{+a} dx\, x\alpha \sim \frac{1}{a} \gg 1 \quad \text{or} \quad \int_{-a}^{+a} dx\, \alpha\beta \sim \frac{1}{a} \gg 1. \tag{40}$$

In this case the potentials ϕ, ψ have large gradients inside the transitional layer. That introduces large mathematical difficulties since there are no convenient small parameters that would permit the use of the surface-wave formalism. As the current sheet with such sharp edges is also physically relatively difficult to realize, from the practical point of view it is less relevant and we will not study it further in this paper.

2.4 Solution inside the current layer

It is worth noting that Eqs. (21) and (22) are singular at the resonant point x_{res}, defined by $\alpha(x_{res}) = 0$. Consequently, their general solution is also singular at x_{res}, with a possible exception of one particular solution that is regular. For simplicity, we assume that α and β are odd functions of x, and thus the resonant point is located at the origin, $x_{res} = 0$.

In the region $x \in (-a, a)$ Eqs. (21) and (22) are solved approximately using the small parameter $\epsilon = k_y a \ll 1$. We expand the solution as

$$\phi = \sum_{n=0}^{\infty} \phi^{(n)}, \quad \psi = \sum_{n=0}^{\infty} \psi^{(n)}, \tag{41}$$

and use the fact that the integrals I_i are small quantities $\sim k_y a$. Then, Eq. (29) breaks down to the following sequence

$$\left[G_i\left(x, \phi^{(0)}, \psi^{(0)}\right)\right]_{-a}^{x} = 0, \quad i = 1, 2, 3, 4., \tag{42}$$

$$\left[G_i\left(x, \phi^{(1)}, \psi^{(1)}\right)\right]_{-a}^{x} = I_i\left(x, \phi^{(0)}, \psi^{(0)}\right), \tag{43}$$

$$\left[G_i\left(x, \phi^{(2)}, \psi^{(2)}\right)\right]_{-a}^{x} = I_i\left(x, \phi^{(1)}, \psi^{(1)}\right), \tag{44}$$

....

where, without the loss of generality, we adopted that the higher-order perturbations, $\phi^{(n)}, \psi^{(n)}$, $n = 1, 2, ..$ and their derivatives reduce to zero at $x = \pm a$.

The singularity $x = 0$ in the integrals $I_1 - -I_4$ is resolved by including a small imaginary part of the frequency, $\omega = \omega' + i\gamma$, with $\gamma \ll \omega'$, which yields

$$\frac{1}{\alpha} \approx \frac{p}{\alpha} + i\pi\,\delta\left(\alpha\right), \tag{45}$$

where p denotes the principal value.

From Eqs. (42)-(44) we can easily calculate all the components $\phi^{(n)}, \psi^{(n)}$ in the expansion (41). We conveniently separate the even and odd parts, in order to distinguish the tearing- and bending-mode components. From the even parts, and after some algebra we obtain

$$\psi_{tear}^{(0)} = C_{\psi,2}G_2(-a) + C_{\psi,4}G_4(-a), \tag{46}$$

$$\phi_{bend}^{(0)} = C_{\phi,1}G_1(-a) + C_{\phi,3}G_3(-a), \tag{47}$$

$$\psi_{tear}^{(n)} = \left[\sum_{i=1}^{4} C_{\psi,i}\cdot\left(\frac{x}{|x|}\right)^{i} I_i\left(x,\phi^{(n-1)},\psi^{(n-1)}\right)\right]_{\text{even part}}, \tag{48}$$

and

$$\phi_{bend}^{(n)} = \left[\sum_{i=1}^{4} C_{\phi,i}\cdot\left(\frac{x}{|x|}\right)^{i-1} I_i\left(x,\phi^{(n-1)},\psi^{(n-1)}\right)\right]_{\text{even part}}, \quad n=1,2,.. \tag{49}$$

The functions ϕ_{tear} and ψ_{bend}, that are odd in x and singular at $x=0$, are then readily expressed as

$$\phi_{tear}^{(n)} = \hat{t}\,\psi_{tear}^{(n)}, \quad \psi_{bend}^{(n)} = \hat{b}\,\phi_{bend}^{(n)}, \quad n=0,1,2,..., \tag{50}$$

where the operators $\hat{t}$ and $\hat{b}$ are defined in Eqs. (25) and (28).

In the case of a current sheet with gentle slopes the coefficients $C_{\psi,i}$ and $C_{\phi,i}$ are determined from the expressions (30)-(33) for the functions G_i, $i = 1--4$, which yields

$$C_{\psi,1}^{G} = C_{\phi,2}^{G} = |x|, \quad C_{\psi,3}^{G} = C_{\phi,4}^{G} = 1, \quad C_{\psi,2}^{G} = C_{\psi,4}^{G} = C_{\phi,1}^{G} = C_{\phi,3}^{G} = 0. \tag{51}$$

Similarly, for a current sheet with sharp edges these coefficients are found from Eqs. (36)-(39). We have

$$C_{\psi,1}^{S} = C_{\phi,2}^{S} = \frac{|x|}{\Delta}\left(1 - x\beta\frac{d\beta}{dx}\right), \quad C_{\psi,2}^{S} = -C_{\phi,1}^{S} = -\frac{x\beta}{\Delta},$$

$$C_{\psi,3}^{S} = C_{\phi,4}^{S} = -\frac{|x|\,\beta^2}{\Delta}\frac{d\beta}{dx}, \quad C_{\psi,4}^{S} = -C_{\phi,3}^{S} = \frac{1}{\Delta}\left(1+\beta^2 - x\beta\frac{d\beta}{dx}\right), \tag{52}$$

where

$$\Delta = \frac{1}{1+\beta^2}\left[\left(1+\beta^2 - x\beta\frac{d\beta}{dx}\right)^2 + \left(x\beta^2\frac{d\beta}{dx}\right)^2\right].$$

2.5 Dispersion relation

Since Eqs. (21) and (22) are fourth order differential equations, to find the leading-order estimate for the growth rate we need to solve Eqs. (42)-(44) with the accuracy to the second-order terms, $\phi^{(2)}$ and $\psi^{(2)}$. Then, setting the upper boundary in Eq.

(29) to $+a$ and substituting the expressions (46)-(50) in the integrals $I_1 - -I_4$, with the accuracy to the second order we have

$$[G_i(x,\phi,\psi)]_{-a}^{+a} = pI_i(x,\phi,\psi) + i\pi \operatorname*{Res}_{x=0}\left[\frac{\partial}{\partial x} I_i(x,\phi,\psi)\right] +$$

$$\frac{i\pi}{2}\sum_{j=1}^{4} \operatorname*{Res}_{x=0}\left[\frac{\partial}{\partial x} I_j(x,\phi,\psi)\right] \times$$

$$\left\{pI_i(x,C_{\phi,j},C_{\psi,j}) + i\pi \operatorname*{Res}_{x=0}\left[\frac{\partial}{\partial x} I_i(x,C_{\phi,j},C_{\psi,j})\right]\right\}. \tag{53}$$

The potentials ϕ, ψ are written as the sums of the linearly independent tearing and bending components, $\phi = \phi_{tear} + \phi_{bend}$, $\psi = \psi_{tear} + \psi_{bend}$, where

$$\phi_{bend} = \Phi_1 f_1 + \Phi_2 f_2, \quad \psi_{tear} = \Psi_1 g_1 + \Psi_2 g_2,$$

$$\psi_{bend} = \Phi_1 \hat{b} f_1 + \Phi_2 \hat{b} f_2, \quad \phi_{tear} = \Psi_1 \hat{t} g_1 + \Psi_2 \hat{t} g_2, \tag{54}$$

where Ψ_l, Φ_l are constants of integration introduced in Eqs. (24) and (27). For $|x| > a$, the x-even functions f_l and g_l are exponential functions

$$f_l = g_l = \exp\left[-\kappa_l\left(|x| - a\right)\right], \quad l = 1, 2.$$

Inside the current sheet, $|x| < a$, within the second-order accuracy, we may adopt f_l and g_l as arbitrary smooth and well-behaved functions that satisfy the appropriate boundary conditions at $x = \pm a$. As there are four different boundary conditions, the simplest interpolating function is the fourth-order polynomial.

Using the expressions (46)-(50) and the linear combinations (54), carefully separating the x-odd and even parts and after some algebra, the continuity conditions (53) yield the following linear system for the constants of integration $\Phi_{1,2}, \Psi_{1,2}$

$$\sum_{l=1}^{2}\left[\left(a_{1,l}^{(0)} + i\epsilon^2 a_{1,l}^{(2)}\right)\Psi_l + \left(i\epsilon b_{1,l}^{(1)} - \epsilon^2 b_{1,l}^{(2)}\right)\Phi_l\right] = 0, \tag{55}$$

$$\sum_{l=1}^{2}\left[\left(i\epsilon a_{2,l}^{(1)} - \epsilon^2 a_{2,l}^{(2)}\right)\Psi_l + \left(b_{2,l}^{(0)} + i\epsilon^2 b_{2,l}^{(2)}\right)\Phi_l\right] = 0, \tag{56}$$

$$\sum_{l=1}^{2} a_{3,l}^{(0)}\Psi_l = 0, \tag{57}$$

$$\sum_{l=1}^{2} b_{4,l}^{(0)}\Phi_l = 0, \tag{58}$$

where, with the accuracy to the second order in the small parameter $\epsilon = k_y a$, the coefficients are given by

$$a_{1,l}^{(0)} = D_{Q,l} + pI_1\left(a, \hat{b}g_l, g_l\right) - \frac{\pi^2}{2\epsilon^2}\left(k_y^2+1\right)\operatorname*{Res}_{x=0}\left(\hat{t}C_{\phi,2}\right)\left(k_y^2+V_*\right)\operatorname*{Res}_{x=0}\left(\hat{b}g_l\right), \quad (59)$$

$$a_{1,l}^{(2)} = \frac{\pi}{2\epsilon^2}\left(k_y^2+V_*\right)\operatorname*{Res}_{x=0}\left(\hat{b}g_l\right)pI_1\left(a, C_{\phi,2}, C_{\psi,2}\right), \quad (60)$$

$$b_{1,l}^{(1)} = \frac{\pi}{\epsilon}\left(k_y^2+1\right)\operatorname*{Res}_{x=0}\left(\hat{t}f_l\right)\left[1+\frac{1}{2}pI_1\left(a, C_{\phi,1}, C_{\psi,1}\right)\right], \quad (61)$$

$$b_{1,l}^{(2)} = \frac{\pi^2}{\epsilon^2}\left(k_y^2+1\right)^2\operatorname*{Res}_{x=0}\left(\hat{t}f_l\right)\operatorname*{Res}_{x=0}\left(\hat{t}C_{\phi,1}\right), \quad (62)$$

$$a_{2,l}^{(1)} = \frac{\pi}{\epsilon}\left(k_y^2+V_*\right)\operatorname*{Res}_{x=0}\left(\hat{b}g_l\right)\left[1+\frac{1}{2}pI_2\left(a, C_{\phi,2}, C_{\psi,2}\right)\right], \quad (63)$$

$$a_{2,l}^{(2)} = \frac{\pi^2}{\epsilon^2}\left(k_y^2+V_*\right)^2\operatorname*{Res}_{x=0}\left(\hat{b}g_l\right)\operatorname*{Res}_{x=0}\left(\hat{b}C_{\psi,2}\right), \quad (64)$$

$$b_{2,l}^{(0)} = D_{P,l} + pI_2\left(a, f_l, \hat{t}f_l\right) - \frac{\pi^2}{2\epsilon^2}\left(k_y^2+V_*\right)\operatorname*{Res}_{x=0}\left(\hat{b}C_{\psi,1}\right)\left(k_y^2+1\right)\operatorname*{Res}_{x=0}\left(\hat{t}f_l\right), \quad (65)$$

$$b_{2,l}^{(2)} = \frac{\pi}{2\epsilon^2}\left(k_y^2+V_*\right)\operatorname*{Res}_{x=0}\left(\hat{t}f_l\right)pI_2\left(a, C_{\phi,1}, C_{\psi,1}\right), \quad (66)$$

$$b_{3,l}^{(0)} = 2P_l\left(1+\kappa_l a\right) + pI_3\left(a, f_l, \hat{t}f_l\right) \quad (67)$$

$$a_{4,l}^{(0)} = 2Q_l\left(1+\kappa_l a\right) + pI_4\left(a, \hat{b}g_l, g_l\right) \quad (68)$$

where $P_l = [1/\beta(a)][\kappa_l^2 - k_y^2 - V_* + \beta^2(a)]$ and $Q_l = [-1/\beta(a)](\kappa_l^2 - k_y^2 - 1)$. For the gentle profile we have $D_{Q,l}^G = D_{P,l}^G = 2\kappa_l$ and $D_{Q,l}^S = 2\kappa_l(1+\beta(a)Q_l)$, and for the steep profile we have $D_{P,l}^S = 2\kappa_l(1+\beta(a)P_l)$. In the above expressions the lowest-order contributions originate from the left-hand-sides of Eq. (53), $[G_i(x)]_{-a}^{+a}$ calculated from the outer solutions (24) and (27), while the corrections proportional to ϵ and ϵ^2 come from the single and double integrals, respectively. It should also be noted that Eqs. (67) and (68) do not contain any contributions of singularities.

Using the definitions (51) and (52) for the coefficients $C_{\phi,l}$ and $C_{\psi,l}$ we find that both for the gentle and steep profiles we have

$$\operatorname*{Res}_{x=0}\left(\hat{t}C_{\phi,l}\right) = \operatorname*{Res}_{x=0}\left(\hat{b}C_{\psi,l}\right) = 0, \quad l = 1, 2. \quad (69)$$

Consequently, in both regimes we have

$$b_{1,l}^{(2)} = a_{2,l}^{(2)} = 0. \quad (70)$$

For the gentle profile, the system (55)-(58) is further simplified since we have

$$pI_1^G\left(a, C_{\phi,2}^G, C_{\psi,2}^G\right) = pI_2^G\left(a, C_{\phi,1}^G, C_{\psi,1}^G\right) = 0, \quad (71)$$

which yields

$$a_{1,l}^{(2)G} = b_{2,l}^{(2)G} = 0. \tag{72}$$

Setting the determinant of the system (55)-(58) to zero and using (70), we obtain the dispersion relation

$$D_{tear}^{(0)} D_{bend}^{(0)} - \\ i\epsilon^2 \left[D_{tear}^{(0)} \left(b_{2,2}^{(2)} b_{4,1}^{(0)} - b_{2,1}^{(2)} b_{4,2}^{(0)}\right) + D_{bend}^{(0)} \left(a_{1,2}^{(2)} a_{3,1}^{(0)} - a_{1,1}^{(2)} a_{3,2}^{(0)}\right)\right] = \mathcal{O}\left(\epsilon^2\right), \tag{73}$$

where the right-hand-side is a second-order small quantity that us purely real. Th expressions $D_{tear}^{(0)}$ and $D_{bend}^{(0)}$ are, respectively, the dispersion functions of the tearing and bending modes in the limit of a zero-width transitional layer

$$D_{tear}^{(0)} = \kappa_1 \left[\kappa_1^2 - k_y^2 - V_* + 2\beta^2(a)\right] - \kappa_2 \left[\kappa_2^2 - k_y^2 - V_* + 2\beta^2(a)\right], \tag{74}$$

$$D_{bend}^{(0)} = \kappa_1 \left[\kappa_1^2 - k_y^2 - 1 + \beta^2(a)\right] - \kappa_2 \left[\kappa_2^2 - k_y^2 - 1 + \beta^2(a)\right]. \tag{75}$$

Obviously, the dispersion relation (73) is *purely real* in the case of a gentle slope, due to the condition (72). Thus, for a gentle slope the possibility of linear instability of this type does not exist.

For a steep profile, a careful analysis of the leading-order dispersion relation $D_{tear}^{(0)S} = 0$ reveals that the tearing mode cannot propagate as an eigenmode. This is illustrated by Fig. 1 which shows that for a broad range of the parameters v_* and k_y the tearing mode is not well-localized, since the corresponding e-folding lengths κ_1 and κ_2 possess large imaginary parts and the above surface-mode analysis is not applicable. Conversely, the dispersion relation $D_{bend}^{(0)S} = 0$ has real roots and the bending mode is an appropriate eigenmode, provided $V_* \neq 0$. Expanding the full dispersion relation (73) in the vicinity of $D_{bend}^{(0)S}(\omega_{bend}^{(0)S}, k_y) = 0$ and using $\gamma \ll \omega$, the growth rate is readily calculated as

$$\gamma_{bend}^S = \epsilon^2 \left(b_{2,2}^{(2)S} b_{4,1}^{(0)S} - b_{2,1}^{(2)S} b_{4,2}^{(0)S}\right) \left[\frac{\partial D_{bend}^{(0)S}}{\partial \omega}\right]_{\omega=\omega_{bend}^{(0)S}}^{-1}. \tag{76}$$

As we discussed before, instability arises from the resonant mode-coupling between negative- and positive energy waves at the point x_{res} where the wave vector is parallel to the local magnetic field $\vec{B}_0(x_{res})$. The tearing mode is a forced quasimode, driven by the eigenmode associated with the bending of the current sheet. These modes are coupled through the lowest order singularities and their relative amplitudes are estimated from Eq. (55) as $\Psi_l/\Phi_l \sim -i\epsilon$. Thus, the tearing mode has a smaller amplitude than its bending counterpart and they are relatively shifted along y by $\pi/2$. It should be noted that due to the presence of dust the symmetry between the electrons and ions is broken, which gives rise to a finite perpendicular current associated with the $\vec{E} \times \vec{B}$ drift, $\vec{j}_\perp = (1/B_{z,0})(\vec{e}_z \times \nabla\phi)\, e z_d n_d$ and, consequently, to the perturbation of the z-component of the magnetic field. The relative amplitude of this magnetic field compared to the perpendicular magnetic field of the tearing mode, is readily estimated as

$$\frac{\delta B_z}{\delta B_{\perp tear}} \sim \frac{\Phi}{\Psi}\left(\frac{m_e}{m_i}\right)^{1/2} \frac{z_d n_d}{(n_{0,e} n_{0,i})^{1/2}}.$$

The above value may be close to that observed in the magnetospheric storms [*Deng and Matsumoto* 2001].

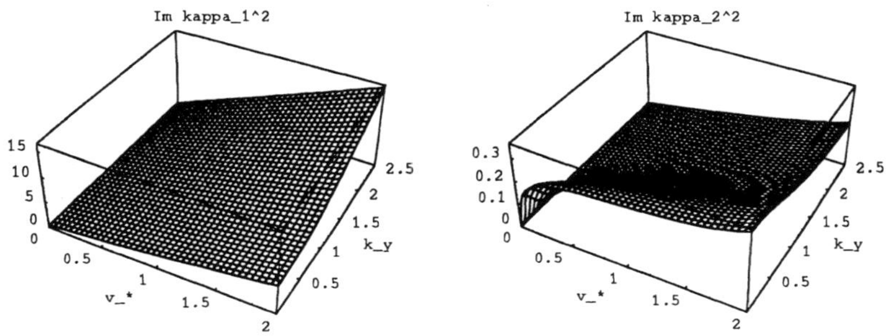

Figure 1: Imaginary parts of the e-folding lengths κ_1^2 and κ_1^2 for the tearing mode on a current sheet with a steep profile. The figure is plotted for $B_{y,0}(a)/B_{z,0} = 0.75\, k_z/k_y$. The e-folding lengths $\kappa_{1,2}$ and the wavenumber k_y are normalized by d^{-1}, and the dust-drift speed v_* by $c_A\, k_z/k_y$

3 Resistive reconnection

3.1 Governing equations

Let us consider a partially ionized dusty plasma in a sheared magnetic field $B_0\vec{e}_z + B_{0y}(x)\vec{e}_y$, where $\vec{e}_z$ ($\vec{e}_y$) is the unit vector along the z (y) axis. The plasma constituents are the electrons, ions, and negatively charged dust grains. At equilibrium, the overall quasi-neutrality condition is $n_{i0} = n_{e0} + z_d n_{d0}$, where n_{j0} is the unperturbed particle number density of the particle species j (j equals e for electrons, i for ions, and d for negatively charged dust grains) and z_d is the number of the electrons residing on the dust grain surface. We consider the resistive tearing instability of intermediate-frequency ($\Omega_d \ll |\partial/\partial t| \ll \Omega_i, \nu_{en} \ll \Omega_e$; where Ω_d, Ω_i, and Ω_e are the dust, ion and electron gyrofrequencies, respectively, and ν_{en} is the electron-neutral collision frequency; the latter is much larger than the electron-ion collision frequency in a partially ionized astrophysical plasmas), low phase speed (in comparison with the speed of light c) electromagnetic waves whose electric and magnetic fields are $\vec{E} = -\nabla_\perp\phi - \vec{e}_z(\partial/\partial t)A_z$ and $\vec{B}_\perp = \nabla A_z \times \vec{e}_z$. The relevant velocities of the electrons, ions, and dust grains are given by, respectively,

$$\vec{v}_{e\perp} \approx \frac{1}{B_0}\vec{e}_z \times \nabla\phi + v_{ez}\frac{\vec{B}_{\perp 0}}{B_0} + v_{e0}\frac{\vec{B}_{\perp 1}}{B_0}, \tag{77}$$

$$\vec{v}_{i\perp} \approx \frac{1}{B_0}\vec{e}_z \times \nabla\phi - \frac{1}{B_0\Omega_i}\left(\frac{\partial}{\partial t} + \nu_{in}\right)\nabla_\perp\phi, \tag{78}$$

and

$$\left(\frac{\partial}{\partial t} + \nu_{dn}\right)\vec{v}_{d\perp} = \frac{z_d e}{m_d}\nabla_\perp\phi - \nabla_\perp\varphi. \tag{79}$$

Here the subscript $\perp$ stands for the component perpendicular to $\vec{e}_z$, and the gravitational potential φ is determined from

$$\nabla_\perp^2\varphi = 4\pi G n_{d1}, \tag{80}$$

in which the dust number density is obtained by means of the dust continuity equation

$$\frac{\partial}{\partial t}n_{d1} + n_{d0}\nabla\cdot v_{d\perp} = 0, \tag{81}$$

where $v_{e0} = (c^2\epsilon_0/en_0)\nabla_\perp^2 A_{z,0}$, $v_{ez} = (c^2\epsilon_0/en_0)\nabla_\perp^2 A_z$, $\vec{B}_{\perp 0} = \nabla A_{z,0} \times \vec{e}_z$, and $\vec{B}_{\perp 1} = \nabla A_z \times \vec{e}_z$, with $A_{z,0}$ being the unperturbed part of the parallel magnetic vector potential, $v_{j\perp}$ is the perpendicular component of the fluid velocity, ν_{in} is the ion-neutral collision frequency, and G is the gravitational constant. The parallel component of the electron fluid velocity v_{ez} is obtained from

$$\nu_{en}v_{ez} = \frac{e}{m_e}\frac{\partial}{\partial t}A_z + \frac{e}{m_e B_0}\vec{B}_{\perp 0}\cdot\nabla\phi, \tag{82}$$

and

$$v_{ez} \approx \frac{c^2\epsilon_0}{en_{e0}}\nabla_\perp^2 A_z, \tag{83}$$

which are the parallel components of the inertialess electron equation of motion and Ampère's law, respectively. The compressional magnetic fields and the parallel ion dynamics are ignored in view of the low-β^* approximation.

We subtract the ion continuity equation from the electron continuity equation, use $n_{i1} = n_{e1} + z_d n_{d1}$ and Eqs. (77), (78) and (83) to derive the modified (by the presence of charged dust) ion vorticity equation

$$\frac{\partial}{\partial t}n_{d1} - \frac{n_{i0}}{z_d B_0\Omega_i}\left(\frac{\partial}{\partial t} + \nu_{in}\right)\nabla_\perp^2\phi$$
$$-\frac{c^2\epsilon_0}{ez_d B_0}\left(\vec{B}_{\perp 0}\cdot\nabla\nabla_\perp^2 A_z + \vec{B}_{\perp 1}\cdot\nabla\nabla_\perp^2 A_{z,0}\right) = 0, \tag{84}$$

where n_{i1} and n_{e1} are small perturbations in the equilibrium ion and electron number densities, respectively. Combining Eqs. (79), (80) and (81) we obtain

$$\left[\left(\frac{\partial}{\partial t}+\nu_{dn}\right)\frac{\partial}{\partial t}-\omega_J^2\right]n_{d1}=-\frac{n_{d0}z_d e}{m_d}\nabla_\perp^2\phi, \tag{85}$$

where $\omega_J=(4\pi G n_{d0})^{1/2}$ is the usual Jeans frequency. On the other hand, from Eqs. (82) and (83) we have

$$\left(\frac{\partial}{\partial t}-\eta\nabla_\perp^2\right)A_z+\frac{1}{B_0}\vec{B}_{\perp 0}\cdot\nabla\phi=0, \tag{86}$$

where $\eta=\nu_{en}\lambda_e^2$ represents the plasma resistivity and λ_e is the electron skin depth.

Equations (84)-(86) are the desired set for studying the resistive tearing mode instabilities in a three-component self-gravitating dusty magnetoplasma including the dust particle dynamics. In the absence of the dust density fluctuation (viz. $n_{d1}=0$), Eqs. (84) and (85) describe the physics of the resistive tearing mode instability in a collisional dusty magnetoplasma (where $n_{i0}\neq n_{e0}$) containing stationary neutrals and stationary charged dust grains [*Birk et al.* 1996].

In order to study the tearing mode instabilities in the sheared magnetic field $B_{0y}(x)=-\partial A_{z,0}/\partial x$, the latter is due to an equilibrium parallel current $J_0=c^2\epsilon_0(\partial B_{0y}/\partial x)$, we suppose that ϕ, A_z and n_{d1} vary as $\exp(\gamma t+iky)$, where γ is the growth rate and k is the wavenumber. Accordingly, Eqs. (84)-(86) can be expressed as

$$\gamma n_{d1}-\frac{n_{i0}(\gamma+\nu_{in})}{z_d B_0\Omega_i}\left(\frac{\partial^2}{\partial x^2}-k^2\right)\phi$$
$$-ik\frac{c^2\epsilon_0}{ez_d B_0}\left[B_{0y}(x)\left(\frac{\partial^2}{\partial x^2}-k^2\right)A_z+J_0'A_z\right]=0, \tag{87}$$

$$\left[(\gamma+\nu_{dn})\gamma-\omega_J^2\right]n_{d1}=-\frac{n_{d0}z_d e}{m_d}\left(\frac{\partial^2}{\partial x^2}-k^2\right)\phi, \tag{88}$$

and

$$\gamma A_z-\eta\left(\frac{\partial^2}{\partial x^2}-k^2\right)A_z+ik\frac{B_{0y}(x)}{B_0}\phi=0. \tag{89}$$

where $J_0'\equiv(\partial^2/\partial x^2)B_{0y}$.

3.2 Solution inside the resistive layer

Following the standard procedure [*Furth et al.* 1963, *White* 1986] for investigating the tearing mode instability, we define an inner region where $B_{0y}(x)\approx(\partial B_{0y}/\partial x)x\equiv B_{0y}'x$, $J_0'\approx 0$ and $\partial^2/\partial x^2\gg k^2$. Thus, in this inner region, Eqs. (87) and (88) are combined to obtain

$$\frac{\partial^2}{\partial x^2}\phi-\frac{c^2\epsilon_0}{en_{i0}(1+\alpha)}\frac{\Omega_i}{(\gamma+\nu_{in})}ikB_{0y}'x\frac{\partial^2}{\partial x^2}A_z, \tag{90}$$

where

$$\alpha = \frac{z_d n_{d0}}{n_{i0}} \frac{\Omega_i}{(\gamma + \nu_{in})} \frac{\gamma \Omega_d}{[(\gamma + \nu_{dn})\gamma - \omega_J^2]} \tag{91}$$

represents the contribution associated with the dust density perturbation. Furthermore, in the inner region, Eq. (89) takes the form

$$\gamma A_z - \eta \partial_x^2 A_z + ik \frac{B'_{0y}}{B_0} x\phi = 0. \tag{92}$$

Using the constant-A_z approximation, we obtain from Eqs. (90) and (92)

$$\xi^{-1}\partial_\xi^2 F - \xi F = -1, \tag{93}$$

where $\xi = x/\delta$ and

$$\delta = \left[\frac{(1+\alpha)(\gamma + \nu_{in})\nu_{en}\lambda_e^2 L_s^2}{k^2 c_A^2}\right]^{1/4} \tag{94}$$

is the width of the inner layer, $F = -ik\delta\phi/\gamma L_s A_z$, $L_s = B_0/B'_{0y} > 0$ is the scale length of the sheared magnetic field, and c_A is the Alfvén speed. The constant-A_z approximation can be justified a posteriori, and has been discussed by *Furth et al.* [1963] and *White* [1986].

3.3 Dispersion relation

Integrating Eq. (92) across the inner layer, we obtain by defining $\Delta' = (1/A_z)(\partial A_z/\partial x)$,

$$\Delta' = \frac{\delta\gamma}{\eta} I, \tag{95}$$

where $I = -\int_{-\infty}^{\infty} d\xi \xi^{-1}\partial_\xi^2 F$ is a number of order unity if we use F as given by the solution of Eq. (93). Inserting the definition of the width δ, we thus readily obtain from Eq. (95) a new dispersion relation

$$\gamma^4(\gamma + \nu_{in})(1+\alpha) = \nu_{en}^3 k^2 c_A^2 \left(\frac{\Delta'}{I}\right)^4 \frac{\lambda_e^6}{L_s^2}. \tag{96}$$

Equation (96) can be analyzed in several limiting cases. First, we note that for $\nu_{in} \ll \gamma$ and $\alpha = 0$, we obtain from (96) the growth rate of the resistive tearing mode [*Birk et al.* 1996] in a dusty plasma. We have

$$\gamma = \left(\frac{kc_A}{L_s}\right)^{2/5} \left(\frac{\Delta'}{I}\right)^{4/5} \eta^{3/5}. \tag{97}$$

Second, for $\nu_{in} \gg \gamma$ and $\alpha = 0$, we obtain from Eq. (96)

$$\gamma = \left(\frac{kc_A}{L_s}\right)^{1/2} \nu_{in}^{-1/4} \frac{\Delta'}{I} \eta^{3/4}. \tag{98}$$

Third, for $\alpha \gg 1$, we obtain from (19)

$$\frac{z_d n_{d0}}{n_{i0}} \frac{\gamma^5 \omega_g^2}{\gamma(\gamma + \nu_{dn}) - \omega_J^2} = \frac{k^2 c_A^2}{L_s^2} \left(\frac{\Delta'}{I}\right)^4 \eta^3, \tag{99}$$

where $\omega_g = \sqrt{\Omega_i \Omega_d}$. For $\gamma \gg \nu_{dn}, \omega_J$, we have from Eq. (99)

$$\gamma = \left(\frac{n_{i0}}{z_d n_{d0}}\right)^{1/3} \left(\frac{k c_A}{L_s \omega_g}\right)^{2/3} \left(\frac{\Delta'}{I}\right)^{4/3} \eta. \tag{100}$$

Fourth, we consider a collision-dominated case in which $\nu_{in}, \nu_{dn} \gg \gamma \gg \omega_J^2/\nu_{dn}$, Here, we obtain from (19)

$$\gamma = \left(\frac{k c_A}{L_s}\right)^{1/2} \left[\left(1 + \frac{z_d n_{d0} \Psi}{n_{i0}}\right) \nu_{in}\right]^{-1/4} \frac{\Delta'}{I} \eta^{3/4}, \tag{101}$$

where $\Psi = \omega_g^2/\nu_{in}\nu_{dn}$. Expressions (100) and (101) give the growth rates of our new resistive tearing mode instability in a partially ionized dusty magnetoplasma.

4 Summary

In this paper, we have investigated the instability of a thin current sheet in a dusty plasma with an inhomogeneous dust density. A new linear surface mode is found in the form of a propagating kink (bending) of the current sheet, which is unstable only in the case of a sheet with sharp boundaries. The bending mode is destabilized by the second-order effects in the singularity $x = 0$, due to the coupling with the local convective cell mode [with $\vec{k} \| \vec{B}_0(x)$]. The small-amplitude, forced tearing mode develops due to the coupling with unstable bending mode in the same singularity. The growth rates and the relative amplitudes of these two surface modes are estimated using the surface-wave formalism.

Furthermore, we have presented a theoretical investigation of the resistive tearing mode instability in a partially ionized dusty magnetoplasma, taking into account the dust particle dynamics. For this purpose, we have derived a set of equations comprising the modified (due to the presence of mobile charged dust grains) ion vorticity equation, the dynamical dust density response to the scalar wave potential, and the parallel component of Ohm's law. The governing equations include collisional interactions between the plasma particles and stationary neutral atoms. Standard techniques [*Furth et al.* 1963, *White* 1986] are then employed to derive a general dispersion relation for the resistive tearing modes. The analysis of the dispersion relation reveals new resistive tearing mode instabilities when the dust particle inertia and collisions between the plasma particles and neutrals play dominant roles. It is expected that the present instabilities, which occur on very long timescales, may be responsible for magnetic field reconnection and subsequent particle energization in astrophysical dusty plasmas such as those in protoplanetary nebulae, cometary tails, interstellar medium, and dark molecular clouds.

Acknowledgements

The authors acknowledge the support of the International Institute for Space Science of Bern (Switzerland) for the team "Complex Plasmas in Space: Application to Solar System and Near Earth Space". The research was partially supported also by the European Commission (Brussels) through the contract No. HPRN-CT-2000-00140. One of the authors (D. J.) acknowledges the financial support from the Deutsche Forschungsgemeinschaft for his three months stay at the University of Bayreuth, F. R. Germany.

References

Birk G. T., Kopp A. and Shukla P. K. 1996, *Phys. Plasmas* **3**, 3564.

Birk G. T. 1998, *Astron. Astrophys.* **330**, 1070.

Birk G. T. and Wiechen H. 2001, *Proceedings of the Sixth International School/Symposium for Space Plasma Simulations (ISSS-6)*, Garching, Germany, edited by Büchner J., Dum C. T. and Scholer M. (Copernicus Gesselschaft, Katlenburg-Lindau), pp. 222-225.

Calfaro E., Grasso D., Porcelli F. and Saluzzi A. 1998, *Phys. Rev. Lett.* **80**, 4430.

Coppi B., Laval G. and Pellat R. 1966 *Phys. Rev. Lett.* **16**, 1207.

Deng X. H. and Matsumoto H. 2001, *Nature* **410**, 557.

Drake J. F., Biskamp D. and Zeiler A. 1997, *Geophyshys. Res. Lett.* **24**, 2921.

Edwards A. W., Campbell D. J., Engelhart W. W., Fahrbach H. U., Gill R. D., Granetz R. S., Tsuii S., Tubbing B. J. D., Weller A., Wesson J. and Zasche D. 1986, *Phys. Rev. Lett.* **57**, 210.

Furth H. P., Kileen J. and Rosenbluth M. N. 1963, *Phys. Fluids* **6**, 659.

Grasso D., Califano F., Pegoraro F. and Porcelli F. 2001, *Phys. Rev. Lett.* **86**, 5051.

Kopp A., Schröer A., Birk G. T. and Shukla P. K. 1997, *Phys. Plasmas* **4**, 4414.

Lee G. and Russell C. T. 1994, *Geophys. Res. Lett.* **21**, 2454.

Levy E. H. and Araki S. 1989, *Icarus* **81**, 74.

Mahajan S. M.,. Hazeltine R. D, Strauss H. R. and Ross D. W. 1978, *Phys. Rev. Lett.* **41**, 1375.

Matsumoto H., Nagano I., Anderson, R. R. *et al.* 1994 *J. Geomagn. Geoelectr.* **46**, 59.

Ness N. F. J. 1965, *Geophys. Res.* **70**, 2989.

Ottaviani M. and Porcelli F. 1993, *Phys. Rev. Lett.* **71**, 3802.

Paschmann G. 1979, *Nature* **282**, 243.

Russel C. T., Saunders M. A., Phillips J. L. and Fedder J. A. 1986, *J. Geophys. Res.* **91**, 1417.

Rutherford P. H. 1973, *Phys. Fluids* **16**, 1903.

Sonnet C. P. 1979 , *Geophys. Res. Lett.* **6**, 677.

Vishniak E. T. and Lazarian A. 1999, *Astrophys. J.* **511**, 193.

White R. B. 1986, *Rev. Mod. Phys.* **58**, 183.

Zweibel E. G. 1989, *Astrophys. J.* **340**, 550.

Zweibel E. G. 1999 , *Phys. Plasmas* **6**, 1725.

Nonlinear Electrostic Waves in Dusty Plasmas

A. A. Mamun*, P. K. Shukla
Fakultät für Physik & Astronomie
Ruhr-Universität Bochum, D-44780 Bochum, Germany

F. Verheest
Sterrenkundig Observatorium, Universiteit Gent
Krijgslaan 281, B-9000 Gent, Belgium

Abstract

We present a rigorous theoretical investigation of different types of nonlinear electrostatic waves that may exist in dusty plasmas. We have focused on dust ion-acoustic solitary and shock waves by considering static charged dust grains, and dust-acoustic solitary and shock waves as well as dust double layers by considering the dynamics of dust particles. We have also studied dust lattice solitary waves in a strongly coupled dusty plasma system. We have investigated the effects of a nonplanar geometry and dust grain charge fluctuations, which have been found to introduce some important new features on and solitary waves. It has been found that the effects of the dust fluid temperature, non-isothermal (trapped and nonthermal) ion distributions, strong dust correlation, etc. do not only significantly modify the dust-acoustic solitary structures, but also introduce some important new features. We have also briefly discussed the implications of these theoretical results in experimental observations of solitary and shock waves in dusty plasmas.

1 Introduction

The physics of charged dust particles has become an outstanding and challenging research topic not only because dust particles are ubiquitous in most space [Goertz 1989; Mendis 1991, 1997; Mendis and Horányi 1991; de Angelis 1992; Mendis and Rosenberg 1994; Shukla 1996; Verheest 1996; Nakano 1998; Zweibel 1999; Verheest 2000; Shukla 2001; Shukla and Mamun 2002] and laboratory plasmas [Winter 1998; Bouchoule 1999; Hollenstein 2000; Shukla and Mamun 2002], but also because it has introduced a great variety of new phenomena associated with waves and instabilities [Rao *et al.* 1990; Shukla and Silin 1992; Verheest 1992, 2000; Varma *et al.* 1993;

*Permanent address: Department of Physics, Jahangirnagar University, Savar, Dhaka, Bangladesh.

Melandsø *et al.* 1993; Winske *et al.* 1995; Barkan *et al.* 1995, 1996; Shukla 1996; Melandsø 1996; Rosenberg 1996; Salimullah 1996; Rosenberg and Kalman 1997; Homann *et al.* 1997; Morfill *et al.* 1997; Kaw and Sen 1998; Merlino *et al.* 1998; Farokhi *et al.* 1999; Thompson *et al.* 1999; Luo *et al.* 2001; Shukla 2001; Shukla and Mamun 2002] and it plays a vital role in understanding different interesting phenomena in astrophysical and space environments, such as interplanetary space, interstellar medium, interstellar or molecular clouds, comets, planetary rings, earth's environments, etc. [Hill and Mendis 1981; Smith *et al.* 1981, 1982; Morfill and Goertz 1982; Goertz 1989; Horányi and Goertz 1990; Mendis 1991, 1997; Mendis and Horányi 1991; de Angelis 1992; Mendis and Rosenberg 1994; Nakano 1998; Zweibel 1999; Verheest 2000; Shukla and Mamun 2002].

The consideration of charged dust grains in a plasma does not only modify the existing plasma wave spectra [Bliokh *et al.* 1985; de Angelis *et al.* 1988; Shukla and Stenflo 1992] but also introduces a number of new novel eigenmodes, such as dust ion-acoustic (DIA) waves [Shukla and Silin 1992; Barkan *et al.* 1996; Merlino *et al.* 1998], dust-acoustic (DA) waves [Rao *et al.* 1990; Barkan *et al.* 1995; Thompson *et al.* 1999], dust lattice (DL) waves [Melandsø 1996; Homann *et al.* 1997; Morfill *et al.* 1997; Farokhi *et al.* 1999] etc.

Shukla and Silin [1992] have first theoretically shown that due to the conservation of equilibrium charge density $n_{e0}e + n_{d0}Z_{d0}e - n_{i0}e = 0$ and the strong inequality $n_{e0} \ll n_{i0}$ [where n_{s0} is the particle number density of the species s with $s = e$, i, d for electrons, ions and dust particles, Z_{d0} is the number of electrons residing onto the dust grain surface at equilibrium and e is the magnitude of the electronic charge] a dusty plasma (with negatively charged static dust grains) supports low-frequency DIA waves with phase speed much smaller (larger) than electron (ion) thermal speed. The dispersion relation (a relation between the wave frequency ω and the wave number k) of the linear DIA waves is [Shukla and Silin 1992] $\omega^2 = (n_{i0}/n_{e0})k^2C_i^2/(1 + k^2\lambda_{De}^2)$, where $C_i = (k_BT_e/m_i)^{1/2}$ is the ion-acoustic speed (with T_e being the electron temperature and m_i being the ion mass) and $\lambda_{De} = (k_BT_e/4\pi n_{e0}e^2)^{1/2}$ is the electron Debye radius. When we consider a long wavelength limit (viz. $k\lambda_{De} \ll 1$), the dispersion relation for the DIA waves becomes $\omega = (n_{i0}/n_{e0})^{1/2}kC_i$. This form of spectrum is similar to the usual ion-acoustic wave spectrum [Lonngren 1983] for a plasma with $n_{i0} = n_{e0}$ and $T_i \ll T_e$ (where T_i is the ion temperature). However, in dusty plasmas we usually have $n_{i0} \gg n_{e0}$ and $T_i \simeq T_e$. Therfore, a dusty plasma cannot support the usual ion-acoustic waves, but can do the DIA waves of Shukla and Silin [1992]. The phase speed ω/k of the DIA waves is larger than C_i because of $n_{i0} \gg n_{e0}$ for negatively charged dust grains. The increase in the phase velocity is attributed to the electron density depletion in the background plasma, so that the electron Debye radius becomes larger. As a result, there appears a stronger space charge electric field which is responsible for the enhanced phase velocity of the DIA waves. The DIA waves have been observed in laboratory experiments [Barkan *et al.* 1996; Merlino *et al.* 1998].

Rao *et al.* [1990] theoretically predicted the existence of extremely low phase velocity (in comparison with the electron and ion thermal speeds) DA waves in an unmagnetized dusty plasma whose constituents are an inertial charged dust fluid and Boltzmann ions and electrons. Thus, in the DA waves the inertia is provided by the dust particle mass and the restoring force comes from the pressures of electrons and ions. The dispersion relation for the DA waves with the phase speed ω/k much smaller (larger) than the ion (dust) thermal speed is given by [Rao *et al.* 1990] $\omega = kC_d(1+k^2\lambda_D^2)^{-1/2}$, where $\lambda_D = (\lambda_{De}^{-2}+\lambda_{Di}^{-2})^{-1/2}$ is the global screening length and $C_d = \omega_{pd}\lambda_D$ is is the dust-acoustic speed. Here $\lambda_{Di} = (k_BT_i/4\pi n_{i0}e^2)^{1/2}$ is the ion Debye radius and $\omega_{pd} = (4\pi n_{i0}e^2/m_d)^{1/2}$ is dust the plasma frequency, where m_d is the mass of the dust particle. It is obvious that one can obtain the DA mode without the consideration of the dynamics of the dust grains. The theoretical prediction of Rao *et al.* [1990] has been conclusively verified by a number of laboratory experiments [Barkan *et al.* 1995; Thompson *et al.* 1999].

It is now well established from both the theoretical and experimental points of view that when the electrostatic energy of the dust particle interaction exceeds the kinetic energy, charged dust grains often form crystal-like structures [Ikezi 1986; Chu and I 1994; Thomas *et al.* 1994; Hayashi and Tachibana 1994; Melzer *et al.* 1994; Morfill *et al.* 1997] in a strongly coupled dusty plasma. The waves associated with such strongly correlated dust crystals, which represent a unique object connecting plasma physics and condensed matter physics, cannot be explained in terms of DIA or DA wave theories. Melandsø [1996] has presented a consistent theory for the DL waves. He obtained the dispersion relation for DL waves by using the analytical expansion technique, previously known from solid state physics [Kittel 1956]. When $a/\lambda_D \gg 1$, where a is the intergrain distance, it is sufficient to include only the interaction between nearest neighbour particles, and the dispersion relation for the DL waves is [Melandsø 1996] is given by $\omega = C_L\sin(ka)/a$, where $C_L = C_0(1+k_Da+k_D^2a^2/2)^{1/2}\exp(-k_Da/2)$, $C_0 = (2q_d^2/m_da)^{1/2}$ is the DL wave speed and $k_D = 1/\lambda_D$. The theory of Melandsø [1996], limited to only the interaction between nearest dust grains, is then improved by Farokhi *et al.* [1999] by taking into account the short- and long-range interactions between the dust particles. The modified dispersion relation of Farokhi *et al.* [1999] can be represented by that of Melandsø [1996] if we replace C_L by $C_L' = C_0k_Da[a/r_d\cosh(k_Da)]^{1/2}$. The DL waves have also been observed in laboratory experiments [Homann *et al.* 1997; Morfill *et al.* 1997].

The linear properties of the DIA, DA and DL waves in dusty plasmas are now well understood and have been reported by a large number of regular and review articles or books during last few years [Rao *et al.* 1990; Shukla and Silin 1992; Varma *et al.* 1993; Melandsø *et al.* 1993; Barkan *et al.* 1995; Barkan *et al.* 1996; Melandsø 1996; Rosenberg 1996; Rosenberg and Kalman 1997; Homann *et al.* 1997; Morfill *et al.* 1997; Merlino *et al.* 1998; Kaw and Sen 1998; Shukla and Rosenberg 1999; Thompson *et al.* 1999; Farokhi *et al.* 1999, 2000; Verheest 2000; Shukla 2001;

Shukla and Mamun 2002]. Linear theory is valid only when the wave amplitude is so small that one may neglect the nonlinearities. However, there are numerous processes via which unstable modes can saturate and attain large amplitudes. When the amplitudes of the waves are sufficiently large, nonlinearities can no longer be ignored. Nonlinearities come from the harmonic generation involving the fluid advection, the nonlinear Lorentz force, trapping of particles in the wave potential, ponderomotive force, etc. Nonlinearities in plasmas contribute to the localization of waves, leading to different types of interesting nonlinear coherent structures (viz. solitary structures, shock waves, double layers, vortices, etc.) which are important from both theoretical and experimental points of view, and have received a great deal of attention for understanding the basic properties of localized electrostatic perturbations in space and laboratory dusty plasmas.

Recently, a number of theoretical attempts have been made in order to study the properties of DIA solitary [Bharuthram and Shukla 1992; Popel and Yu 1995; Shukla and Mamun 2002; Mamun and Shukla 2002a, 2002b] and shock [Popel *el.* 1996; 2000, 2001; Shukla and Mamun 2001, 2002; Mamun and Shukla 2002b; 2002c] waves, DA solitary [Rao *et al.* 1990; Mamun *et al.* 1996a, 1996b; Mamun 1998, 1999; Ma and Liu 1997; Singh and Rao 1997; Mendoza-Briceño *et al.* 2000; Mamun and Shukla 2001; Ivlev and Morfill 2001; Gupta *et al.* 2001; Shukla and Mamun 2002] and shock [Melandsø and Shukla 1995; Shukla and Mamun 2001, 2002; Mamun and Shukla 2002b] waves, and DL solitary waves [Melandsø 1996; Farokhi *et al.* 1999]. The DIA solitary [Nakamura and Sharma 2001] and shock waves [Nakamura *et al.* 1999; Luo *et al.* 1999; Nakamura 2002] and DL solitary waves [Samsonov *et al.* 2002] have been experimentally observed. To the best of our knowledge, there is no review article where detailed information about DIA, DA and DL solitary waves are systematically presented. Therefore, we review here the nonlinear propagation of DIA, DA and DL waves and show how dust particles introduce new features to the nonlinear structures in an unmagnetized dusty plasma. We consider a number of important effects, such as nonplanar (spherical/cylindrical] geometries and dust charge fluctuations, etc. on DIA and DA solitary waves. We also include the effects of trapped and nonthermal ion distributions, strong correlation among dust particles, etc. on DA solitary waves. We show that these effects do not only significantly modify the basic properties of DIA and DIA solitary waves, but also introduce new features/phenomena in dusty plasmas, viz. shock structures, dust double layers (DDL), etc.

We have organized our present chapter in the following fashion. We present our model equations in §2. We first consider the dynamics of ions, Boltzmann electrons and stationary dust particles, and study the properties of DIA solitary waves including the effects of adiabatic ion temperature and nonplanar geometries and dust charge fluctuations (cf. §3). We study DIA shock waves (which are formed due to dust grain charge fluctuations) in §4. We then switch our attention to the dynamics of dust particles (where electrons and ions are Boltzmannean), and studied the properties of DA solitary waves including the effects of adiabatic dust fluid temperature, trapped ion distribution, nonplanar geometries and dust charge fluctuations (cf. §5). We also focus on DDL (which are formed due to trapped ion distribution) and DA shock structures (which are formed due to the strong correlations among dust particles) in §6 and §7, respectively. We then consider strongly correlated dust particles and study the DL solitary waves in §8. We finally conclude with a brief discussion in §9.

2 Governing Equations

We consider the propagation of electrostatic waves in a dusty plasma containing electrons, ions and negatively charged dust grains. The dynamics of low phase velocity waves, whose phase speed V_p is much less than the electron thermal speed V_{Te}, is governed by

$$\frac{\partial N_j}{\partial t} + \nabla \cdot N_j \mathbf{U}_j = 0, \tag{1}$$

$$N_j m_j \left(\frac{\partial N_j}{\partial t} + \mathbf{U}_j \cdot \nabla \mathbf{U}_j \right) = -q_j N_j \nabla \Phi - \nabla P_j, \tag{2}$$

$$\frac{\partial P_j}{\partial t} + \mathbf{U}_j \cdot \nabla P_j + \gamma P_j \nabla \cdot \mathbf{U}_j = 0, \tag{3}$$

$$\nabla^2 \Phi = 4\pi e \left[n_{e0} \exp\left(\frac{e\Phi}{k_B T_e} \right) - N_i + Z_d N_d \right], \tag{4}$$

where N_j represents the ion (for $j = i$) or dust (for $j = d$) number density. U_j is the fluid velocity, P_j is the fluid pressure, q_j is the ion or dust charge (i.e. $q_i = e$ and $q_i = -Z_d e$, where Z_d is number of electrons residing on to the dust grain surface), γ is the adiabatic index (which is 3 for one-dimensional case and 5/3 in three-dimensional case) and Φ is the electrostatic wave potential. We have assumed a Boltzmannean electron density response, which is valid as long as $V_p \ll V_{Te}$.

3 Dust Ion-Acoustic Solitary Waves

We confine ourself, in this section, to one dimensional DIA solitary waves in dusty plasma in which dust particles are stationary and provide only the background charge-neutrality. The basic equations (1)–(4) for one dimensional DIA waves can be expressed in terms of normalized variables as

$$\frac{\partial n_i}{\partial t} + \frac{\partial}{\partial x}(n_i u_i) = 0, \tag{5}$$

$$\frac{\partial u_i}{\partial t} + u_i \frac{\partial u_i}{\partial x} = -\frac{\partial \phi}{\partial x} - \frac{\sigma_i}{n_i} \frac{\partial p_i}{\partial x}, \tag{6}$$

$$\frac{\partial p_i}{\partial t} + u_i \frac{\partial p_i}{\partial x} + 3 p_i \frac{\partial u_i}{\partial x} = 0, \tag{7}$$

$$\frac{\partial^2 \phi}{\partial x^2} = \mu \exp(\phi) - n_i + (1-\mu), \tag{8}$$

where n_i is the ion number density normalized by n_{i0}, u_i is the ion fluid velocity normalized by the ion-acoustic speed C_i, $\sigma_i = T_i/T_e$, p_i is the ion pressure normalized by $n_{i0}k_B T_i$, ϕ is the ion-acoustic wave potential normalized by $K_B T_e/e$ and $\mu = n_{e0}/n_{i0}$. The space variable x is normalized by the modified electron Debye radius $\lambda_{Dem} = (K_B T_e/4\pi n_{i0} e^2)^{1/2}$ and the time variable t is normalized by the ion plasma period $\omega_{pi}^{-1} = (m_i/4\pi n_{i0} e^2)^{1/2}$.

To obtain a solitary wave solution, we assume that all the dependent variables depend on a single independent variable $\xi = x - Mt$, where M is the Mach number (the velocity of the solitary wave normalized by the ion-acoustic speed C_i). Considering the steady state condition, i.e. $\partial/\partial t = 0$, we can write Eqs. (5)–(8) as

$$M\frac{\partial n_i}{\partial \xi} - \frac{\partial}{\partial \xi}(n_i u_i) = 0, \tag{9}$$

$$M\frac{\partial u_i}{\partial \xi} - u_i\frac{\partial u_i}{\partial \xi} - \frac{\sigma_i}{n_i}\frac{\partial p_i}{\partial \xi} = \frac{\partial \phi}{\partial \xi}, \tag{10}$$

$$M\frac{\partial p_i}{\partial \xi} - u_i\frac{\partial p_i}{\partial \xi} - 3p_i\frac{\partial u_i}{\partial \xi} = 0, \tag{11}$$

$$\frac{\partial^2 \phi}{\partial \xi^2} = \mu \exp(\phi) - n_i + (1-\mu). \tag{12}$$

Now, under the appropriate boundary conditions, viz. $\phi \to 0$, $u \to 0$, $p_i \to 1$ and $n \to 1$ at $\xi \to \pm\infty$, Eqs. (9) and (11) can be integrated to give

$$n_i = \frac{1}{1 - u_i/M}, \tag{13}$$

$$p_i = n_i^3. \tag{14}$$

Substituting Eq. (13) into Eq. (10) we obtain

$$M\frac{\partial u_i}{\partial \xi} - u_i\frac{\partial u_i}{\partial \xi} - \sigma_i\frac{\partial p_i}{\partial \xi} + \frac{\sigma_i}{M}u_i\frac{\partial p_i}{\partial \xi} = \frac{\partial \phi}{\partial \xi}. \tag{15}$$

Multiplying Eq. (11) by σ_i/M one can write

$$\sigma_i\frac{\partial p_i}{\partial \xi} - \frac{\sigma_i}{M}u_i\frac{\partial p_i}{\partial \xi} - 3p_i\frac{\sigma_i}{M}\frac{\partial u_i}{\partial \xi} = 0. \tag{16}$$

Now, subtracting twice Eq. (15) from Eq. (16), one can get a differential equation which has the form

$$3\sigma_i\frac{\partial p_i}{\partial \xi} - 3\frac{\sigma_i}{M}\frac{\partial}{\partial \xi}(p_i u_i) - 2M\frac{\partial u_i}{\partial \xi} + 2u_i\frac{\partial u_i}{\partial \xi} + 2\frac{\partial \phi}{\partial \xi} = 0. \tag{17}$$

The integration of this equation yields

$$3\frac{\sigma_i}{M}p_i u_i - 3\sigma_i(p_i - 1) + 2Mu_i - u_i^2 - 2\phi = 0, \tag{18}$$

where we have used the same boundary conditions, viz. $\phi \to 0$, $u_i \to 0$, $p_i \to 1$ and $n_i \to 1$ at $\xi \to \pm\infty$. Substituting u_i and p_i into Eq. (18) one can obtain a quadratic equation for n_i^2

$$3\sigma_i n_i^4 - (3\sigma_i + M^2 - 2\phi)n_i^2 + M^2 = 0. \tag{19}$$

The solution of Eq. (19) is given by

$$n_i = \frac{\sigma_{i1}}{\sqrt{2}\sigma_{i0}}\left[1 - \frac{2\phi}{M^2\sigma_{i1}^2} - \sqrt{\left(1 - \frac{2\phi}{M^2\sigma_{i1}^2}\right)^2 - 4\frac{\sigma_{i0}^2}{\sigma_{i1}^4}}\right]^{1/2}, \tag{20}$$

where $\sigma_{i0} = \sqrt{3\sigma_i/M^2}$ and $\sigma_{i1} = \sqrt{1+\sigma_{i0}^2}$. Substituting Eq. (20) into Eq. (12) and integrating the resultant equation, we have an energy integral [Bernstein *et al.* 1957; Sagdeev 1966]

$$\frac{1}{2}\left(\frac{d\phi}{d\xi}\right)^2 + V(\phi) = 0, \tag{21}$$

where the pseudo-potential $V(\phi)$ is given by

$$\begin{aligned} V(\phi) &= -\mu\exp(\phi) - -(1-\mu)\phi \\ &\quad - \frac{M^2\sigma_{i1}}{\sqrt{2}}\left[1 - \frac{2\phi}{M^2\sigma_{i1}^2} + \sqrt{\left(1 - \frac{2\phi}{M^2\sigma_{i1}^2}\right)^2 - 4\frac{\sigma_{i0}^2}{\sigma_1^4}}\right]^{1/2} \\ &\quad - \frac{2\sqrt{2}\sigma}{\sigma_{i1}^3}\left[1 - \frac{2\phi}{M^2\sigma_{i1}^2} + \sqrt{\left(1 - \frac{2\phi}{M^2\sigma_{i1}^2}\right)^2 - 4\frac{\sigma_0^2}{\sigma_1^4}}\right]^{-3/2} + C_1. \end{aligned} \tag{22}$$

Here C_1 is the integration constant which we will choose in such a manner that $V(\phi) = 0$ at $\phi = 0$. It is important to mention that in our study the condition for ion density to be real, $|1 - 2\phi/M^2\sigma_1^2| \geq 2\sigma_0/\sigma_1^2$, must always be satisfied.

To find the condition for the existence of DIA solitary waves analytically, we first consider the cold ion limit $\sigma_i = 0$ for which Eq. (22) reduces to

$$V(\phi) = \mu[1 - \exp(\phi)] - (1-\mu)\phi + M^2\left(1 - \sqrt{1 - \frac{2\phi}{M^2}}\right). \tag{23}$$

It is obvious from Eq. (23) that $V(\phi) = dV(\phi)/d\phi = 0$ at $\phi = 0$. Therefore, solitary wave solutions of Eq. (23) exist if (i) $(d^2V/d\phi^2)_{\phi=0} < 0$ so that the fixed point at the origin is unstable and (ii) $(d^3V/d\phi^3)_{\phi=0} > (<)\ 0$ for solitary waves with

$\phi > (<) \, 0$. The nature of these solitary waves, whose amplitude tends to zero as the Mach number M tends to its critical value, can be found by expanding the Sagdeev potential $V(\phi)$ to third order in a Taylor series in ϕ. The critical Mach number is that which corresponds to the vanishing of the quadratic term. At the same time, if the cubic term is negative, there is a potential well on the negative side and if the cubic term is positive, there is a potential well on the positive side. Therefore, by expanding the Sagdeev potential $V(\phi)$ around the origin, the critical Mach number at which the second derivative changes sign can be found as $M_c = \mu^{-1/2}$. At this critical value of M the cubic term of $V(\phi)$ can be expressed as $(3\mu - 1)\mu^{-1/2}/2$ which reveals that the cubic term is positive (negative) for $\mu > (<) 1/3$. This means that depending on the value of μ, DIA solitary waves with both positive and negative potential can exist with $\phi > (<) 0$ when $\mu > (<) 1/3$.

To examine how the ion fluid temperature (σ_i) modifies the properties of the arbitrary amplitude DIA solitary waves, we have analyzed the pseudo-potential potential $V(\phi)$ given by Eq. (22), and found that as we increase σ_i, (i) we need higher values of the critical Mach number (corresponding to a lower value of μ) in order to have the DIA solitary waves, and (ii) the amplitude of both the positive and negative solitary waves decreases, but their width increases. We have also found the same results by numerical analysis of the energy integral (21). Since in most space and laboratory dusty plasmas $\mu < 1/3$, unlike the usual ion-acoustic solitary waves (associated with a positive potential) in a two component electron-ion plasma [Lonngren 1983] DIA solitary waves have a new feature in that these are associated with a negative potential which is due to the presence of negatively charged dust grains.

We have discussed the properties of the DIA solitary waves by assuming a planar geometry and constant dust and grain charge. However, it is shown that the effects of nonplanar geometry and dust grain charge fluctuations introduce new features or significantly modify the properties of DIA solitary waves [Shukla and Mamun 2002, Mamun and Shukla 2002a, 2002b, 2002c]. W now consider the cold ion limit ($\sigma_i = 0$) just to avoid mathematical complexities, and investigate the effects of nonplanar geometry and dust grain charge fluctuations on the properties of the DIA solitary waves.

3.1 Nonplanar Geometry

The nonlinear dynamics of the DIA waves, whose phase speed is much smaller (larger) than the electron (ion) thermal speed (viz. $V_{Ti} \ll V_p \ll V_{Te}$), in nonplanar cylindrical and spherical geometries is governed by

$$\frac{\partial n_i}{\partial t} + \frac{1}{r^\nu}\frac{\partial}{\partial r}(r^\nu n_i u_i) = 0, \tag{24}$$

$$\frac{\partial u_i}{\partial t} + u_i\frac{\partial u_i}{\partial r} = -\frac{\partial \phi}{\partial r}, \tag{25}$$

$$\frac{1}{r^\nu}\frac{\partial}{\partial r}\left(r^\nu\frac{\partial\phi}{\partial r}\right) = \mu\exp(\phi) - n_i + (1-\mu), \tag{26}$$

where $\nu = 0$ for a one-dimensional geometry and $\nu = 1$ (2) for a nonplanar cylindrical (spherical) geometry, We assumed here a constant dust grain charge (i.e. $Z_d = Z_{d0}$). To investigate ingoing solutions of Eqs. (24)–(26), we introduce stretched coordinates [Washimi and Taniuti 1966; Maxon and Viecelli 1974] $\zeta = -\epsilon^{1/2}(r+v_0t)$ and $\tau = \epsilon^{3/2}t$, expand n_i, u_i, and ϕ in a power series of ϵ

$$n_i = 1 + \epsilon n_i^{(1)} + \epsilon^2 n_i^{(2)} + \cdots, \tag{27}$$

$$u_i = \epsilon u_i^{(1)} + \epsilon^2 u_i^{(2)} + \cdots, \tag{28}$$

$$\phi = \epsilon\phi^{(1)} + \epsilon^2\phi^{(2)} + \cdots, \tag{29}$$

and develop equations in various powers of ϵ. To lowest order in ϵ, Eqs. (24)–(29), give $n_i^{(1)} = -u_i^{(1)}/v_0$, $u_i^{(1)} = -\phi^{(1)}/v_0$ and $v_0 = \mu^{-1/2}$. To next higher order in ϵ, we obtain a set of equations

$$\frac{\partial n_i^{(1)}}{\partial\tau} - v_0\frac{\partial n_i^{(2)}}{\partial\zeta} - \frac{\partial u_i^{(2)}}{\partial\zeta} - \frac{\partial}{\partial\zeta}[n_i^{(1)}u_i^{(1)}] - \frac{\nu u_i^{(1)}}{v_0\tau} = 0, \tag{30}$$

$$\frac{\partial u_i^{(1)}}{\partial\tau} - v_0\frac{\partial u_i^{(2)}}{\partial\zeta} - u_i^{(1)}\frac{\partial u_i^{(1)}}{\partial\zeta} - \frac{\partial\phi^{(2)}}{\partial\zeta} = 0, \tag{31}$$

$$\frac{\partial^2\phi^{(1)}}{\partial\zeta^2} - \mu\phi^{(2)} + n_i^{(2)} - \frac{1}{2}\mu[\phi^{(1)}]^2 = 0. \tag{32}$$

Combining Eqs. (30)–(32) we deduce a modified Korteweg-de Vries (KdV) equation

$$\frac{\partial\phi^{(1)}}{\partial\tau} + \frac{\nu}{2\tau}\phi^{(1)} + A_g\phi^{(1)}\frac{\partial\phi^{(1)}}{\partial\zeta} + B_g\frac{\partial^3\phi^{(1)}}{\partial\zeta^3} = 0, \tag{33}$$

where $A_g = (3\mu-1)/\sqrt{2\mu}$ and $B_g = 1/2\mu^{3/2}$. We have already mentioned that $\nu = 0$ corresponds to a one-dimensional geometry. Thus, for a one-dimensional geometry ($\nu = 0$) and for a moving frame moving with a speed u_0, the stationary solitary wave solution of Eq. (33) is

$$\phi^{(1)}(\nu = 0) = \left(\frac{3u_0}{A_g}\right)\mathrm{sech}^2\left[\sqrt{\frac{u_0}{4B_g}}(\zeta - u_0\tau)\right]. \tag{34}$$

It is obvious from Eq. (34) that for $\mu > (<)\ 1/3$, a dusty plasma supports compressive (rarefactive) DIA solitary waves which are associated with a positive (negative) potential. We now turn to Eq. (33) with the term $(\nu/2\tau)\phi^{(1)}$ which is due to the effect of the nonplanar (cylindrical or spherical) geometry. An exact analytic solution of Eq. (33) is not possible. Therefore, we have numerically solved Eq. (33) and have

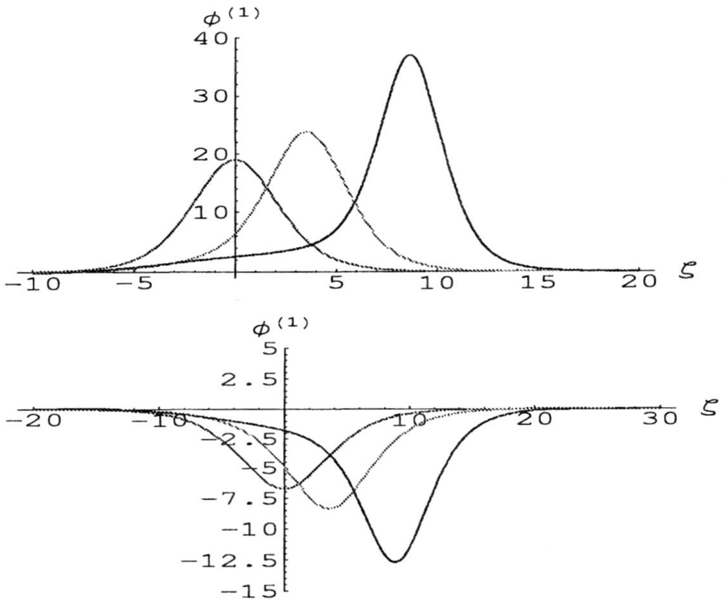

Figure 1: Time evolution of cylindrical ($\nu = 1$) DIA solitary waves: $\phi^{(1)}$ versus spatial coordinate ζ at times $\tau = -9$ (left), $\tau = -6$ (middle), and $\tau = -3$ (right) for $\mu = 0.4$ (upper plot) and $\mu = 0.2$ (lower plot). We choose our initial pulse at $\tau = -9$ and show how its amplitude increases with decreasing the value of τ (after Mamun and Shukla 2002a).

studied the effects of cylindrical and spherical geometries on time-dependent DIA solitary waves. The results are depicted in figures 1 and 2. The initial condition that we have used in our numerical analysis is in the form of the stationary solution of Eq. (34) without the term $(\nu/2\tau)\phi$, i.e. in the form $\phi^{(1)} = (3/A_g)\text{sech}^2(\zeta/\sqrt{4B_g})$. Figure 1 shows how the effect of a cylindrical geometry modifies the DIA solitary waves associated with both positive ($\mu = 0.4 > 1/3$) and negative ($\mu = 0.2 < 1/3$) potentials. Figure 2 shows how the effect of a spherical geometry modifies the DIA solitary waves associated with both positive ($\mu = 0.4 > 1/3$) and negative ($\mu = 0.2 < 1/3$) potentials.

The numerical solutions of (6) (displayed in figures 1 and 2) reveal that for a large value of τ (e.g. $\tau = -9$) the spherical and cylindrical solitary waves are similar to one-dimensional solitary waves. This is because for a large value of τ the term $(\nu/2\tau)\phi^{(1)}$, which is due to the effect of the cylindrical or spherical geometry, is no longer dominant. However, as the value of τ decreases, the term $(\nu/2\tau)\phi^{(1)}$ becomes dominant and both spherical and cylindrical solitary waves differ from one-dimensional solitary waves. It is found that as the value of τ decreases, the amplitude of these localized pulses increases. It is also found that the amplitude of cylindrical solitary waves is larger than that of the one-dimensional solitary waves

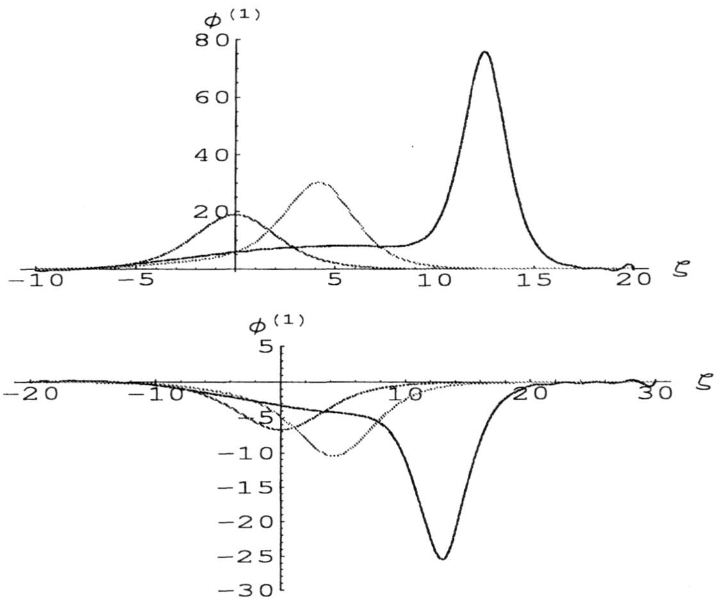

Figure 2: Time evolution of spherical ($\nu = 2$) DIA solitary waves: $\phi^{(1)}$ versus spatial coordinate ζ at times $\tau = -9$ (left), $\tau = -6$ (middle), and $\tau = -3$ (right) for $\mu = 0.4$ (upper plot) and $\mu = 0.2$ (lower plot). We choose our initial pulse at $\tau = -9$ and show how its amplitude increases with decreasing the value of τ (after Mamun and Shukla 2002a).

but smaller than that of the spherical ones.

3.2 Dust Charge Fluctuations

We study the propagation of DIA solitary waves in an unmagnetized dusty plasma where the dust charge is not constant but varies with space and time. The nonlinear dynamics of one-dimensional DIA waves, whose phase speed is much smaller (larger) than the electron (ion) thermal speed, is governed by Eq. (5) and

$$\frac{\partial u_i}{\partial t} + u_i \frac{\partial u_i}{\partial x} = -\frac{\partial \phi}{\partial x}, \tag{35}$$

$$\frac{\partial^2 \phi}{\partial x^2} = \mu \exp(\phi) - n_i + (1-\mu) z_d, \tag{36}$$

where z_d is the number of electrons residing onto the dust grain surface normalized by its equilibrium value Z_{d0}. We note that z_d is not constant but varies with space and time. Thus, Eqs. (5), (35) and (36) are completed by the normalized dust grain charging equation (Shukla and Mamun 2002)

$$\eta \frac{\partial z_d}{\partial t} = \mu\beta \exp(\phi - \alpha z_d) - \beta_i n_i u_i \left(1 + \frac{2\alpha z_d}{u_i^2}\right), \tag{37}$$

where $\eta = \sqrt{\alpha m_e(1-\mu)/2m_i}$, $\alpha = Z_{d0}e^2/k_B T_e r_d$, $\beta = (r_d/a)^{3/2}$, $a = n_{d0}^{-1/3}$, and $\beta_i = \beta\sqrt{\pi m_e/8m_i}$. We note that at equilibrium $\mu\beta\exp(-\alpha) = \beta_i u_0(1+2\alpha/u_0^2)$, where u_0 is the ion streaming speed normalized by C_i.

To study small but finite amplitude DIA solitary waves, we first introduce stretched coordinates [Washimi and Taniuti 1966] $\xi = \epsilon^{1/2}(x - v_0 t)$, $\tau = \epsilon^{3/2}t$, and $\eta = \epsilon\eta_0$, where ϵ is the expansion parameter, measuring the amplitude of the wave or the weakness of the wave dispersion. We then expand n_i, u_i, ϕ and z_d in a power series of ϵ

$$n_i = 1 + \epsilon n_i^{(1)} + \epsilon^2 n_i^{(2)} + \cdots, \tag{38}$$

$$u_i = u_0 + \epsilon u_i^{(1)} + \epsilon^2 u_i^{(2)} + \cdots, \tag{39}$$

$$\phi = \epsilon\phi^{(1)} + \epsilon^2\phi^{(2)} + \cdots, \tag{40}$$

$$z_d = 1 + \epsilon z_d^{(1)} + \epsilon^2 z_d^{(2)} + \cdots. \tag{41}$$

We develop the equations in powers of ϵ. To lowest order in ϵ, Eqs. (5), (35)–(37) give

$$w_0 n_i^{(1)} = u_i^{(1)}, \tag{42}$$

$$w_0 u_i^{(1)} = \phi^{(1)}, \tag{43}$$

$$0 = \mu\phi^{(1)} - n_i^{(1)} + (1-\mu)z_d^{(1)}, \tag{44}$$

$$0 = \beta_e\phi^{(1)} - \alpha u_\beta z_d^{(1)} - \beta_i u_1 u_i^{(1)} - u_0\beta_i u_2 n_i^{(1)}, \tag{45}$$

where $w_0 = v_0 - u_0$, $u_1 = 1 - 2\alpha/u_0$, $u_2 = 1 + 2\alpha/u_0^2$, $u_\beta = \beta_e + 2\beta_i/u_0$, and $\beta_e = \mu\beta\exp(-\alpha)$. Now, substituting $n_i^{(1)}$, $u_i^{(1)}$, and $z_d^{(1)}$ [obtained from Eqs. (42), (43) and (45)] into Eq. (44), we obtain the dispersion relation

$$a w_0^2 - b w_0 - c = 0, \tag{46}$$

where

$$a = \mu + \frac{(1-\mu)\beta_e}{\alpha u_\beta}, \tag{47}$$

$$b = \frac{(1-\mu)u_1\beta_i}{\alpha u_\beta}, \tag{48}$$

$$c = 1 + \frac{(1-\mu)u_2 u_0\beta_i}{\alpha u_\beta}. \tag{49}$$

To next higher order in ϵ, from Eq. (5) and Eqs. (35)–(37) we obtain a set of equations

$$\frac{1}{w_0^2}\frac{\partial\phi^{(1)}}{\partial\tau} - w_0\frac{\partial n_i^{(2)}}{\partial\xi} + \frac{\partial u_i^{(2)}}{\partial\xi} + \frac{2}{w_0^3}\phi^{(1)}\frac{\partial\phi^{(1)}}{\partial\xi} = 0, \tag{50}$$

$$\frac{1}{w_0}\frac{\partial\phi^{(1)}}{\partial\tau} - w_0\frac{\partial u_i^{(2)}}{\partial\xi} + \frac{1}{w_0^2}\phi^{(1)}\frac{\partial\phi^{(1)}}{\partial\xi} = \frac{\partial\phi^{(2)}}{\partial\xi}, \tag{51}$$

$$\frac{\partial^2\phi^{(1)}}{\partial\xi^2} = \mu\phi^{(2)} + \frac{1}{2}\mu[\phi^{(1)}]^2 - n_i^{(2)} + (1-\mu)z_d^{(2)}, \tag{52}$$

$$0 = \beta_e\phi^{(2)} - \alpha u_\beta z_d^{(2)} - \beta_i u_1 u_i^{(2)} - u_0\beta_i u_2 n_i^{(2)} + \beta_1\phi^{(1)}\frac{\partial\phi^{(1)}}{\partial\xi}, \tag{53}$$

where

$$\beta_1 = \beta_e[1 + (\alpha\beta_0 - 2)\alpha\beta_0] - \frac{2\beta_i}{w_0^3}\left[1 + \frac{w_\alpha}{u_0^3}\right], \tag{54}$$

$$\beta_0 = \frac{1}{\alpha u_\beta}\left[\beta_e - \frac{\beta_i w_2}{w_0} + \frac{2\alpha\beta_i}{w_0 u_0}\left(1 - \frac{1}{w_0}\right)\right], \tag{55}$$

$w_1 = 1 - u_0/w_0$, $w_2 = 1 + u_0/w_0$, and $w_\alpha = 2\alpha w_0 w_1(1 - \beta_0 w_0 u_0)$. Now, combining equations Eqs. (50)–(53) we deduce a KdV equation

$$\frac{\partial\phi^{(1)}}{\partial\tau} + A_c\phi^{(1)}\frac{\partial\phi^{(1)}}{\partial\xi} + B_c\frac{\partial^3\phi^{(1)}}{\partial\xi^3} = 0, \tag{56}$$

where

$$A_c = \frac{3c + bw_0 - \beta_2 w_0^4}{w_0(2c + bw_0)}, \tag{57}$$

$$B_c = \frac{w_0^3}{2c + bw_0}, \tag{58}$$

$$\beta_2 = \mu + \frac{\beta_1(1-\mu)}{\alpha u_\beta}. \tag{59}$$

The stationary solution of the KdV equation (56) is obtained by transforming the independent variables ξ to $\zeta = \xi - U_0\tau$, where U_0 is a constant speed normalized by C_i, and imposing the appropriate boundary conditions for localized perturbations, viz. $\phi^{(1)} \to 0$, $d\phi^{(1)}/d\zeta \to 0$, $d^2\phi^{(1)}/d\zeta^2 \to 0$ at $\zeta \to \pm\infty$. Accordingly, the stationary solitary wave solution of Eq. (56) is

$$\phi^{(1)} = \psi\ \text{sech}^2\left[(\xi - U_0\tau)/\Delta\right], \tag{60}$$

where $\psi = 3U_0/A_c$ and $\Delta = \sqrt{4B_c/U_0}$ represent the amplitude and the width of the solitary waves, respectively. It is obvious from (60) that there exists compressive (rarefactive) solitary waves if $A_c > 0$ ($A_c < 0$). We have numerically analyzed ψ and Δ for the parameters corresponding to space dusty plasma parameters, $n_{d0} = 10^{-7}$ cm^{-3}, $r_d = 1$ μm, $k_BT_e = 50$ eV, and $Z_{d0} = 10^3$ [Mendis and Rosenberg 1994; Shukla and Mamun 2002], which correspond to $\alpha = 0.0288$ and $\beta = 3\times10^{-10}$, as well as laboratory dusty plasma parameters, $n_{d0} = 10^5$ cm^{-3}, $r_d = 5$ μm, $k_BT_e = 0.2$ eV and $Z_{d0} = 10^3$ [Barkan *et al.* 1996; Merlino *et al.* 1998], which correspond to $\alpha = 1.44$ and $\beta = 3\times10^{-4}$. We found that ψ is always positive (which is obvious from the upper plot of figure 3, since $U_0 > 0$). This means that our present dusty plasma model can support only compressive solitary waves (solitary waves with

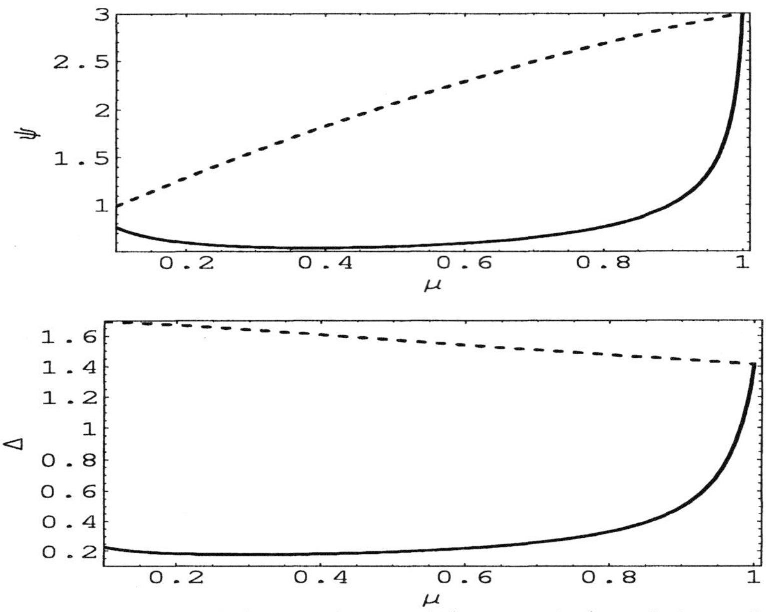

Figure 3: Variation of the amplitude ψ (upper plot) and the width Δ of the solitary waves with μ for $U_0 = 1$. The solid curves correspond to space dusty plasma parameters: $\alpha = 0.0288$ and $\beta = 3 \times 10^{-10}$, whereas the dotted curves correspond to laboratory dusty plasma parameters: $\alpha = 1.44$ and $\beta = 3 \times 10^{-4}$ (after Mamun and Shukla 2002c).

$\phi > 0$). The results of our numerical analyses (cf. figure 3) show how the amplitude and the width of these compressive solitary waves vary with μ for both the space [Mendis and Rosenberg 1994; Shukla and Mamun 2002] and laboratory [Barkan *et al.* 1996; Merlino *et al.* 1998] dusty plasma situations.

4 Dust Ion-Acoustic Shock Waves

We have studied DIA solitary waves which arise because of the balance between the effects of nonlinearity and dispersion when the effect of dissipation is negligible in comparison with that of nonlinearity and dispersion. However, when the dissipative effect is comparable to or more dominant than the dispersive effect, one encounters shock waves. We have designed the present section to present a theoretical model of DIA shock waves [Shukla 2000, Shukla and Mamun 2002; Mamun and Shukla 2002c] and to describe their experimental observations [Nakamura *et al.* 1999].

4.1 Theoretical Model

Shukla [2000] presented an analytical model for DIA shock waves in an unmagnetized dusty plasma with a constant dust grain charge, and derived the KdV-Burger's equation, which admits shock waves solutions [Shukla and Mamun 2002] by including an arbitrary dissipative (kinematic viscosity) term on the right hand side of the momentum balance equation. There is an important question regarding the source/mechanism of this dissipative term. Motivated by this question, recently, we have considered an unmagnetized dusty plasma with charge fluctuating dust grains, and have proposed that dust grain charge fluctuations can arise this dissipative term and can be responsible for the formation of DIA shock waves in a dusty plasma [Mamun and Shukla 2002c]. The theoretical model for DIA waves that we recently proposed in a dusty plasma with charge fluctuating dust grains can be summerized as follows.

We have seen in the preceeding section that the parameter η does not play any role in our analysis of the DIA solitary waves. This is because of the scaling that we have used. We now consider a situation in which we can scale the parameter η as [Shukla 2000] $\eta = \epsilon^{1/2}\eta_0$. This additional scaling, remaining stretching/scaling of other variables being as before, introduces a new term, namely $-u_0\eta_0\partial z_d^{(1)}/\partial\xi$, in the right-hand side of Eq. (53). Therefore, with $\eta = \epsilon^{1/2}\eta_0$, Eq. (53) can be rewritten as

$$-u_0\eta_0\frac{\partial z_d^{(1)}}{\partial\xi} = \beta_e\phi^{(2)} - \alpha u_\beta z_d^{(2)} + \beta_1\phi^{(1)}\frac{\partial\phi^{(1)}}{\partial\xi} - \beta_i u_1 u_i^{(2)} - u_0\beta_i u_2 n_i^{(2)}. \tag{61}$$

Replacing Eq. (53) by Eq. (61) and performing all mathematical steps as we did in order to derive Eq. (56), we obtain a KdV-Burgers equation

$$\frac{\partial\phi^{(1)}}{\partial\tau} + A_c\phi^{(1)}\frac{\partial\phi^{(1)}}{\partial\xi} + B_c\frac{\partial^3\phi^{(1)}}{\partial\xi^3} = C_c\frac{\partial^2\phi^{(1)}}{\partial\xi^2}, \tag{62}$$

where

$$C_c = \frac{B_c v_0\eta_0\beta_0(1-\mu)}{\alpha(\beta_e + 2\beta_i/u_0)}. \tag{63}$$

An exact analytical solution of Eq. (62) is not possible. However, we can deduce some approximate analytical solutions. Let us try to find an analytical solution which is valid for both space and laboratory dusty plasmas. We first transform the independent variables ξ to $\zeta = \xi - U_0\tau$. Under the steady state condition we then find from Eq. (62) a third order ordinary differential equation for $\phi^{(1)}(\zeta) = \phi$. The latter can be integrated once, giving

$$B_c\frac{d^2\phi}{\partial\zeta^2} - C_c\frac{d\phi}{d\zeta} + \frac{A_c}{2}\phi^2 - U_0\phi = 0, \tag{64}$$

where we have imposed the appropriate boundary conditions, viz. $\phi \to 0$, $d\phi/d\zeta \to 0$, $d^2\phi/d\zeta^2 \to 0$ at $\zeta \to \infty$. For analysis, we can use a simple mechanical analogy [Karpman 1975] based on the fact that it has the form of an equation of motion for a pseudo particle of mass B_c, of pseudo time $-\zeta$, and pseudo position ϕ in a force field with potential

$$V(\phi) = \frac{A_c}{6}\phi^3 - \frac{U_0}{2}\phi^2, \tag{65}$$

and a frictional force with coefficient C_c. The nature of the pseudo potential $V(\phi)$ for typical space (upper plot) and laboratory (lower plot) dusty plasma parameters is depicted in figure 4. If the frictional force were absent, the quasi-particle entering

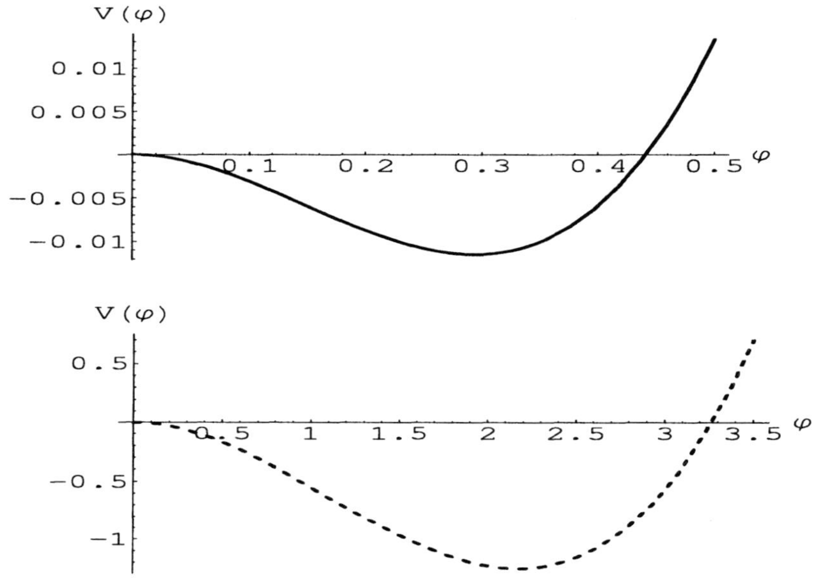

Figure 4: The behaviour of the potential $V(\phi)$ for $\mu = 0.5$. The upper plot corresponds to space dusty plasma parameters: $\alpha = 0.0288$ and $\beta = 3 \times 10^{-10}$, whereas the lower plot corresponds to laboratory dusty plasma parameters: $\alpha = 1.44$ and $\beta = 3 \times 10^{-4}$ (after Mamun and Shukla 2002c).

from the left will go the right-hand-side of the well ($\phi < 0$), reflect, and return to $\phi = 0$, making a single transit. This corresponds to the DIA solitary waves defined by Eq. (60). However, since a frictional force is present, i.e. the particle suffers a loss of energy, it will never return $\phi = 0$ but will oscillate about some negative value of ϕ corresponding to the minimum of $V(\phi)$. We assume that at pseudo time $-\zeta = -\infty$, i.e. at $\zeta = \infty$, the quasi-particle is located at the coordinate origin, i.e. $\phi(\infty) = 0$ and at pseudo time $-\zeta = \infty$, i.e. at $\zeta = -\infty$, the quasi-particle is located at a point corresponding to the minimum of $V(\phi)$, i.e. $\phi(-\infty) = 2U_0/A_c$. Thus, the solution of Eq. (63) describes a shock wave whose speed U_0 is related to the extreme values $\phi(\infty) = 0$ and $\phi(-\infty) = 2U_0/A_c$ by $\phi(-\infty) - \phi(\infty) = 2U_0/A_c$.

The nature of these shock structures depends on the relative values between the dispersive and dissipative coefficients B_c and C_c. If the value of C_c is very small, the energy of the particle decreases very slowly and the first few oscillations at the wave front will be close to solitary waves as defined by Eq. (60). However, if the value of C_c is larger than a certain critical value, the motion of the particle will be aperiodic and we obtain a shock wave with a monotonic structure. We now determine the condition

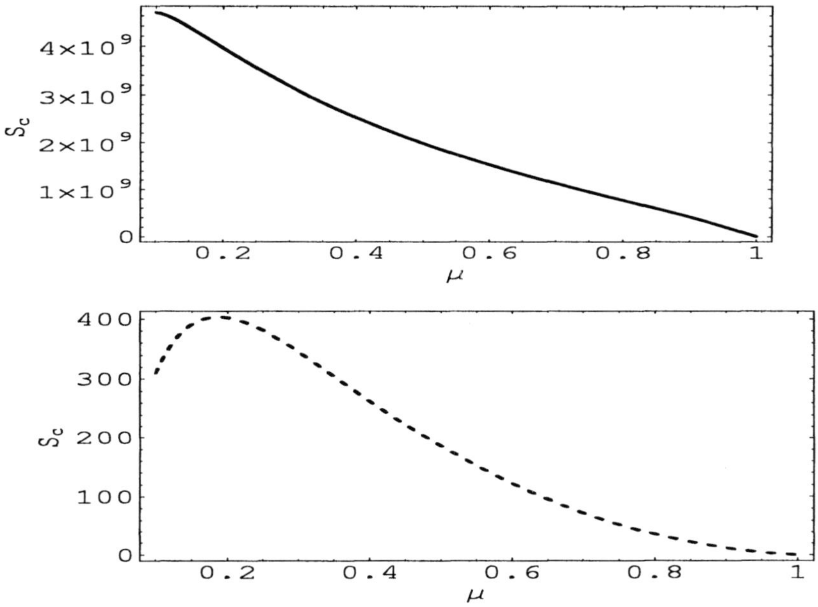

Figure 5: Variation of S_c with μ for $U_0 = 1$ and $\epsilon = 10^{-2}$. The upper plot corresponds to space dusty plasma parameters: $\alpha = 0.0288$ and $\beta = 3 \times 10^{-10}$, whereas the lower plot corresponds to laboratory dusty plasma parameters: $\alpha = 1.44$ and $\beta = 3 \times 10^{-4}$ (after Mamun and Shukla 2002c).

for monotonic or oscillating shock profiles by investigating the asymptotic behaviour of the solutions of Eq. (64) for $\zeta \to -\infty$. We first substitute $\phi(\zeta) = \phi_0 + \tilde{\phi}(\zeta)$, where $\tilde{\phi} \ll \phi_0$, into Eq. (64), then linearize it with respect to $\tilde{\phi}$ and obtain

$$B_c \frac{d^2\tilde{\phi}}{\partial\zeta^2} - C_c \frac{\partial\tilde{\phi}}{\partial\zeta} + U_0\tilde{\phi} = 0. \tag{66}$$

The solutions of Eq. (66) are proportional to $\exp(p\zeta)$, where p is given by

$$p = \frac{C_c \pm \sqrt{C_c^2 - 4B_cU_0}}{2B_c}. \tag{67}$$

It implies that the shock wave has a monotonic profile for $S_c = C_c/2\sqrt{B_cU_0} > 1$ and an oscillating profile for $S_c < 1$. We have numerically analyzed S_c for space ($\alpha = 0.0288$ and $\beta = 3 \times 10^{-10}$) and laboratory ($\alpha = 1.44$ and $\beta = 3 \times 10^{-4}$) dusty plasma parameters and have found that $S_c \gg 1$ is always valid for $0 < \mu < 1$ (cf. figure 5). This means that for typical space and laboratory dusty plasma parameters

Eq. (66) [or Eq. (64) under steady state condition] exhibits a monotonic shock wave solution which is given by

$$\phi^{(1)} \simeq \psi_{sh} - \psi_{sh} \tanh\left[(\xi - U_0\tau)/\Delta_{sh}\right], \tag{68}$$

where $\psi_{sh} = U_0/A_c$ and $\Delta_{sh} = 2C_c/U_0$ represent the amplitude and the width of the shock wave, respectively. We note that since $S_c \gg 1$, we have neglected

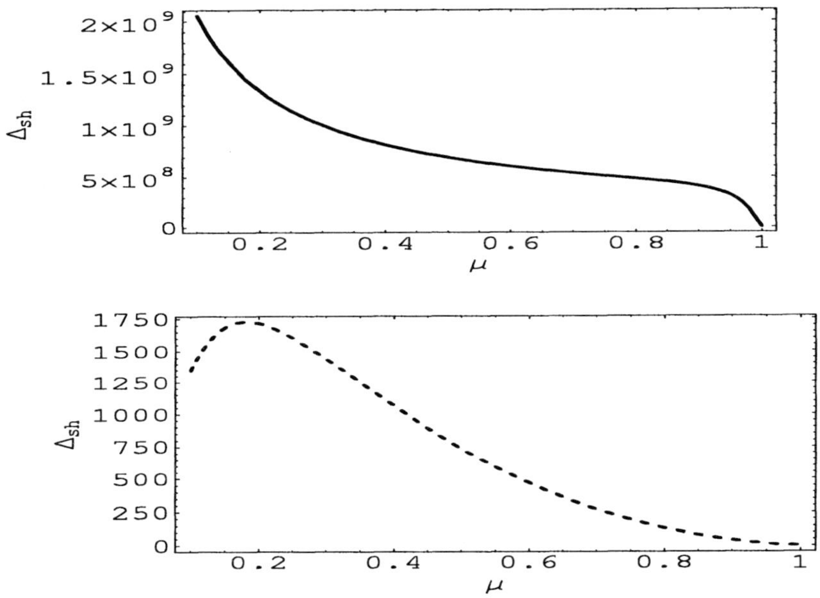

Figure 6: Variation of the shock width Δ with μ for $U_0 = 1$ and $\epsilon = 10^{-2}$. The upper plot corresponds to space dusty plasma parameters: $\alpha = 0.0288$ and $\beta = 3 \times 10^{-10}$, whereas the lower plot corresponds to laboratory dusty plasma parameters: $\alpha = 1.44$ and $\beta = 3 \times 10^{-4}$ [after Mamun and Shukla 2002c].

the dispersive term just to avoid the complexity of the mathematics which does not affect the properties of the shock structure that may exist in our space and laboratory dusty plasmas. Since $\psi_{sh} = \psi/3$, the variation of the shock amplitude ψ_{sh} with μ for both space and laboratory dusty plasmas can be represented by the upper plot of figure 3. We have also graphically shown how the width Δ_{sh} of the shock wave varies with μ for both space ($\alpha = 0.0288$ and $\beta = 3 \times 10^{-10}$) and laboratory ($\alpha = 1.44$ and $\beta = 3 \times 10^{-4}$) dusty plasmas. This is depicted in figure 6. We note that the shock structures predicted by Popel *et al.* [2000] do not follow our shock solution (21), and look like the ion density steepening as observed by Luo *et al.* [1999]. However, the shock structures found in our present investigation are very close to those observed by Nakamura et al. [1999].

4.2 Experimental Observation

DIA shock waves were experimentally excited in a dusty double plasma (DP) device by Nakamura *et al.* [1999]. We here briefly illustrate the formation of these experimentally excited DIA shock waves by summarizing the experimental work of Nakamura *et al.* [1999]. The inner diameter of the dusty DP device is 40 cm and its length is 90 cm. The device is separated into a source and a target section by a fine mesh grid which is kept electrically floating. The chamber is evacuated down to 5×10^{-7} torr with a turbo-molecular pump. The argon gas is bled into the chamber at a partial pressure of about 5×10^{-4} torr. A dust dispersing setup fitted at the target section consisted of a dust reservoir coupled to an ultrasonic vibrator. The dust reservoir consists of a fine stainless steel mesh (118 lines per cm) of 10 cm (width) $\times$ 16 cm (axial length) area at the bottom end and is placed horizontally closer to the anode wall. An ultrasonic vibrator is tuned at 27 kHz to vibrate the dust reservoir by using a signal generator and a power amplifier. Glass powder of average diameter 8.8 μm is used. The dust number density is easily controlled by adjusting the power of the signal applied to the vibrator and is measured from the intensity of the laser light which passes through the dust column and is collected by a photodiode array. A maximum dust density of the order of 10^5 cm^{-3} is obtained in this setup. The plasma parameters measured by a plane Langmuir probe of 6 mm diameter and a retarding potential analyzer are as follows: $n_e = 10^8 - 10^9$ cm^{-3}, $T_e = (1 - 1.5) \times 10^4$ K, $T_i \simeq 0.1\ T_e$, $Z_d \simeq 10^5$ for $n_d < 10^3$ cm^{-3} and $Z_d \simeq 10^2$ for $n_d < 10^5$ cm^{-3}. The shock waves are excited in the plasma by applying a ramp signal with an amplitude of 2 V and a rise time of approximate 10 μs. The Langmuir probe is biased above the plasma potential in order to detect the signal as fluctuations in the electron saturation currents. Oscillatory shock waves are first excited in the plasma without the dust and the dust density is then increased in small steps, keeping the probe fixed at 12 cm measured from the grid. The corresponding time normalized by the ion plasma period (ω_{pi}^{-1}) for the signal at 12 cm is about 150. A number of these signals are shown in figure 7.

It reveals that the oscillatory wave structure behind the shock becomes less in number with increasing dust particle number density and finally completely disappears at a sufficiently high dust particle number density, leaving only the laminar shock front. The shock speed also increases with increasing dust particle number density. It is also noted that the particle density behind the shock remains constant, although the amplitude of the shock front (steepened part) seems to decrease when the dust particle number density is increased.

The effect of the dust particle number density on the ion acoustic compressional pulses has also been experimentally studied by Luo *et al.* [1999, 2000] who observed a steepening of the ion-acoustic pulses as they propagated through a dusty plasma if the percentage of the negative charge in the plasma on the dust grains was about 75% or more.

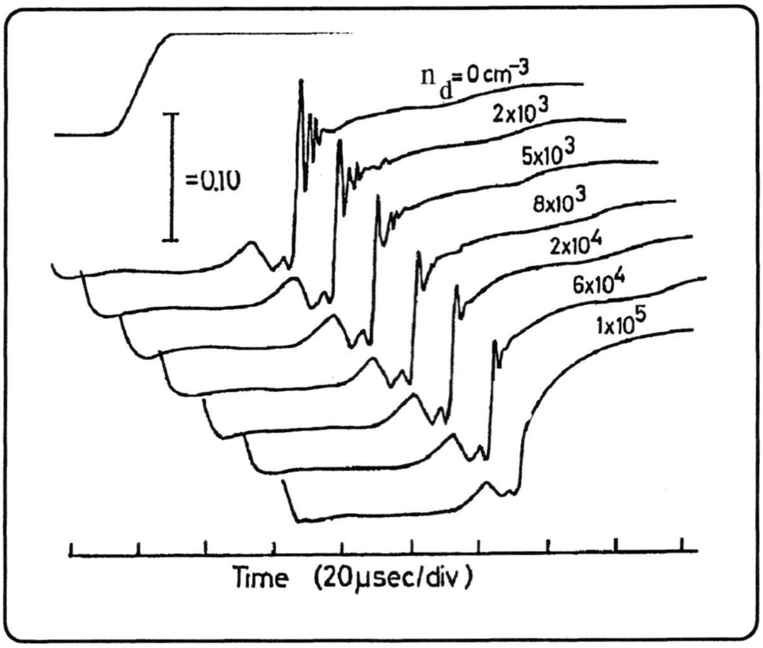

Figure 7: Variation of the plasma number density with time at a fixed probe position (12 cm) showing the transition of an oscillatory shock to a monotonic shock when the dust particle number density is increased (after Nakamura *et al.* 1999).

5 Dust-Acoustic Solitary Waves

We consider here the dynamics of negatively charged dust particles, and study the nonlinear behaviour of the DA waves whose phase speed is much smaller than the electron and ion thermal speeds, viz. $V_p \ll V_{Te}, V_{Ti}$. The nonlinear dynamics of DA solitary waves in a one-dimensional unmagnetized dusty plasma system with a constant dust grain charge is thus governed by [Rao *et al.* 1990]

$$\frac{\partial n_d}{\partial t} + \frac{\partial}{\partial z}(n_d u_d) = 0, \tag{69}$$

$$\frac{\partial u_d}{\partial t} + u_d \frac{\partial u_d}{\partial z} = \frac{\partial \varphi}{\partial z} - \frac{\sigma_d}{n_d}\frac{\partial p_d}{\partial z} \tag{70}$$

$$\frac{\partial p_d}{\partial t} + u_d \frac{\partial p_d}{\partial z} + 3 p_d \frac{\partial u_d}{\partial z} = 0, \tag{71}$$

$$\frac{\partial^2 \varphi}{\partial z^2} = n_d + \mu_e \exp(\sigma_i \varphi) - \mu_i \exp(-\varphi), \tag{72}$$

where n_d is the particle number density normalized by n_{d0}, u_d is the dust fluid velocity normalized by $C_d = (Z_{d0} k_B T_i / m_d)^{1/2}$ and φ is the electrostatic wave potential normalized by $k_B T_i / e$, p_d is the dust fluid pressure normalized by $n_{d0} k_B T_d$,

$\sigma_d = T_d/Z_dT_i$, $\mu_e = n_{e0}/Z_{d0}n_{d0} = \mu/(1-\mu)$, and $\mu_i = n_{i0}/Z_{d0}n_{d0} = 1/(1-\mu)$. The time ($t$) and space ($z$) variables are in units of the dust plasma period ${\omega_{pd}}^{-1} = (m_d/4\pi n_{d0}Z_{d0}^2e^2)^{1/2}$ and the Debye length $\lambda_{Dim} = (k_BT_i/4\pi Z_{d0}n_{d0}e^2)^{1/2}$, respectively.

To study arbitrary amplitude DA solitary waves in such a dusty plasma system, we assume that all the dependent variables depend on a single independent variable $\xi = z - Mt$ and reduce Eq. (69)–(72) to an energy integral (by following the mathematical procedure presented in §3)

$$\frac{1}{2}\left(\frac{d\varphi}{d\xi}\right)^2 + V(\varphi) = 0, \tag{73}$$

where the pseudo-potential potential $V(\varphi)$ is

$$V(\varphi) = -\left(\frac{\mu_e}{\sigma_i}\right)\exp(\sigma_i\varphi) - \mu_i\exp(-\varphi) - M^2\sqrt{\sigma_{d0}}\left[e^{\theta/2} + \frac{1}{3}e^{-3\theta/2}\right] + C_1, \tag{74}$$

with

$$\theta = \cosh^{-1}\left[\frac{\sigma_{d1}^2}{2\sigma_{d0}}\left(1 + \frac{2\varphi}{M^2\sigma_{d1}^2}\right)\right], \tag{75}$$

$\sigma_{d0} = \sqrt{3\sigma_d/M^2}$, $\sigma_{d1} = \sqrt{1+\sigma_{d0}^2}$ and C_1 is an integration constant. The latter is determined from $V(\varphi) = 0$ at $\varphi = 0$. It is important to note that we cannot consider the limit $\sigma_d \to 0$ in the Sagdeev potential $V(\varphi)$ in its present form. To consider the limit $\sigma_d \to 0$, we express θ as

$$\theta = \ln\left[\frac{\sigma_{d1}^2}{2\sigma_{d0}}\left(1 + \frac{2\varphi}{M^2\sigma_{d1}^2}\right) + \sqrt{\frac{\sigma_{d1}^4}{4\sigma_{d0}^2}\left(1 + \frac{2\varphi}{M^2\sigma_{d1}^2}\right)^2 - 1}\right]. \tag{76}$$

We note that in our study the condition for the dust density to be real requires $|1+2\varphi/M^2\sigma_{d1}^2| \geq 2\sigma_{d0}/\sigma_{d1}^2$. To find the condition for the existence of DA solitary waves analytically, we first consider the cold ion limit $\sigma_d = 0$ for which Eq. (76) reduces to the form [Mamun 1999]

$$V(\varphi) = \mu_i[1 - \exp(-\varphi)] + \frac{\mu_e}{\sigma_i}[1 - \exp(\sigma_i\varphi)] + M^2\left[1 - \left(1 + \frac{2\varphi}{M^2}\right)^{1/2}\right]. \tag{77}$$

It is obvious from Eq. (77) that $V(\varphi) = dV(\varphi)/d\varphi = 0$ at $\varphi = 0$. Therefore, solitary wave solutions of Eq. (73) exist if (i) $(d^2V/d\varphi^2)_{\varphi=0} < 0$ so that the fixed point at the origin is unstable and (ii) $(d^3V/d\varphi^3)_{\varphi=0} > (<)\ 0$ for solitary waves with $\varphi > (<)\ 0$. The nature of these solitary waves, whose amplitude tends to zero as the Mach number M tends to its critical value, can be found by expanding the Sagdeev potential $V(\varphi)$ to third order in a Taylor series in φ. The critical Mach number is that which corresponds to the vanishing of the quadratic term. At the

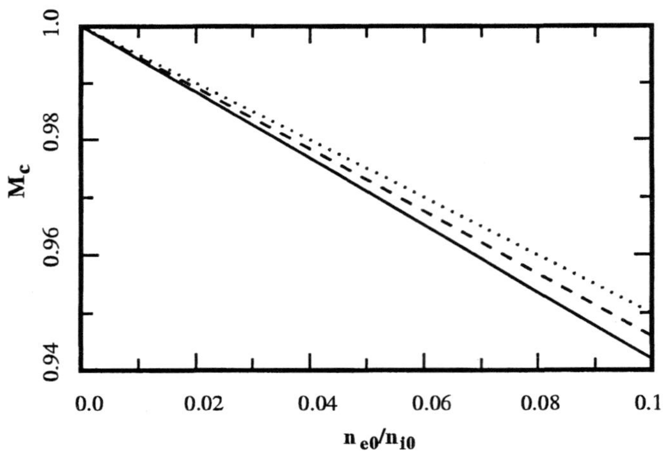

Figure 8: Variation of the critical Mach number M_c against n_{e0}/n_{i0} for $\sigma_i = 0.01$ (solid curve), $\sigma_i = 0.05$ (dashed curve) and $\sigma_i = 0.1$ (dotted curve) (after Mamun 1999).

same time, if the cubic term is negative, there is a potential well on the negative side, and if the cubic term is positive, there is a potential well on the positive side. Therefore, by expanding the Sagdeev potential $V(\varphi)$ around the origin, the critical Mach number at which the second derivative changes sign can be found as $M_c = \sqrt{(1-\mu)/(1+\sigma_i\mu)}$. Figure 8 depicts the variation of the critical Mach number M_c against n_{e0}/n_{i0} ($=\mu$) for different values of σ_i. The figure shows that the critical Mach number increases with σ_i, but decreases with μ. At the critical value of M the cubic term of $V(\varphi)$ can be expressed as $-[1+(3+\sigma_i\mu)\sigma_i\mu+\mu(1+\sigma_i^2)/2]/[3(1-\mu)^2]$. This clearly reveals that the cubic term is always negative for any value of σ_i and μ, i.e. DA solitary waves with $\varphi < 0$ can only exist.

It is of interest to examine whether or not there exists an upper limit of M for which DA solitary waves can exist. This upper limit of M can be found by the condition $V(\varphi_c) \geq 0$, where $\varphi_c = -M^2/2$ is the minimum value of φ for which the dust number density n_d is real. Thus, the upper limit of M is that maximum value of M for which $S_m \geq 0$, where $S_m = \mu_i + \mu_e/\sigma_i + M^2 - \mu_i \exp(M^2/2) - (\mu_e/\sigma_i)\exp(-\sigma_i M^2/2)$. Figure 9 shows the variation of S_m against M for different values of μ. We notice that as we increase μ, the upper limit of M decreases. We note that for $\sigma_i = 0.05$ and $\mu = 0.1$, there exist DA solitary waves with $\varphi < 0$ for $0.95 < M < 1.52$. We have also numerically analyzed the Sagdeev potential $V(\varphi)$ and have found the minimum and maximum values of M for which the DA solitary waves exist. These are incorporated in figures 10 and 11, which show that for $\sigma_i = 0.05$ and $\mu = 0.1$ there exists a potential well on the negative φ-axis for $0.95 < M < 1.52$, i.e. there exists DA solitary waves with $\varphi < 0$ for $0.95 < M < 1.52$.

To examine how the dust fluid temperature (σ_d) modifies the properties of the

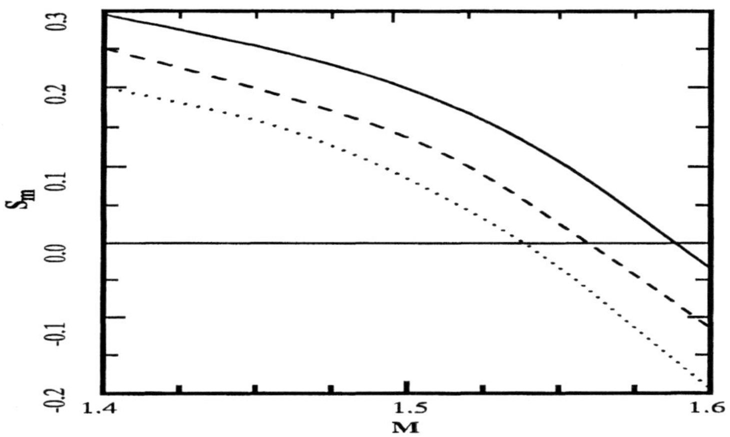

Figure 9: Variation of S_m against M for $\sigma_i = 0.05$, $\mu = 0$ (solid curve), $\mu = 0.05$ (dashed curve) and $\mu = 0.1$ (dotted curve). The upper limit of M is that value for which $S_m = 0$ (after Mamun 1999).

arbitrary amplitude DA solitary waves, we analyze the pseudo-potential potential $V(\varphi)$ given by Eq. (74), and found that as we increase σ_d, (i) we need higher value of the Mach number in order to have solitary wave solutions of Eq. (73), and (ii) the amplitude of the solitary waves decreases, but their width increases. We have also found the same results by numerical analysis of the energy integral (73). The potential profiles, as solutions of Eq. (73), are displayed in figure 12, showing the effects of dust temperature on their amplitude and width,.

We have discussed the properties of the DA solitary waves by assuming a planar geometry, Maxwellian ion distribution, and constant dust and grain charge. However, it can be shown shown that the effects of nonplanar geometry and dust grain charge fluctuations introduce new features or significantly modify the properties of the DA solitary waves [Mamun *et al.* 1996b; Ma and Liu 1997; Mamun and Shukla 2001; Shukla and Mamun 2002; Mamun and Shukla 2002c]. We now consider the cold dust limit ($\sigma_d = 0$) just to avoid the mathematical complexities, and investigate the effects of a non-Maxwellian (vortex-like and non-thermal) ion distribution, nonplanar geometry, and dust grain charge fluctuations on the properties of the DA solitary waves.

5.1 Vortex-Like Ion Distribution

It is well known [Schamel 1972, 1986, 1995, 2000] that the electron and ion distribution functions can be significantly modified in the presence of large amplitude waves that are excited by the two-stream instability [Winske *et al.* 1995]. Accordingly, the electron and ion number densities depart from a Boltzmann distribution when a phase space vortex distribution appears in a plasma. For the DA waves, the

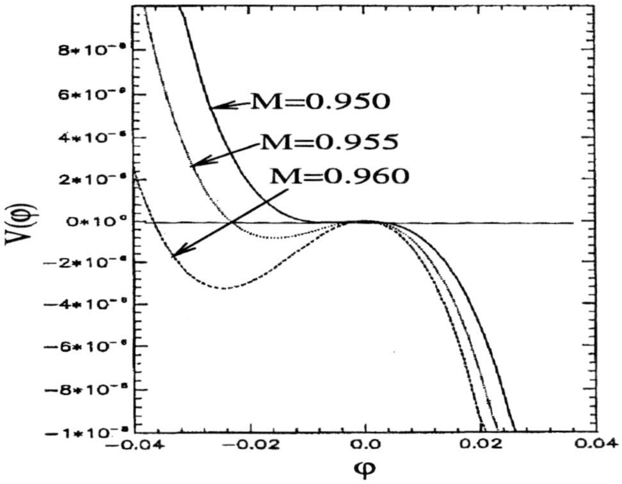

Figure 10: Behaviour of $V(\varphi)$ for $\sigma_i = 0.05$ and $\mu = 0.1$. It reveals that DA solitary waves with $\varphi < 0$ exist if $M > 0.95$ (after Mamun 1999).

ion trapping in the wave potential is of our interest. To study the effects of non-isothermal ions on the DA solitary waves, we consider the trapped or vortex-like [Schamel 1972, 1975] ion distribution $f_i = f_{if} + f_{it}$, where

$$f_{if} = \frac{1}{\sqrt{2\pi}} \exp\left[-\frac{1}{2}(v_i^2 + 2\varphi)\right] \tag{78}$$

for $|v_i| > \sqrt{-2\varphi}$ and

$$f_{it} = \frac{1}{\sqrt{2\pi}} \exp\left[-\frac{1}{2}\sigma_{it}(v_i^2 + 2\varphi)\right] \tag{79}$$

for $|v_i| \leq \sqrt{-2\varphi}$. We note that the ion distribution function, as prescribed above, is continuous in velocity space and satisfies the regularity requirements for an admissible BGK solution [Bernstein *et al.* 1957]. Here the ion velocity v_i in Eqs. (78) and (79) is normalized by the ion thermal speed V_{Ti}, and $\sigma_{it} = T_i/T_{it}$, the ratio of the free ion temperature T_i to the trapped ion temperature T_{it}, is a parameter determining the number of trapped ions. Integrating the ion distributions over velocity space we readily obtain the ion number density n_i as [Schamel 1986]

$$n_i = I(-\varphi) + \frac{1}{\sqrt{\sigma_{it}}} \exp(-\sigma_{it}\varphi)\, \mathrm{erf}(\sqrt{-\sigma_{it}\varphi}) \tag{80}$$

for $\sigma_{it} > 0$ and

$$n_i = I(-\varphi) + \frac{1}{\sqrt{\pi |\sigma_{it}|}} W_D(\sqrt{\sigma_{it}\varphi}) \tag{81}$$

for $\sigma_{it} < 0$, where

$$I(z_0) = [1 - \mathrm{erf}(\sqrt{z_0})] \exp(z_0), \tag{82}$$

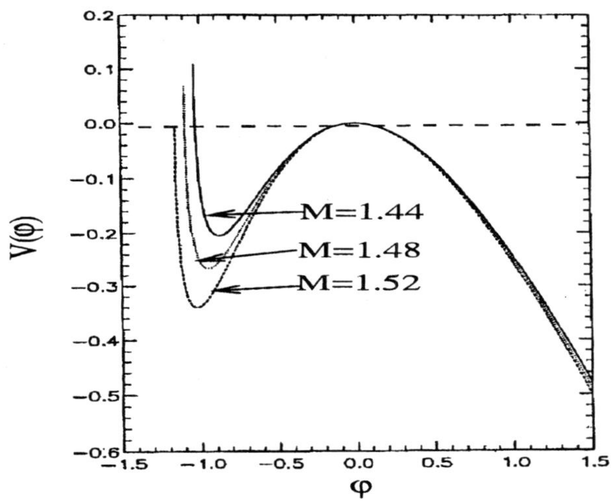

Figure 11: Behaviour of $V(\varphi)$ for $\sigma_i = 0.05$ and $\mu = 0.1$. It shows that DA solitary waves with $\varphi < 0$ exist if $M \leq 1.52$ (after Mamun 1999).

$$\text{erf}(z_0) = \frac{2}{\sqrt{\pi}} \int_0^{z_0} \exp(-y^2) dy, \tag{83}$$

$$W_D(z_0) = \exp(-z_0^2) \int_0^{z_0} \exp(y^2) dy. \tag{84}$$

If we expand n_i in the small amplitude limit (viz. $\varphi \ll 1$) and keep terms up to φ^2, it is found that n_i is the same for both $\sigma_{it} < 0$ and $\sigma_{it} > 0$. It is expressed as

$$n_i \simeq 1 - \varphi - \frac{4(1-\sigma_{it})}{3\sqrt{\pi}}(-\varphi)^{3/2} + \frac{1}{2}\varphi^2. \tag{85}$$

We now follow the reductive perturbation technique of Schamel and construct a weakly nonlinear theory for DA solitary waves by introducing the stretched coordinates $\zeta = \epsilon^{1/4}(z - v_0 t)$ and $\tau = \epsilon^{3/4} t$. Using Eq. (69), Eq. (72) with $\sigma_d = 0$, and Eq. (78) with the replacement of $\mu_i \exp(-\varphi)$ by right hand side of Eq. (85), and applying the reductive perturbation technique of Schamel [1975] we can derive

$$\frac{\partial \varphi^{(1)}}{\partial \tau} + a_t \sqrt{-\varphi^{(1)}} \frac{\partial \varphi^{(1)}}{\partial \zeta} + b_s \frac{\partial^3 \varphi^{(1)}}{\partial \zeta^3} = 0, \tag{86}$$

where

$$a_t = \frac{v_0^3 (1 - \sigma_{it})}{\sqrt{\pi}(1 - \mu)}. \tag{87}$$

Equation (86) is a modified KdV equation exhibiting a stronger nonlinearity, smaller width and larger propagation speed of the DA solitary waves.

The stationary soliton-like solution of the modified KdV equation (86) can be obtained by transforming the space variable ζ to $\eta = \zeta - u_0 \tau$ and by imposing the

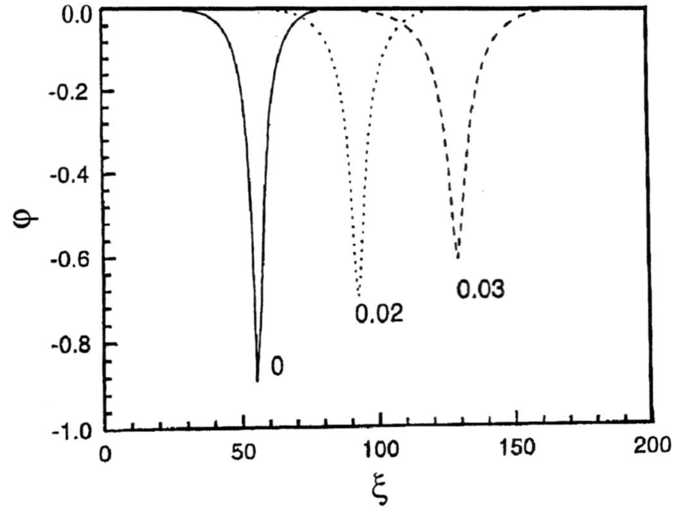

Figure 12: Potential profiles for $\sigma_i = 0.2$, $\mu = 0.1$, $M = 1.455$ and $\sigma_d = 0$ (solid curve), $\sigma_d = 0.02$ (dotted curve) and $\sigma_d = 0.03$ (dashed curve) (after Shukla and Mamun 2002).

appropriate boundary conditions, viz. $\varphi \to 0$, $d\varphi^{(1)}/d\eta \to 0$, $d^2\varphi^{(1)}/d\eta^2 \to 0$ at $\eta \to \pm\infty$. Thus, the stationary solution of Eq. (86) can be expressed as

$$\varphi^{(1)} = -\varphi_m^{(1)}\mathrm{sech}^4[(\zeta - u_0\tau)/\Delta_t], \tag{88}$$

where the amplitude $\varphi_m^{(1)}$ and the width Δ_t are given by $\varphi_m^{(1)} = (15u_0/8a_t)^2$ and $\Delta_t = \sqrt{16b_s/u_0}$, respectively. As $u_0 > 0$ and $\mu < 1$, Eq. (88) reveals that there exist solitary waves with negative potential only. It is also observed that as u_0 increases, the amplitude increases while the width decreases.

5.2 Nonthermal Ion Distribution

To study the effect of a non-thermal ion distribution on the properties of DA solitary waves, we choose a more general class of the ion distribution which includes the population of nonthermal (fast) ions. Thus, we take [Mamun *et al.* 1996b]

$$f_i(v_i) = \frac{1 + \alpha_i v_i^4}{(1 + 3\alpha_i)\sqrt{2\pi}} \exp(-v_i^2/2), \tag{89}$$

where v_i is the ion speed normalized by the ion thermal speed V_{Ti} and α_i is a parameter determining the population of nonthermal (fast) ions in our dusty plasma model. The effect of electrostatic disturbance on the equilibrium ion distribution can easily be introduced by replacing v_i^2 with $v_i^2 + 2\varphi$. The resulting distribution function is then integrated over velocity space, yielding [Mamun *et al.* 1996b]

$$n_i = (1 + \alpha_0\varphi + \alpha_0\varphi^2)\exp(-\varphi), \tag{90}$$

where $\alpha_0 = 4\alpha_i/(1+3\alpha_i)$. We now make all the dependent variables depend only on a single variable $\xi = z - Mt$, use the steady state condition, impose the appropriate boundary conditions (namely $n_d \to 1$, $u_d \to 0$, $\varphi \to 0$ and $d\varphi/d\xi \to 0$ at $\xi \to \pm\infty$) and reduce to an energy integral $(1/2)(d\varphi/d\xi)^2 + V(\varphi) = 0$, where $V(\varphi)$ is given by [Mamun *et al.* 1996b]

$$\begin{aligned} V(\varphi) &= \mu_i[1 + 3\alpha_0 - (1 + 3\alpha_0 + 3\alpha_0\varphi + \alpha_0\varphi^2)\exp(-\varphi)] \\ &+ \frac{\mu_e}{\sigma_i}[1 - \exp(\sigma_i\varphi)] + M^2\left[1 - \left(1 + \frac{2\varphi}{M^2}\right)^{1/2}\right]. \end{aligned} \tag{91}$$

Now, following the analytical steps or numerical analysis of the Sagdeev potential $V(\varphi)$ described as before, we can show that when $\alpha_i > 0.155$ and $M > 1.41$, the potential well develops on both the positive and negative φ-axis [Mamun *et al.* 1996b]. This means that the presence of nonthermal ions ($\alpha_i > 0.155$) supports the coexistence of compressive and rarefactive DA solitary waves (DA solitary waves with $\varphi < 0$ and $\varphi > 0$).

5.3 Nonplanar Geometry

It is well known that in laboratory devices one may encounter multi dimen- sional DA solitary waves. We are, therefore, interested in examining radially ingoing DA solitary waves in nonplanar (cylindrical and spherical) geometries. The dynamics of low phase velocity nonlinear DA waves in cylindrical and spherical geometries are governed by [Maxon and Viecelli 1974]

$$\frac{\partial n_d}{\partial t} + \frac{1}{r^\nu}\frac{\partial}{\partial r}(r^\nu n_d u_d) = 0, \tag{92}$$

$$\frac{\partial u_d}{\partial t} + u_d\frac{\partial u_d}{\partial r} = \frac{\partial \varphi}{\partial r}, \tag{93}$$

$$\frac{1}{r^\nu}\frac{\partial}{\partial r}\left(r^\nu\frac{\partial\varphi}{\partial r}\right) = n_d + \mu_e\exp(\sigma_i\varphi) - \mu_i\exp(-\varphi), \tag{94}$$

where $\nu = 1, 2$ for cylindrical and spherical geometries, respectively. The space variable r is normalized by the Debye radius λ_{Ddm}.

To investigate ingoing solutions of Eqs. (92)–(94), we introduce the stretched coordinates [Maxon and Viecelli 1974] $\zeta = -\epsilon^{1/2}(r + v_0 t)$ and $\tau = \epsilon^{3/2}t$, expand n_d, u_d and φ in powers of ϵ, and develop equations in various powers of ϵ. Thus, performing all the mathematical steps as in the case of cylindrical or spherical dust ion-acoustic waves, one can deduce a modified Korteweg-de Vries equation

$$\frac{\partial\varphi^{(1)}}{\partial\tau} + \frac{\nu}{2\tau}\varphi^{(1)} - A_{gd}\varphi^{(1)}\frac{\partial\varphi^{(1)}}{\partial\zeta} + B_{gd}\frac{\partial^3\varphi^{(1)}}{\partial\zeta^3} = 0, \tag{95}$$

where

$$A_{gd} = \frac{v_0^3}{(1-\mu)^2}[1 + (3 + \sigma_i\mu)\sigma_i\mu + \frac{1}{2}\mu(1 + \sigma_i^2)] \tag{96}$$

and $B_{gd} = v_0^3/2$. If we compare Eq. (95) with equation (13) of Mamun [1999], it is obvious that the term $(\nu/2\tau)\varphi^{(1)}$ in Eq. (95) is due to the effect of the cylindrical or spherical geometry. We numerically solve Eq. (95) and study the effects of spherical and cylindrical geometries on time-dependent DA solitary waves. The results for spherical and cylindrical solitary waves are shown in figure 13. The initial condition

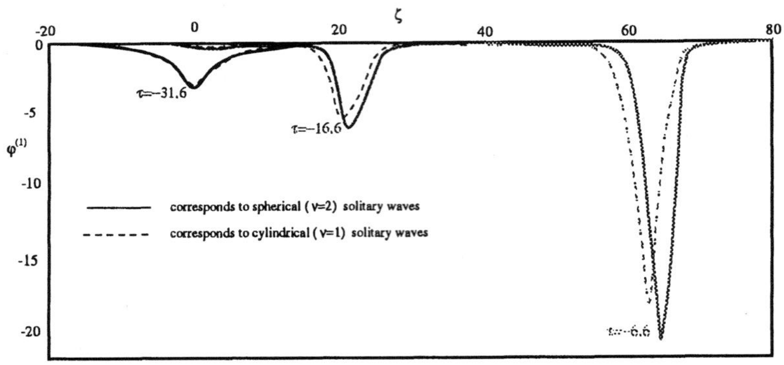

Figure 13: Time evolution of spherical and cylindrical solitary waves: $\varphi^{(1)}$ versus spatial coordinate ζ at times $\tau = -31.6$, $\tau = -16.6$ and $\tau = -6.6$ for $\mu = 0.1$ and $\sigma_i = 0.1$. The solid line corresponds to spherical ($\nu = 2$) solitary waves and the dashed line corresponds to cylindrical ($\nu = 1$) solitary waves (after Mamun and Shukla 1999).

that we have used in our numerical analysis is in the form of the stationary solution of Eq. (95) without the term $(\nu/2\tau)\varphi$, i.e. $\varphi^{(1)} = -\varphi_m^{(1)}\mathrm{sech}^2(\zeta/\Delta_{gd})$, where $\varphi_m^{(1)} = 3/A_{gd}$ and $\Delta_{gd} = \sqrt{4B_{gd}}$. Numerical solutions of Eq. (95), shown in figure 13, reveal that for a large value of τ (e.g. $\tau = -31.6$) the spherical and cylindrical solitary waves are similar to one-dimensional solitary waves. This is because for a large value of τ the term $(\nu/2\tau)\varphi^{(1)}$, which is due to the effect of the cylindrical or spherical geometry, is no longer dominant. However, as the value of τ decreases, the term $(\nu/2\tau)\varphi^{(1)}$ becomes dominant and both spherical and cylindrical solitary waves differ from one-dimensional solitary waves. It is also found that as the value of τ decreases, the amplitude of these localized pulses increases. Furthermore, cylindrical solitary waves travel faster than one-dimensional solitary waves but slower than spherical ones, and that the amplitude of cylindrical solitary waves is larger than that of the one-dimensional solitary waves but slightly smaller than that of the spherical ones. In conclusion, we mention that the present results may help to understand the salient features of multi-dimensional DA solitary waves when data for space and laboratory observations become available.

5.4 Dust Charge Fluctuation

We now investigate the effect of the dust grain charge fluctuation [Rao and Shukla 1994; Ma and Liu 1997] on arbitrary amplitude DA solitary waves. We use Eq. (69)

and

$$\frac{\partial u_d}{\partial t} + u_d \frac{\partial u_d}{\partial z} = z_d \frac{\partial \varphi}{\partial z} \tag{97}$$

$$\frac{\partial^2 \varphi}{\partial z^2} = z_d n_d + \mu_e \exp(\sigma_i \varphi) - \mu_i \exp(-\varphi), \tag{98}$$

where z_d is normalized by its equilibrium value Z_{d0}. Eqs. (97) and (98) are completed by the charging equation, which can be expressed in terms of the normalized variables as

$$\frac{\partial z_d}{\partial t} + u_d \frac{\partial z_d}{\partial z} = -(I_e + I_i) \tag{99}$$

where the normalized electron and ion currents I_e and I_i are

$$I_e = -\sqrt{\frac{1}{2\pi}} \frac{V_{Te}}{\omega_{pd}} 4\pi r_d^2 n_{d0} \mu_e n_e \exp(-\sigma_i S_d z_d), \tag{100}$$

$$I_i = \sqrt{\frac{1}{2\pi}} \frac{V_{Ti}}{\omega_{pd}} 4\pi r_d^2 n_{d0} \mu_i n_i (1 + S_d z_d), \tag{101}$$

where $S_d = Z_{d0} e^2 / r_d k_B T_i$. Using the current balance equation ($I_e + I_i = 0$) as well as substituting n_e and n_i we have

$$(1 + S_d z_d) \frac{\delta V_{Ti}}{V_{Te}} = \exp[-\sigma_i S_d z_d + (1 + \sigma_i)\varphi] \tag{102}$$

which can be simplified to

$$\varphi = \frac{1}{1 + \sigma_i} \left[\sigma_i S_d z_d + \ln\left(\frac{\delta V_{Ti}}{V_{T_e}}\right) + \ln(1 + S_d z_d) \right]. \tag{103}$$

To reduce Eqs. (69), (97) and (98) into a single equation, we introduce a dimensionless function [Ma and Liu 1997]

$$\Psi = -\int_0^{\varphi} z_d d\varphi \tag{104}$$

which, after using Eq. (103), can be expressed as

$$\Psi = -\frac{\varphi}{S_d} - (z_d - 1) - \frac{\sigma_i S_d}{2(1 + \sigma_i)} (z_d^2 - 1). \tag{105}$$

To study arbitrary amplitude time-independent DA solitary waves, we make all the dependent variables depend only on a single variable $\xi = z - Mt$, use the steady state condition and impose the appropriate boundary conditions for localized perturbations. Hence from Eqs. (69), (97) and (104) we obtain

$$n_d = \left(1 - \frac{2\Psi}{M^2}\right)^{-1/2}. \tag{106}$$

Substituting Eq. (106) into Eq. (98) and integrating the resultant equation, we obtain the energy integral

$$\frac{1}{2}\left(\frac{d\varphi}{d\xi}\right)^2 + V(\varphi, z_d) = 0, \tag{107}$$

where the Sagdeev potential $V(\varphi, z_d)$ is of the form [Ma and Liu 1997]

$$V(\varphi, z_d) = \mu_i[1 - \exp(-\varphi)] + \frac{\mu_e}{\sigma_i}[1 - \exp(\sigma_i\varphi)] + M^2\left[\left(1 - \frac{2\Psi}{M^2}\right)^{1/2} - 1\right]. \tag{108}$$

It is obvious from Eq. (108) that the effect of dust grain charge fluctuations is contained in Ψ. Now analyzing the Sagdeev potential $V(\varphi, z_d)$ as before, we can

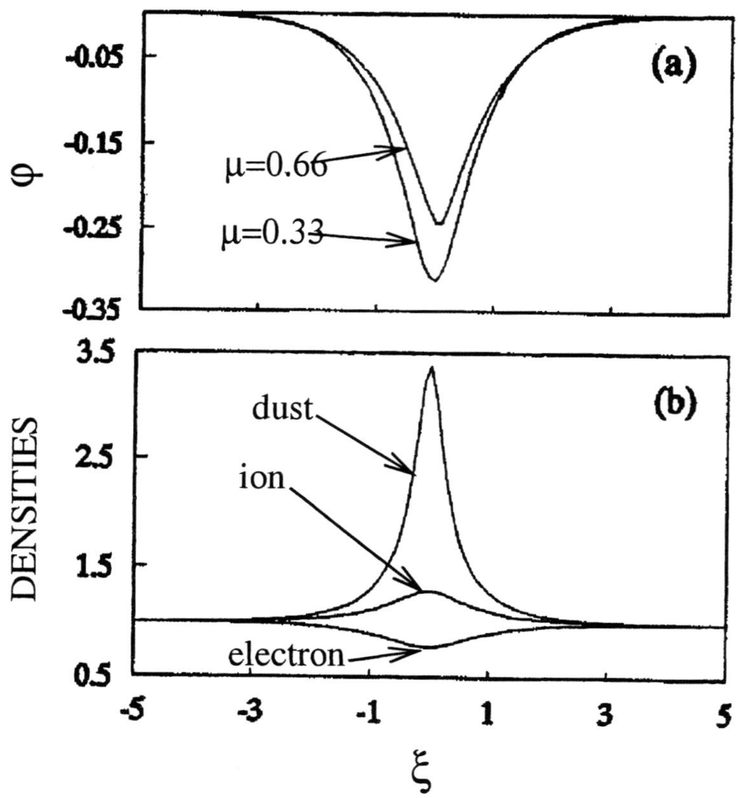

Figure 14: Solitary wave solutions of Eq. with $\sigma_i = 1$ and $M = S_d^{-1/2}$: a) potential profiles for $\mu = 0.66$ and $\mu = 0.66$, and b) electron, ion and dust density profiles for $\mu = 0.33$ (after Ma and Liu 1997).

easily investigate the properties of the DA solitary waves accounting for dust charge fluctuations. One can also directly solve Eqs. (105) and (108) numerically and examine the structures of the DA solitary profiles (cf figure 14) that may exist in a dusty plasma with fluctuating charges on the dust grain surface.

6 Dust-Acoustic Shock Waves

The nonlinear propagation of DA waves in a strongly coupled dusty plasma can be investigated by means of the generalized hydrodynamic (GH) model, namely Eqs. (69) and (72), and

$$(1+\tau_m D_t)\left[n_d\left(D_t u_d + \nu_{dn} u_d - \frac{\partial\varphi}{\partial z}\right)\right] = \eta_d \frac{\partial^2 u_d}{\partial z^2}, \tag{109}$$

where $D_t = \partial/\partial t + u_d \partial/\partial z$, ν_{dn} is normalized by the dust plasma frequency ω_{pd}, the viscoelastic relaxation time τ_m is normalized by the dust plasma period ω_{pd}^{-1} and $\eta_d = (\tau_d/m_d n_{d0}\lambda_{Dm}^2)[\eta_b + (4/3)\zeta_b]$ is the normalized longitudinal viscosity coefficient with η_b and ζ_b being the shear and bulk transport coefficients [Kaw and Sen 1998].

To derive a dynamical equation for DA shock waves from our basic equations (69), (72) and (109), we employ the reductive perturbation technique. Thus, as before, we introduce the stretched coordinates $\xi = \epsilon^{1/2}(z - u_0 t)$ and $\tau = \epsilon^{3/2} t$, and use the expansion of n_d, u_d and φ, and finally derive the KdV-Burgers equation

$$A_d^{-1}\frac{\partial\varphi^{(1)}}{\partial\tau} + \varphi^{(1)}\frac{\partial\varphi^{(1)}}{\partial\xi} + \beta_d\frac{\partial^3\varphi^{(1)}}{\partial\xi^3} = \mu_d\frac{\partial^2\varphi^{(1)}}{\partial\xi^2}, \tag{110}$$

where $A_d = (v_0^3 a_d/2)(1+\nu_{dn}\tau_m/2)^{-1}$, $\beta_d = 1/a_d$, $\mu_d = \eta_{d0}/a_d v_0^3$, $a_d = (\nu_{dn}\tau_m - a_{\mu\sigma})/v_0^4$ and $a_{\mu\sigma} = 2v_0^4[1+(3+\sigma_i\mu)\sigma_i\mu + (1+\sigma_i^2)\mu/2]/(1-\mu)^2$. Thus, for a weakly coupled or collisionless dusty plasma ($\nu_{dn}\tau_m \to 0$) we have $a_{\mu\sigma} < \nu_{dn}\tau_m$, i.e. $a_d < 0$, i.e. all coefficients (A_d, β_d and μ_d) are negative, whereas for a strongly coupled highly collisional dusty plasma satisfying $\nu_{dn}\tau_m > a_{\mu\sigma}$ we have $a_d > 0$, i.e. all coefficients (A_d, β_d and μ_d) are positive. We note that for the usual dusty plasma parameters (viz. $\mu = 0.1$ and $\sigma_i = 1$) we have $a_{\mu\delta} \simeq 2.0$. This means that for $T_i \le T_e$ and $n_{e0} \le 0.1 n_{i0}$ all coefficients (A_d, β_d and μ_d) can be positive if $\nu_{dn}\tau_m \ge 2$.

We now find the shock solutions of the KdV-Burgers equation (110) which, after transformation of the space variable ξ to $\zeta = \xi - U_0\tau$, where U_0 is the normalized velocity of the shock waves, can be integrated once, yielding

$$\beta_d\frac{d^2\varphi^{(1)}}{d\zeta^2} - \mu_d\frac{d\varphi^{(1)}}{d\zeta} + \frac{1}{2}[\varphi^{(1)}]^2 - \frac{U_0}{A_d}\varphi^{(1)} = 0, \tag{111}$$

where we have used the steady state condition and have imposed the appropriate boundary conditions, viz. $\varphi^{(1)} \to 0$, $d\varphi^{(1)}/d\zeta \to 0$, $d^2\varphi^{(1)}/d\zeta^2 \to 0$ at $\zeta \to \infty$. We can now easily show [Karpman 1975; Hasegawa 1975] that Eq. (111) describes a shock wave whose speed U_0 is related to the extreme values $\varphi^{(1)}(-\infty) - \varphi^{(1)}(\infty) = Y$ by $U_0/A_d = Y/2$. Thus, in the rest frame the normalized speed of the shock waves is $(1 + A_d Y/2)$. The nature of these shock waves depends on the relative values of the dispersive and dissipative parameters β_d and μ_d.

We first consider a situation where the dissipative term is dominant over the dispersive term, i.e. we can express Eq. (111) as

$$\left(\varphi^{(1)} - \frac{U_0}{A_d}\right)\frac{d\varphi^{(1)}}{d\zeta} = \mu_d \frac{d^2\varphi^{(1)}}{d\zeta^2}. \tag{112}$$

Using the condition that y is bounded as $\zeta \to \pm\infty$, Eq. (112) can be integrated to obtain [Karpman 1975]

$$\varphi^{(1)} = \frac{U_0}{A_d}\left[1 - \tanh\left(\frac{U_0}{2\mu_d}(\xi - U_0\tau)\right)\right]. \tag{113}$$

Equation (113) represents a monotonic shock solution with the shock speed U_0, the shock height U_0/A_d and the shock thickness $2\mu_d/U_0$. The shock solution appears because of the dissipative term, which is proportional to the viscosity coefficient.

Now, we discuss the combined effects of dissipation (β_d) and dispersion (μ_d). When μ_d is extremely small, the shock waves will have an oscillatory profile in which the first few oscillations at the wave front will be close to solitons moving with the speed U_0 [Karpman 1975]. If μ_d is increased and it is larger than a critical value μ_{dc}, the shock wave will have a monotonic behaviour. To determine the value of the dissipation coefficient μ_d corresponding to monotonic or oscillating shock profiles, we investigate the asymptotic behaviour of the solution of Eq. (111) for $\zeta \to -\infty$. We first substitute $\varphi^{(1)}(\zeta) = y_0 + y_1(\zeta)$, where $y_1 \ll y_0$, into equation Eq. (111) and then linearize it with respect to y_1 in order to obtain

$$\beta_d \frac{d^2 y_1}{\partial\zeta^2} - \mu_d \frac{\partial y_1}{\partial\zeta} + \frac{U_0}{A_d} y_1 = 0. \tag{114}$$

The solution of Eq. (114) is proportional to $\exp(p_s x)$, where p_s is

$$p_s = \frac{\mu_d}{2\beta_d} \pm \left(\frac{\mu_d^2}{4\beta_d^2} - \frac{U_0}{A_d\beta_d}\right)^{1/2}. \tag{115}$$

It turns out that the shock wave has a monotonic profile for $\mu_d > \mu_{dc}$ and an oscillatory profile for $\mu_d < \mu_{dc}$, where $\mu_{dc} = (4\beta_d U_0/A_d)^{1/2}$.

7 Dust Double Layers

A double layer is a monotonic transition of the electric potential connecting smoothly two differently biased plasmas. This is achieved by a bipolar charge distribution, as shown in figure 15 where the electric and classical potentials [φ and $V(\varphi)$] are displayed together with the charge density. According to Poisson's equation, a positively charged layer gives rise to a region of negative curvature of the potential and a negatively charged layer gives rise to a region of positive curvature of the

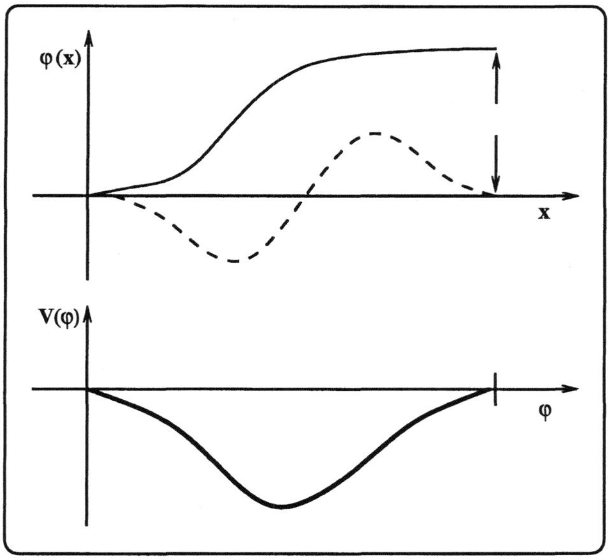

Figure 15: The electric and classical potentials [φ and $V(\varphi)$] of a double layer. The dotted line represents the total charge density (after Schamel 1986).

potential, and hence two oppositely charged layers are needed to built-up the double layer structure [Schamel 1986]. To have the double layers, we take into account the trapped particle effect. The reason is that if only streaming (i.e. non-reflected) particles would be present, the spatial constancy of the electron and ion currents would imply that the required charge neutrality cannot hold simultaneously on both sides of the double layer structure. Due to the different acceleration of electrons and ions in the double layer, their densities are affected. We assume that in the double layer the electron density response is Boltzmannean, while the ion density response is non-Maxwellian and is given by Eq. (81). Thus, the structure of a small amplitude dust double layer is deduced from

$$\frac{\partial^2\varphi}{\partial\xi^2} = \lambda_1\varphi + \frac{4}{3}b_{ti}\,(-\varphi)^{3/2} - \lambda_2\varphi^2, \tag{116}$$

where $\lambda_1 = -M^{-2}+\mu_e\sigma_i+\mu_i$, $b_{ti} = \mu_i(1-\sigma_i)/\sqrt{\pi}$, and $\lambda_2 = (3/2M^4)+(\mu_e\sigma_i^2-\mu_i)/2$. Introducing $-\varphi = \psi$, multiplying Eq. (116) by $\partial\psi/\partial\xi$ and integrating once, we obtain

$$\frac{1}{2}\left(\frac{\partial\psi}{\partial\xi}\right)^2 + U_{sd}(\psi) = 0, \tag{117}$$

where the classical potential is

$$U_{sd}(\psi) = -\frac{\lambda_1}{2}\psi^2 + \frac{2}{5}b_{ti}\psi^{5/2} + \frac{\lambda_2}{3}\psi^3. \tag{118}$$

The dust double layers must satisfy two conditions, namely $U_{sd}(\psi_m) = 0$ and $(\partial U_{sd}/\partial \psi)_{\psi=\psi_m} = 0$, which express the vanishing of the electric field and the charge density on the high potential side. Accordingly, we obtain

$$\lambda_1 - \frac{4}{5} b_{ti} \psi_m^{1/2} - \frac{2}{3} \lambda_2 \psi_m = 0, \tag{119}$$

$$\lambda_1 - b_{ti} \psi_m^{1/2} - \lambda_2 \psi_m = 0, \tag{120}$$

which can be solved for λ_1 and b_{ti}, resulting in

$$\lambda_1 = -\frac{2}{3} \lambda_2 \psi_m, \tag{121}$$

$$b_{ti} = -\frac{5}{3} \lambda_2 \psi_m^{1/2}. \tag{122}$$

Subsequently, we can express Eq. (117) as

$$\left(\frac{\partial \psi}{\partial \xi}\right)^2 + \frac{2}{3} \lambda_2 \psi^2 \left(\sqrt{\psi} - \sqrt{\psi_m}\right)^2 = 0, \tag{123}$$

which admits a double layer solution [Kim 1983]

$$\psi(\xi) = \frac{1}{4} \psi_m \left[1 + \tanh\left(\frac{\xi}{\lambda_{dl}}\right)\right]^2, \tag{124}$$

where the width of the dust double layers is

$$\lambda_{dl} = \left(\frac{\lambda_2}{24} \psi_m\right)^{-1/2}. \tag{125}$$

We observe that the dust double layers require $\lambda_2 > 0$, which yields $M^4(\mu_e \sigma_i^2 - \mu_i) + 1.5 > 0$.

8 Dust Lattice Solitary Waves

Melandsø [1996] has first presented a consistent theory for longitudinal DL solitary waves. The theory of Melandsø [1996], which is limited to only the interaction between nearest dust grains (to be discussed later), is then improved by Farokhi *et al.* [1999] who have taken into account the short- and long-range interactions between the dust particles. We now study the properties of small but finite amplitude DL solitary waves by summarizing the works of Farokhi *et al.* [1999, 2000].

We consider a multi-component dusty plasma in which extremely massive, negatively charged dust grains are considered as discrete particles, and electrons and the ions are assumed to be Boltzmannean (i.e. $N_e = n_{e0}\exp(e\Phi/k_BT_e)$ and $N_i = n_{i0}\exp(-e\Phi/k_BT_i)$]. The ambipolar potential Φ satisfies Poisson's equation

$$\nabla^2\Phi - 4\pi e[n_{e0}\exp(e\Phi/k_BT_e) - n_{i0}\exp(-e\Phi/k_BT_i)] = -\sum_n q_n(\mathbf{r},\Phi), \tag{126}$$

where the space charge contribution from all dust particles are included by the right-hand side term. The space charge density, corresponding to the nth spherical grain of radius r_{dn}, can be represented as

$$q_n(\mathbf{r},\Phi) = \frac{Q_n}{4\pi r_{dn}^2}\int_{S_n} dS\hat{\mathbf{n}}\delta(\mathbf{r} - [\mathbf{r_n} + r_{dn}\hat{\mathbf{n}}]), \tag{127}$$

where S_n is the surface area of the nthe grain, $\mathbf{r_n}$ is the radius-vector of the grain, $\hat{\mathbf{n}}$ a unit vector perpendicular to the surface element dS, and Q_n denotes the total charge of the grain. Assuming that the grains are conductors, we have $Q_n = r_{dn}\phi(\mathbf{r_n})$. When the dust grain radius r_d is assumed to be much smaller than the distance of the intergrain distance a, (i.e. $r_d \ll a$), we can neglect the last term in the argument of the δ-function, and from Eq. (127) we then obtain

$$q_n(\mathbf{r},\Phi) = Q_n\delta(\mathbf{r} - \mathbf{r_n}) = r_{dn}\delta(\mathbf{r} - \mathbf{r_n})\Phi(\mathbf{r}). \tag{128}$$

Thus, substituting Eq. (128) into Eq. (126), and assuming $\Phi/k_BT_e \ll 1$ and $\Phi/k_BT_i \ll 1$, we have a Schrödinger type equation

$$\nabla^2\Phi + [-k_D^2 - V(\mathbf{r})]\Phi = 0, \tag{129}$$

$$V(\mathbf{r}) = -\sum_n r_{dn}\delta(\mathbf{r} - \mathbf{r_n}). \tag{130}$$

We note that Eq. (129) is often used in solid state physics for the description of lattice waves in the one-dimensional approximation [Kittel 1956]. Equation (129) for one dimensional case can be written as

$$\frac{\partial^2\Phi(z)}{\partial z^2} - k_D^2\Phi(z) + \sum_n b_n\delta(z - z_n)\Phi(z) = 0, \tag{131}$$

where $b_n = N_0 r_{dn}/S$ is the size of the dust grain, averaged over the area S perpendicular to the linear lattice, and N_0 the number of the dust grains embraced by the area S. Equation (131) determines the real potential Φ and the condition of the periodicity (for the ideal lattice) cannot contain an exponential factor (the Bloch coefficient), as happens in solid state physics. The solutions of (131) are

$$\Phi(z) = A_n\sinh[k_D(z - z_n)] + B_n\cosh[k_D(z - z_n)] \tag{132}$$

for $z_n < z < z_{n+1}$ and

$$\Phi(z) = A_{n-1}\sinh[k_D(z - z_{n-1})] + B_{n-1}\cosh[k_D(z - z_{n-1})] \tag{133}$$

for $z_{n-1} < z < z_n$. Obviously, $B_n = \Phi(z_n) = \Phi_n$, where Φ_n is the potential of the n-th grain, and Φ_n is assumed to be known.

The solution of Eq. (131) and its derivative must satisfy two boundary conditions, namely

$$\Phi(z)|_{z_n+0} = \Phi|_{z_n-0}, \tag{134}$$

$$\frac{\partial\Phi}{\partial z}|_{z_n+0} - \frac{\partial\Phi}{\partial z}|_{z_n-0} = -b_n\Phi_n. \tag{135}$$

Using Eqs. (132)–(135) one can easily obtain the coefficients A_{n-1} and A_n:

$$A_{n-1} = \Phi_n \sinh^{-1}[k_D(z_n - z_{n-1})] - \Phi_{n-1}\coth[k_D(z_n - z_{n-1})], \tag{136}$$

$$A_n = \Phi_n \coth\left[k_D(z_n - z_{n-1})\right] - b_n/k_D - \Phi_{n-1}\sinh^{-1}\left[k_D(z_n - z_{n-1})\right]. \tag{137}$$

Equations (136) and (137) must be identical when $n \to n+1$. This condition gives

$$\Phi_n\left\{\tanh\left[\frac{k_D}{2}(z_n - z_{n-1})\right] + \tanh\left[\frac{k_D}{2}(z_{n+1} - z_n)\right] - b_n/k_D\right\}$$
$$= \frac{\Phi_{n-1} - \Phi_n}{\sinh[k_D(z_n - z_{n-1})]} + \frac{\Phi_{n+1} - \Phi_n}{\sinh[k_D(z_{n+1} - z_n)]}. \tag{138}$$

Thus, the solution of Eq. (132) takes the form

$$\Phi(z) = \frac{\Phi_n \sinh[k_D(z_{n+1} - z)] + \Phi_{n+1} \sinh[k_D(z - z_n)]}{\sinh[k_D(z_{n+1} - z_n)]}. \tag{139}$$

Under the condition specified by Eq. (138), the solution of Eq. (133) has an analogous form, which can be found from Eq. (139) by the replacement $n \to n-1$.

We now assume that the dust particles execute small oscillations about their equilibrium position. The dust grains are assumed to have the same potential Φ_0 (and the charge Q_0) with a uniform separation distance a. During the oscillations, distances between the grains change, and according to Eq. (138), the grain potential change as well. However, on a time scale of the dust lattice wave appearance, we assume that the dust particles maintain their equilibrium potential (and the charge). The effect of statistical dust charge fluctuations on dust lattice waves has been examined by Morfill *et al.* [1999]. Equation (139) can then be expressed in the form

$$\Phi(z) = \Phi_0 \frac{\cosh\left[\frac{k_D}{2}(z_{n+1} + z_n - 2z)\right]}{\cosh\left[\frac{k_D}{2}(z_{n+1} - z_n)\right]}, \tag{140}$$

The above expression gives the total electrostatic potential energy of the interaction for the dust grains system, viz.

$$\begin{aligned} U &= \sum_n Q_n\Phi(z_n) \\ &= Q_0\Phi_0 \sum_n \frac{\cosh\left[\frac{k_D}{2}(z_{n+1} + z_{n-1} - 2z_n\right]}{\cosh\left[\frac{k_D}{2}(z_{n+1} - z_{n-1})\right]}. \end{aligned} \tag{141}$$

We note that in Eq. (141) we have excluded the interaction of the charge with its own field. The equation of motion for the n-th dust grain, therefore, becomes

$$m_d \frac{\partial^2 z_n}{\partial t^2} = -\frac{\partial U}{\partial z_n}, \tag{142}$$

where U is given by Eq. (141). We now consider small longitudinal oscillations ($\tilde{z}_n$) of dust grains, about their equilibrium positions z_{n0}, i.e. $z_n = z_{n0} + \tilde{z}_n$, where $\tilde{z}_n \ll z_{n0}$, and derive the nonlinear equation for the DL waves. For this purpose, we need the expanded version of the potential energy U [Eq. (141)] and the corresponding equation of motion [Eq. (142)], which read

$$U = \frac{Q_0\Phi_0}{\cosh(k_D a)} \frac{1 + \frac{k_D^2}{4}(\tilde{z}_{n+1} + \tilde{z}_{n-1} - 2\tilde{z}_n)^2 + \frac{k_D^4}{384}(\tilde{z}_{n+1} + \tilde{z}_{n-1} - 2\tilde{z}_n)^4}{1 + \frac{k_D}{2}\tanh(k_D a)(\tilde{z}_{n+1} - \tilde{z}_{n-1})}, \tag{143}$$

$$\frac{\partial^2 \tilde{z}_n}{\partial t^2} = \frac{Q_0\Phi_0 k_D^2}{m_d \cosh(k_D a)} \frac{(\tilde{z}_{n+1} + \tilde{z}_{n-1} - 2\tilde{z}_n) + \frac{k_D^2}{48}(\tilde{z}_{n+1} + \tilde{z}_{n-1} - 2\tilde{z}_n)^3}{1 + \frac{k_D}{2}\tanh(k_D a)(\tilde{z}_{n+1} - \tilde{z}_{n-1})}. \tag{144}$$

To simplify Eq. (144), we will only include weak non-linearities. We will also consider the long wavelength approximation ($k_D a \ll 1$), where the wave dispersion is small and Eq. (144) can be approximated by the differential equations for a continuum [Melandsø 1996]. Thus, we introduce the Taylor expansions of $\tilde{z}_{n+1}$ and $\tilde{z}_{n-1}$ as $\tilde{z}_n + (\partial\tilde{z}/\partial z)_n a + 0.5(\partial^2\tilde{z}/\partial z^2)_n a^2 + \ldots$ and $\tilde{z}_n - (\partial\tilde{z}/\partial z)_n a + 0.5(\partial^2\tilde{z}/\partial z^2)_n a^2 + \ldots$, respectively. Inserting these expansions in Eq. (144), we obtain [Farokhi *et al.* 1999]

$$\frac{\partial^2 \tilde{z}}{\partial t^2} - C_L^2 \left[\frac{\partial^2 \tilde{z}}{\partial z^2} + \frac{a^2}{12}\frac{\partial^4 \tilde{z}}{\partial z^4} - \frac{k_D a}{2}\tanh(k_D a)\frac{\partial}{\partial z}\left(\frac{\partial \tilde{z}}{\partial z}\right)^2\right] = 0. \tag{145}$$

Here

$$C_L = \left[\frac{Q_0\Phi_0 k_D^2 a^2}{m_d \cosh(k_D a)}\right]^{1/2} \tag{146}$$

is the phase speed of the DL wavers for $k_D a \ll 1$, where the Debye shielding around each dust particle has been neglected. Equation (145) includes the DL waves propagating in both positive and negative z directions with phase velocities close to C_L and $-C_L$, respectively. For the waves with phase velocities close to C_L, Eq. (145)takes the form

$$\left(\frac{\partial}{\partial t} + C_L\frac{\partial}{\partial z}\right)\tilde{z} + C_L\left[\frac{a^2}{12}\frac{\partial^3 \tilde{z}}{\partial z^3} - \frac{k_D a}{2}\tanh(k_D a)\left(\frac{\partial \tilde{z}}{\partial z}\right)^2\right] = 0. \tag{147}$$

Equation (147) is now transformed to a wave-frame moving with a relative speed C_L in a new frame ($Z = z - C_L\tau$ and $\tau = t$). It is then differentiated with respect to Z, yielding the KdV equation

$$\frac{\partial u}{\partial \tau} - a_1 u\frac{\partial u}{\partial Z} + b_1\frac{\partial^3 u}{\partial Z^3} = 0 \tag{148}$$

where $u = \partial\xi/\partial Z$, $a_1 = k_D a \tanh(k_D a)$, and $b_1 = C_L a^2/12$.

The stationary solution of the KdV equation (148) is now obtained by transforming Z to $\xi = Z - u_0\tau$, where u_0 is the solitary wave speed and $\tau = T$, and imposing the appropriate boundary conditions for localized perturbations, viz. $u \to 0$, $du/d\xi \to 0$, $d^2u/d\xi^2 \to 0$ at $\xi \to \pm\infty$. Accordingly, the stationary solution of Eq. (148) is of the form

$$u = -u_m \text{sech}^2[\xi/\Delta_1], \tag{149}$$

where the amplitude u_m and width Δ_{DL} of the DL solitary waves are given by $u_m = 3u_0/a_1$ and $\Delta_1 = \sqrt{4b_1/u_0}$, respectively. As $u_0 > 0$ and $a_1 > 0$, Eq. (149) predicts the existence of small but finite amplitude DLS waves with $u < 0$.

Recently, Samsonov *et al.* [2002] have theoretically and experimentally studied the properties of DL solitary waves and compared the theoretical results with the experimental observations. The theoretical analysis of DL solitary waves presented by Samsonov *et al.* [2002], which is finally the same as presented by Melandsø [1999], can be summarized as follows.

The longitudinal oscillations of dust grains (with a constant charge q_d) in a one dimensional linear chain are governed by the equations of motion

$$m_d\left(\frac{\partial^2 \mathbf{r}_j}{\partial t^2} + \nu_{dn}\frac{\partial \mathbf{r}_j}{\partial t}\right) = -\sum_i \frac{\partial U_{ij}}{\partial \mathbf{r}_j} + \mathbf{F}_{\text{ext}}, \tag{150}$$

$$U_{ij} = \frac{q_d^2}{|\mathbf{r}_i - \mathbf{r}_j|}\exp\left(-\frac{|\mathbf{r}_i - \mathbf{r}_j|}{\lambda_D}\right), \tag{151}$$

where $\mathbf{r}_j$ is position of the jth dust particle, ν_{dn} is the collision frequency of dust particles with background neutral gas, U_{ij} is the Debye-Hückel potential, and $\mathbf{F}_{\text{ext}}$ is the external force consisting of two components, namely a stationary confinement force and a nonstationary exciting force.

We consider a one dimensional problem, and assume that dust particles are sufficiently strongly coupled so that they only obtain small longitudinal oscillations $x_j = x_{j0} + \tilde{x}_j$ around their equilibrium positions x_{j0}. Using the perturbation theory ($\tilde{x}_j \ll x_{j0}$) along with the long wavelength approximation, where dust particle separation a is small in comparison with the wavelength, Eq. (150) can be expressed as [Samsonov *et al.* 2002]

$$\begin{aligned}\frac{\partial^2 \tilde{x}_j}{\partial t^2} + \nu_{dn}\frac{\partial \tilde{x}_j}{\partial t} &= \kappa\frac{\partial}{\partial x}\left\{\frac{C_L^2}{\kappa}\left[\frac{\partial \tilde{x}_j}{\partial x} - \frac{\Lambda}{2}\left(\frac{\partial \tilde{x}_j}{\partial x}\right)^2\right]\right. \\ &\left. + \frac{\partial}{\partial x}\left[\frac{C_L^2}{\kappa}\mathcal{D}^2\frac{\partial^2 \tilde{x}_j}{\partial x^2}\right]\right\} + \frac{F_{\text{ext}}}{m_d},\end{aligned} \tag{152}$$

where $C_L = C_0\sqrt{\kappa^2[G(\kappa)/\kappa]''/2}$ is the dust-lattice speed, $\kappa = a/\lambda_D$ is the lattice parameter, $C_0 = \sqrt{2q_d^2/m_d\lambda_D}$, $\Lambda = -\kappa[G(\kappa)/\kappa]'''/[G(\kappa)/\kappa]''$ is the dimensionless nonlinear coefficient, and $\mathcal{D} = \sqrt{\lambda_D^2\kappa^2[G(\kappa)''/\kappa]''/[12G(\kappa)/\kappa]''}$ is the dispersion length. Here $G(\kappa) = -\ln[\exp(\kappa) - 1]$ and the prime denotes the derivative with respective κ. It is obvious that the kinetic coefficients C_L, Λ and $\mathcal{D}$ are functions of the lattice parameter κ. These are $C_L = C_0(\sqrt{1+\kappa+\kappa^2/2})\exp(-\kappa/2)$, $\Lambda = 3(1+\kappa+\kappa^2/2+\kappa^3/6)/(1+\kappa+\kappa^2/2)$, and $\mathcal{D} = \lambda_D\kappa/\sqrt{12}$ for $\kappa \gg 1$ (nearest neighbour approximation), whereas $C_L = C_0\sqrt{(3-2\ln\kappa)/2}$, $\Lambda = (11-6\ln\kappa)/(3-2\ln\kappa)$, and $\mathcal{D} = \lambda_D/\sqrt{3-2\ln\kappa}$ for $\kappa \ll 1$.

When the weakly inhomogeneous case is considered so that lattice parameter changes on a scale much larger than the size of the solitary waves, i.e. κ, C_L, Λ and $\mathcal{D}$ are independent of x, Eq. (152) can be expressed as [Samsonov *et al.* 2002]

$$\frac{\partial^2 \tilde{x}_j}{\partial t^2} + \nu_{dn}\frac{\partial \tilde{x}_j}{\partial t} = C_L^2\left[\frac{\partial^2 \tilde{x}_j}{\partial x^2} - \frac{\Lambda}{2}\frac{\partial}{\partial x}\left(\frac{\partial \tilde{x}_j}{\partial x}\right)^2 + \mathcal{D}^2\frac{\partial^4 \tilde{x}_j}{\partial x^4}\right]. \tag{153}$$

For experimental conditions the damping length ($C_L/\nu_{dn} \simeq 9-15$ mm) is much larger than the width of the disturbance [Samsonov *et al.* 2002]. Therefore, one can omit the friction term [second term of Eq. (153)] and can obtain Eq. (26) derived by Melandsø [1999], which for waves with phase velocities close to C_L can be reduced to [Melandsø 1999]

$$\frac{\partial \tilde{x}_j}{\partial t} + C_L\frac{\partial \tilde{x}_j}{\partial x} = -\frac{\Lambda}{4}\frac{\partial}{\partial x}\left(\frac{\partial \tilde{x}_j}{\partial x}\right)^2 + \frac{C_L\mathcal{D}^2}{2}\frac{\partial^3 \tilde{x}_j}{\partial x^3}. \tag{154}$$

Transforming Eq. (154) to a frame X moving with the relative speed C_L with respect to the x frame ($X = x - C_L t$, $t = T$), and differentiating it with respect to Z, we obtain the well-known KdV equation [Melandsø 1999]

$$\frac{\partial v}{\partial T} - a_2 v\frac{\partial v}{\partial X} + b_2\frac{\partial^3 v}{\partial X^3} = 0. \tag{155}$$

where $v = \partial\tilde{x}_j/\partial X$, $a_2 = \Lambda/2$ and $b_2 = C_L\mathcal{D}^2/2$ are the nonlinear and the dispersion coefficients, respectively. The steady state solution of is exactly similar to Eq. (149), i.e.

$$v = -v_m \mathrm{sech}^2[(X - u_0\tau)/\Delta_2] \tag{156}$$

where the amplitude v_m and the width Δ_2 are given by $3u_0/a_2$ and $\sqrt{4b_2/u_0}$, respectively. As $u_0 > 0$ and $a_2 > 0$, Eq. (156) predicts the existence of small but finite amplitude DDL waves with $v < 0$.

Samsonov *et al.* [2002] have experimentally studied the properties (amplitude and width) of the DL solitary waves and shown that the theoretical results completely agree with their experimental observations. Their experimental conditions are as follows. The experiment was performed in a capacitively coupled rf discharge. The discharge chamber had a lower disk electrode and an upper ring electrode. The upper electrode and the chamber were grounded. A rf power of 10 W (measured as forward minus reverse) was applied to the lower electrode. The working (argon) gas pressure in the chamber was 1.8 Pa. Mono-dispersive plastic microspheres (dust particles) (8.9 ± 0.1) μm in diameter were levitated in the sheath above the lower electrode forming a mono-layer hexagonal lattice. The monolayer particle cloud was about 6 cm in diameter and levitated at a height of $\simeq$ 9 mm above the lower electrode. The particle separation in the lattice was 650 μm at the excitation edge, 550 μm at the middle, and 720 μm at the outer edge. The particles were illuminated by a horizontal thin (0.2-0.3 mm) sheet of light from a double Nd:YAG diode pumped laser (532 nm).

9 Discussion

We present a rigorous theoretical investigation of different types of nonlinear electrostatic waves that may exist in dusty plasmas. Particularly, we have focused on DIA solitary and shock waves, DA solitary and shock waves as well as DDL. We have also studied DL solitary waves by considering a strongly coupled dusty plasma situation. We have investigated the effects of non-planer geometry and dust grain charge fluctuations on DIA and DA solitary waves. We have also studied the effects of dust fluid temperature, nonisothermal (trapped and nonthermal) ion distributions, and strong dust correlation, etc. on the properties of the DA solitary waves.

We have studied the properties of DIA solitary and shock waves in an unmagnetized dusty plasma composed of inertial ions, Boltzmann electrons, and immobile dust grains. The DIA solitary waves with a positive (negative) potential can exist for $\mu > (<)\ 1/3$. To examine how the ion fluid temperature (σ_i) modifies the properties of the arbitrary amplitude DIA solitary waves, we analyze the pseudo-potential potential $V(\varphi)$ given by Eq. (22), and found that as we increase σ_i, (i) we need higher value of the cortical Mach number (corresponding to a lower value of μ) in order to have DIA solitary waves with negative potential, and (ii) the amplitude of both positive and negative the solitary wave decreases, but their width increases.

To examine nonplanar geometry on DIA solitary waves we found that the term $(\nu/2\tau)\phi^{(1)}$ in Eq. (33) is due to the effect of the nonplanar (cylindrical or spherical) geometry. The numerical solutions of Eq. (33) (cf. figures 1 and 2) reveal that for a large value of τ (e.g. $\tau = -9$) the spherical and cylindrical solitary waves are similar to one-dimensional solitary waves. This is because for a large value of τ the term $(\nu/2\tau)\phi^{(1)}$, which is due to the effect of the cylindrical or spherical geometry, is no longer dominant. However, as the value of τ decreases, the term $(\nu/2\tau)\phi^{(1)}$ becomes dominant and both spherical and cylindrical solitary waves differ from one-dimensional solitary waves. It is found that as the value of τ decreases, the amplitude of these localized pulses increases. It is also found that the amplitude of cylindrical solitary waves is larger than that of the one-dimensional solitary waves but smaller than that of the spherical ones.

We have shown that the effects of dust grain charge fluctuations modify the properties of the DIA solitary waves. It has been found that the effects of dust grain charge fluctuations reduce the speed of the DIA solitary waves. The characteristics of these DIA solitary waves in the space dusty plasma condition are found to be different from those in laboratory dusty plasma condition. It is seen that for the parameters corresponding to space dusty plasma situations [Mendis and Rosenberg 1994; Shukla and Mamun 20002], as we increase μ, both the amplitude and the width of the DIA solitary waves remains constant for $\mu < 0.5$, but increases very rapidly for $\mu > 0.5$. On the other hand, for laboratory dusty plasma parameters [Barkan *et al.* 1996; Merlino *etal.* 1998], as we increase μ, the amplitude increases, but the width decreases.

We have shown that dust grain charge fluctuations are responsible for the formation of DIA shock waves in both space and laboratory dusty plasmas. The shock wave amplitude is one third of the solitary wave amplitude, but their dependence on dusty plasma parameters are exactly the same. Furthermore, the shock width decreases with μ (when $1 > \mu > 0.2$) for both space and laboratory dusty plasma parameters. However, for $0 < \mu < 0.2$ as we increase μ, the shock width increases for laboratory dusty plasma conditions, but decreases for space dusty plasma conditions.

We have studied the properties of DA solitary and shock waves and DDL in an unmagnetized dusty plasma by considering the dynamics of charged dust particles and Boltzmannean electron and ion density responses. We have shown that such a dusty plasma can support DA solitary waves with a negative potential only, corresponding to a hump in the dust number density. We found that as we increase the dust fluid temperature, the amplitude of DA solitary waves decreases, but their width increases.

We have shown that due to the effect of trapped ion distribution, a dusty plasma admits a modified KdV equation, exhibiting a stronger nonlinearity, smaller width and larger propagation speed. On the other hand, the presence of nonthermal (fast) ions may allow compressive and rarefactive solitary waves to coexist. This may be of relevance to electrostatic solitary structures observed by Freja satellite and Viking spacecraft. When we consider a nonplanar geometry (cylindrical or spherical), we find that a dusty plasma admits a modified KdV equation containing $(\nu/2\tau)\varphi^{(1)}$. This extra term is due to the effect of cylindrical ($\nu = 1$) or spherical ($\nu = 2$) geometry. It is also found that as the value of τ decreases, the amplitude of these localized pulses increases. Furthermore, the cylindrical solitary waves travel faster than the one-dimensional solitary waves but slower than the spherical ones, and that the amplitude of cylindrical solitary waves is larger than that of the one-dimensional solitary waves, but slightly smaller than that of the spherical ones.

A strongly coupled (correlated) dusty plasma may support DA shock waves instead of DA solitary structures due to the dissipation effect μ_d. When μ_d is extremely small, shock waves will have an oscillatory profile in which first few oscillations will be close to solitary structures. If μ_d is increased and it is larger than a critical value μ_{dc}, shock waves will have a monotonic behaviour.

We have finally studied the properties of DL solitary waves by constructing a nonlinear evolution equation. which admit a solitary wave solution. We have shown that small but finite amplitude DL solitary waves exist with $\partial\xi/\partial Z < 0$.

Acknowledgements

The authors acknowledge the support of the International Space Science Institute at Bern (Switzerland) for the team "Complex Plasmas in Space: Applications to Solar System and Near Earth Space". A. A. Mamun gratefully acknowledges the financial support of the Alexander von Humboldt-Stiftung (Bonn, Germany). F. Verheest acknowledges research grants from the Bijzonder Onderzoeksfonds (Universiteit Gent) and from the Fonds voor Wetenschappelijk Onderzoek (Vlaanderen).

References

Barkan A., Merlino R. L. and D'Angelo N. 1995, *Phys. Plasmas* **2**, 3563.

Barkan A., D'Angelo N. and Merlino R. L. 1996, *Planet. Space Sci.* **44**, 239.

Bernstein I. B., Greene J. B. and Kruskal M. D. 1957, *Phys. Rev.* **108**, 546.

Bharuthram R. and Shukla P. K. 1992, *Planet. Space Sci.* **40**, 973.

Bliokh P. V. and Yaroshenko V. V. 1985, *Sov. Astron.* **29**, 330.

Bouchoule A. 1999, in *Dusty Plasmas: Physics, Chemistry and Technical Impacts in Plasma Processing* ed A. Bouchoule (New York: John Wiley & Sons) pp 305-396.

Chu J. H. and I. Lin 1994, *Phys. Rev. Lett.* **72**, 4009.

de Angelis U 1992, *Phys. Scripta* **45**, 465.

de Angelis, U. Formisano V. and Giordano M. 1988, *J. Plasma Phys.* **40**, 399.

Farokhi B., Shukla P. K., Tsintsadze N. L. and Tskhakaya D. D. 1999, *Phys. Lett. A* **264**, 318.

Farokhi B., Shukla P. K., Tsintsadze N. L. and Tskhakaya D. D. 2000, *Phys. Plasmas* **7**, 814.

Gupta M. R., Sarkar S., Ghosh S., Debnath M. and Khan M. 2001, *Rev. Rev. E* **63**, 046406.

Goertz C. K. 1989, *Rev. Geophys.* **27**, 271.

Goertz C. K. and Morfill 1983, *Icarus* **53**, 219.

Hasegawa A. 1975, *Plasma Instabilities and Nonlinear Effects* (Berlin: Springer Verlag).

Hayashi H. and Tachibana K. 1994, *Jpn. J. Appl. Phys.* **33**, 804.

Hill J. R. and Mendis D. A. 1981, *Moon and Planets* **24**, 431.

Hollenstein C. 2000, *Plasma Phys. Control. Fusion* **42**, R93.

Homann A., Melzer A., Peters S. and Piel A. 1997, *Phys. Rev. E* **56**, 7138.

Horányi M. and Goertz C. K. 1990, *Astrophys. J.* **361**, 105.

Ikezi H. 1986, *Phys. Fluids* **29**, 1764.

Ivlev, A. V. and Morfill G. 2001, *Phys. Rev. E.* **63**, 0226412.

Karpman V. I. 1975, *Nonlinear Waves in Dispersive Media* (Pergamon Press, Oxford).

Kaw P. K. and Sen A. 1998, *Phys. Plasmas* **5**, 3552.

Kim K. Y. 1983, *Phys. Lett. A* **97**, 45.

Kittel C. 1956, *Introduction to Solid State Physics* (New York: John Wiley & Sons Inc.) Chapter 11.

Lonngren K. E. 1983, *Plasma Phys.* **25**, 943.

Luo Q. Z., D'Angelo N. and Merlino R. L. 1999, *Phys. Plasmas* **6**, 3455.

Luo Q. Z., D'Angelo N. and Merlino R. L. 2000, *Phys. Plasmas* **7**, 2370.

Luo Q. Z., D'Angelo N. and Merlino R. L. 2001, *Phys. Plasmas* **8**, 31.

Ma J. X. and Liu J. 1997, *Phys. Plasmas* **4**, 253.

Mamun A. A. 1998, *Phys. Scripta* **57**, 258.

Mamun A. A. 1999, *Astrophys. Space Sci.* **268**, 443.

Mamun A. A., Cairns R. A. and Shukla P. K. 1996a, *Phys. Plasmas* **3**, 702.

Mamun A. A., Cairns R. A. and Shukla P. K. 1996b, *Phys. Plasmas* **3**, 2610.

Mamun A. A. and Shukla P. K. 2001, *Phys. Lett. A.* **290**, 173.

Mamun A. A. and Shukla P. K. 2002a, *Phys. Plasmas* **9** 1468.

Mamun A. A. and Shukla P. K. 2002b, *Phys. Scritpta* **T98**, 107.

Mamun A. A. and Shukla P. K. 2002c, *IEEE Trans. Plasma Sci.* **30**, in press.

Maxon S. and Viecelli J. 1974, *Phys. Rev. Lett.* **32**, 4.

Melandsø F. 1996, *Phys. Plasmas* **3**, 3890.

Melandsø F., Aslaksen T. and Havnes O. 1993, *Planet. Space Sci.* **41**, 321.

Melandsø F. and Shukla P. K. 1995, *Planet. Space Sci.* **43**, 635.

Melzer A., Trottenberg T. and Piel A. 1994, *Phys. Lett. A* **191**, 301.

Mendis D. A. 1991, *Astrophys. Space Sci.* **176**, 163.

Mendis D. A. 1997, in *Advances in Dusty Plasmas*, eds P. K. Shukla, D. A. Mendis and T. Desai (Singapore: World Scientific) pp 3-19.

Mendis D. A. and Horanyi M. 1991, in *Cometary Plasma Processes*, ed A. J. Johnston (Washington DC: Geophys. Mongr. No. 6) p 17.

Mendis D. A. and Rosenberg M. 1994, *Ann. Rev. Astron. Astrophys.* **32**, 419.

Mendoza-Briceño, C. A., Russel, S. M. and Mamun A. A. 2000, *Planet. Space Sci.* **48**, 599.

Merlino R. L., Barkan A., Thompson C. and D'Angelo N. 1998, *Phys. Plasmas* **5**, 1607.

Morfill G. E., Thomas H. M. and Zuzic M. 1997, in *Advances in Dusty Plasmas* eds P. K. Shukla, D. A. Mendis and T. Desai (Singapore: World Scientific) pp 99-142.

Nakamura Y. 2002, *Phys. Rev. Lett.* **83**, 1602.

Nakamura Y., Bailung H. and Shukla P. K. 1999, *Phys. Plasmas* **9**, 440.

Nakamura Y. and Sharma A. 2001, *Phys. Plasmas* **8**, 3921.

Nakano T. 1998, *Astrophys. J* **494**, 587.

Popel S. I. and Yu M. Y. 1995, *Contrib. Plasma Phys.* **35**, 103.

Popel S. I., Yu M. Y. and Tsytovich V. N. 1996, Phys. Plasmas **3** 4313.

Popel S. I., Gisko A. A., Golub A. P., Bingham R. and Shukla P. K. 2000, *Phys. Plasmas* **3**, 4313.

Popel S. I., Golub A. P. and Losseva T. V. 2001, *JETP Letters* **7**, 396.

Rao N. N. and Shukla P. K. 1994, *J. Plasma Phys.* **49** 375.

Rao N. N., Shukla P. K. and Yu M. Y. 1990, *Planet Space Sci.* **38**, 543.

Rosenberg M. 1996, *J. Vac. Sci. Technol. A* **14**, 631.

Rosenberg M. and Kalman G. 1997, *Phys. Rev. E* **56**, 7166.

Sagdeev R. Z. 1966, in *Reviews of Plasma Physics* ed M. A. Leontovich (Consultants Bureau, New York), Vol 4, p 23.

Salimullah M. 1996, *Phys. Lett. A* **215**, 296.

Samsonov D., Ivlev A. V., Quinn R. A., Morfill G. and Zhdanov S, *Phys. Rev Lett.* **88**, 095004.

Schamel H. 1972, *Plasma Phys.* **14**, 905.

Schamel H. 1975, *J. Plasma Phys.* **13**, 129.

Schamel H. 1986, *Phys. Rep.* **140**, 161.

Schamel, H. 2000, *Phys. Plasmas* 7, 4831.

Shukla P. K. 1996, in *The Physics of Dusty Plasmas* (World Scientific, Singapore) pp 107-121.

Shukla P. K. 2000, Phys. Plasmas **7**, 1044.

Shukla P. K. 2001, *Phys. Plasmas* **8**, 1791.

Shukla P. K. and Mamun A. A. 2001, *IEEE Trans. Plasma Sci.* **29**, 221.

Shukla P. K. and Mamun A. A. 2002, *Introduction to Dusty Plasma Physics* (Institute of Physics Publishing Ltd., Bristol).

Shukla P. K. and Rosenberg M. 1999, *Phys. Plasmas* **6**, 1038.

Shukla P. K. and Silin V. P. 1992, *Phys. Scripta* **45**, 508.

Shukla P. K. and Stenflo L. 1992, Astrophys. Space Sci. **190**, 23.

Singh S. V. and Rao N. N. 1997, *Phys. Lett. A* **235**, 164.

Smith B. A. *et al* 1981, *Science* **212**, 163.

Smith B. A. *et al* 1982, *Science* **215**, 504.

Thomas H., Morfill G. E., Demmel V., Goree J., Feuerbacher B. and Möhlmann D. 1994, *Phys. Rev. Lett.* **73**, 652.

Thompson C., Barkan A., Merlino R. L. and D'Angelo N. 1999, *IEEE Trans. Plasma Sci.* **27**, 146.

Varma R. K., Shukla P. K. and Krishan V. 1993, *Phys. Rev. E* **47**, 3612.

Verheest F. 1992, *Planet. Space Sci.* **40**, 1.

Verheest F. 1996, *Space Sci. Rev.* **77**, 267.

Verheest F. 2000, *Waves in Dusty Space Plasmas* (Dordrecht: Kluwer Academic Publishers).

Washimi H. and Taniuti T. 1966, *Phys. Rev. Lett.* **17**, 996.

Winter J. 1998, *Plasma Phys. Control. Fusion* **40**, 1201.

Winske D., Gary S. P., Jones M. E., Rosenberg M., Mendis D. A. and Chow V. W. 1995, *Geophys. Res. Lett.* **22**, 2069.

Zweibel E. 1999, Phys. Plasmas **6**, 1725.

Coherent Vortices in Dusty Plasmas

D. Jovanović
Institute of Physics, P. O. Box 57, Yu–11001 Belgrade, Yugoslavia
P. K. Shukla
Institut für Theoretische Physik IV,
Ruhr–Universität Bochum, D–44780 Bochum, Germany

Abstract

The nonlinear dynamics is studied of the plasma that consists of electrons, ions and macroscopic dust particles, in the characteristic frequency range below the ion cyclotron and magnetosonic frequencies and above the dust-acoustic and dust charging frequencies. Regarding the dust grains as immobile and with constant charges, the plasma is described by the fluid ions and the first two moments of the drift-kinetic equation for the electrons, accounting for the contributions of the stress-tensor and finite mass to the electron dynamics. In the strongly nonlinear regime the plasma dynamics is governed by the vector nonlinearities arising from the $\vec{E} \times \vec{B}$ convection and the magnetic field lines' bending which leads to the plasma self-organization in the form of coherent vortices. For the strictly two-dimensional, electrostatic perturbations in a cold magnetized plasma with nonuniform dust density, it is shown that purely damped convective cell modes acquire a real frequency which is associated with the balance between the divergence of the non-zero $\vec{E} \times \vec{B}$ and the divergence of the ion polarization current. It is found that the dynamics of weakly interacting finite frequency convective cell modes is governed by the generalized Navier-Stokes equation. The new term arising from the dust inhomogeneity provides the possibility of two-dimensional dipolar vortex solution, in contrast to the dust-free plasmas. Similar localized dipolar vortex solution is also found for the three-dimensional electrostatic perturbations corresponding to the coupled electron-drift and dust ion-acoustic nonlinear modes in a warm plasma. In addition, a vortex-chain solution is found when the translational speed is fixed. In a fully electromagnetic, three-dimensional case the dipole solution is found for a linear dust density profile, while for a parabolic density profiles the stable simple monopolar vortices do not exist and a coherent localized nonlinear solution is constructed in the form of a tripole.

1 Introduction

The dusty plasma has attracted a great deal of interest in the past decade [*Shukla et al.* 1996]. Such plasmas, beside the electron and ions, contain as the third constituent a non-negligible number of macroscopic grains, or dust particles, whose mass is many orders of magnitude (typically $\sim 10^8$ times) larger than the ion mass and which carry large, usually negative, electric charges. Dusty plasmas are often encountered in nature, e. g. in the cometary tails, planetary atmospheres and rings, interstellar clouds, etc [*Spitzer* 1978]. They are also created in large laboratory devices used for the controlled fusion experiments, where the dust grains appear in the plasma as the result of the sputtering of the first wall of the discharge vessel, bombarded by the suprathermal particles and fusion products.

The presence of dust affects the plasma dispersion characteristics in several ways. In the very low-frequency domain, the dust introduces new modes of plasma oscillations that are associated with the motion of dust grains and their charge variation. Conversely, on the timescale that is short compared with the grain related transit times, the presence of dust affects the plasma background quantities, such as the electron and ion number densities, density scalelengths, etc. Particularly important is its role in the magneto-hydrodynamic (MHD) plasma, i. e. in the characteristic frequency regime below the ion cyclotron frequency. In the presence of density and/or temperature inhomogeneities and magnetic and velocity shears, such plasmas are subject to various linear instabilities. In the case when the source of free energy is provided by the magnetic shear (i. e. by the zero-order current parallel to the background magnetic field) the destabilization comes from the reconnection of the magnetic field lines by the magnetic-field aligned electromagnetic perturbations, supported by the electron inertia. Conversely, when the free energy is provided by the perpendicular plasma flow (i. e. by the electric field perpendicular to the magnetic field) the destabilization comes from the oblique modes, supported by the Boltzmann-distributed, inertialess electrons. The presence of the dust component breaks the anti-symmetry between the leading-order electron and ion currents, the plasma ceases to behave as a single conductive fluid, and as the consequence new linear and nonlinear modes are enabled, as well as new types of linear instabilities. While the extreme cases of cold and hot electrons have been the subject of extensive analytical and numerical studies, the interplay of the electron diamagnetic and inertial effects to the plasma stability and nonlinear behavior has not been fully understood.

In order to fully understand the collective processes in dusty plasmas it is necessary to develop the theory of nonlinear waves, as the generalization of the electron-ion plasma results. It is known that coherent nonlinear structures exist in two-species plasmas, that are subject to various instabilities mentioned above. Depending on the properties of the unperturbed plasma state and on the asymptotic boundary conditions, these structures may take the form of either vortex dipoles, tripoles or chains [*Larichev and Reznik* 1976], [*Jovanović et al.* 1987], [*Jovanović et al.* 1998], [*Jovanović and Pegoraro* 2000]. While the dipoles move relative to the surrounding fluid, the tripoles and chains are carried by the fluid and supported by the shear flows. Vortices are considered to play an important role in the spatial self-organization of plasmas and to represent one of the possible sources for the development of turbulence and of the enhanced particle and energy transport in magnetized plasma. In this paper, we construct plasma vortices for different plasma geometries. For a linear density profile we find the standard dipole solution, while in the case of parabolic density profiles a coherent localized nonlinear solution is constructed in the form of a tripole, since stable simple monopolar vortices do not exist. Due to the presence of dust grains, the electron and ion diamagnetic currents do not cancel out, which limits the energy of these structures, provides their good spatial localization and increases the stability, compared to their dust-free plasma counterparts.

The paper is organized as follows. In section 2 we present the linear and nonlinear properties of the finite frequency convective cell mode, which is now known as the Shukla-Varma mode [*Shukla et al.* 1993]. In section 3 we study nonlinearly coupled electron-drift and dust ion-acoustic waves in a warm, nonuniform dusty magnetoplasma and construct coherent nonlinear solutions in the form of a dipolar vortex and a vortex chain. The full set of governing equations for the electromagnetic nonlinear perturbations in a complex plasma geometry is presented in section 4. The nonlinear equations are derived and the analytical solutions in the form of dipolar and tripolar vortices are presented and discussed. Conclusions are given in section 5.

2 Linear and Nonlinear Shukla-Varma Modes

We discuss the linear and nonlinear Shukla-Varma modes in a cold ($T_e = T_i = 0$) nonuniform dusty magnetoplasma whose constituents are electrons, ions, and charged static dust grains. The plasma contains an equilibrium density gradient $\partial n_{j0}/\partial x$ and an external magnetic field $\vec{e_z} B_0$, where n_{j0} is the unperturbed number density of the species j (j equals e for the electrons, i for the ions, and d for the dust grains), $\vec{e_z}$ is the unit vector along the z axis and B_0 is the strength of the magnetic field. The quasi-neutrality condition at the equilibrium is now given by

$$n_{i0}(x) = n_{e0}(x) + z_d(x) n_{d0}(x) , \tag{1}$$

where z_d is the number of electrons residing on the dust grain surface.

Consider now the perturbation of the equilibrium in the presence of two-dimensional electrostatic convective cell motion. Thus, in the electrostatic field ($\vec{E} = -\nabla\phi$, where ϕ is the electrostatic potential), the perpendicular component (to the constant external magnetic field $\vec{e}_z B_0$) of the fluid velocities in the drift-approximation ($|\partial/\partial t| \ll \Omega_i = eB_0/m_i$, where e is the magnitude of the electron charge and m_i is the ion mass) are

$$\vec{v}_{e\perp} \approx \vec{v}_{EB} , \tag{2}$$

and

$$\vec{v}_{i\perp} \approx \vec{v}_{EB} + \vec{v}_{ip} , \tag{3}$$

where $\vec{v}_{EB} = \vec{E} \times \vec{e_z}/B_0$ and $\vec{v}_{ip} = -(1/B_0\Omega_i)(\partial/\partial t + \vec{v}_{EB} \cdot \nabla)\nabla_\perp \phi$ are, respectively, the $\vec{E} \times \vec{B}_0$ and the ion polarization drifts.

The relevant governing equations for our purposes are

$$\frac{\partial}{\partial t}(n_e - n_i + z_d n_d) + \nabla_\perp \cdot (n_e \vec{v}_e - n_i \vec{v}_i) = 0\,, \tag{4}$$

and

$$\nabla_\perp^2 \phi = \frac{e}{\epsilon_0}(n_e - n_i + z_d n_d)\,, \tag{5}$$

where e is the magnitude of the electron charge.

Letting $n_j = n_{j0}(x) + n_{j1}$, where $n_{j1} \ll n_{j0}$, we obtain from Eqs. (1)-(5) the generalized Navier-Stokes equation

$$d_t \nabla_\perp^2 \phi + \frac{\Omega_i}{n_{i0}} \vec{e_z} \times \nabla\,(z_d n_{d0}) \cdot \nabla \phi + \frac{\omega_{pi}^2}{\Omega_i^2} d_t \nabla_\perp^2 \phi = 0\,, \tag{6}$$

where $\omega_{pi} = (n_{i0} e^2/\epsilon_0 m_i)^{1/2}$ is the ion plasma frequency. We have denoted $d_t = \partial/\partial t + \vec{v}_{EB} \cdot \nabla_\perp$, and assumed that $|\partial/\partial t| \gg \omega_{pd}$, where ω_{pd} is the dust plasma frequency.

The origin of the various terms in Eq. (6) is obvious. The first term comes from the deviation from the quasi-neutrality, the second term originates from the $\vec{E} \times \vec{B}_0$ convection of the equilibrium dust number density since $n_{e0} \neq n_{i0}$, whereas the third term arises from the divergence of the ion polarization drift. In the absence of the charged dust grains, the second term vanishes by virtue of the equilibrium quasi-neutrality condition.

In the linear limit, for the perturbations $\propto \exp[i(\vec{k}_\perp \vec{r}_\perp - \omega t)]$, Eq. (6) yields the Shukla-Varma frequency [*Shukla et al.* 1993]

$$\omega = \Omega_i \frac{k_y k_d}{k_\perp^2}\,, \tag{7}$$

where k_y is the component of the wave vector in the y direction, $k_d = n_{i0}^{-1} \partial(z_d n_{d0})/\partial x$, and $k_\perp$ is the wavenumber perpendicular to $\vec{e}_z$. Eq. (7) reveals that the Shukla-Varma frequency, which is directly proportional to $\partial(z_d n_{d0})/\partial x$, is attributed to the presence of the static dust grains in a nonuniform magnetoplasma. Shukla-Varma mode would not arise in an electron-ion plasma without the dust grains.

In the following, we show that the presence of the dust density gradient term in Eq. (6) provides the possibility of a two-dimensional coherent dipolar vortex structure. Accordingly, we look for the stationary solution of the generalized Navier-Stokes equation (6) in the stationary frame $\xi = y - ut$, where u is the translational speed of the vortex. Thus, Eq. (6) takes the form

$$u \frac{\partial}{\partial \xi} \nabla_\perp^2 \phi - u_d \partial_\xi \phi - \frac{1}{B_0} J(\phi, \nabla_\perp^2 \phi) = 0\,, \tag{8}$$

where $u_d = (\Omega_i/n_{i0})[\partial(z_d n_{d0})/\partial x]$, and the Jacobian is denoted by $J(\phi, \nabla_\perp^2 \phi) = (\partial_x \phi \partial_y - \partial_y \phi \partial_x)\nabla_\perp^2 \phi$.

The dipolar vortex solution [*Larichev and Reznik* 1976, *Stenflo* 1987] of Eq. (8) in the outer region ($r > R$) is

$$\phi_o(r,\theta) = \frac{CR}{\lambda_1^2 + \lambda_2^2} \frac{K_1(\lambda_1 r)}{K_0(\lambda_1 R)} \cos\theta\,, \tag{9}$$

whereas the solution in the inner region ($r < R$) reads

$$\phi_i(r,\theta) = \frac{CR}{\lambda_2^2}\left(\frac{r}{R} - \frac{\lambda_1^2}{\lambda_1^2 + \lambda_2^2}\frac{J_1(\lambda_2 r)}{J_0(\lambda_2 R)}\right)\cos\theta\,, \tag{10}$$

where $C = uB_0(\lambda_1^2 + \lambda_2^2), \lambda_1 = (u_d/u)^{1/2}, r = (x^2 + \xi^2)^{1/2}, \theta = \tan^{-1}(\xi/x)$, and R is the vortex radius. Furthermore, $J_1(K_1)$ is the Bessel function (modified Bessel function) of the first order. Clearly, $u_d > 0$ is required for the dipolar vortex to exist.

For a given value of $\lambda_1 > 0$, the value of λ_2 is determined from

$$\frac{J_2(\lambda_2 R)}{J_1(\lambda_2 R)} = -\frac{\lambda_2 K_2(\lambda_1 R)}{\lambda_1 K_1(\lambda_1 R)}\,, \tag{11}$$

where J_2 and K_2 are the Bessel and modified Bessel functions of the second order.

3 Electron-Drift Vortices

In this section, we discuss the properties of nonlinearly coupled electron-drift and dust ion-acoustic waves in a nonuniform dusty magnetoplasma. Similarly to the configuration studied in section 2, the plasma is assumed to contain an equilibrium density gradient $\partial n_{j0}/\partial x$ that is perpendicular to the equilibrium homogeneous magnetic field $B_0\vec{e_z}$. In the presence of low-frequency ($\omega \ll \Omega_i$), long wavelength (in comparison with the ion gyroradius ρ_i) electrostatic field $\vec{E} = -\nabla\phi$ and finite electron and ion pressures, the perpendicular components of the electron and ion fluid velocities in a warm magnetoplasma are given by, respectively,

$$\vec{v}_{e\perp} \approx \frac{1}{B_0}\vec{e_z} \times \nabla\phi - \frac{1}{eB_0 n_e}\vec{e_z} \times \nabla p_e\,, \tag{12}$$

and

$$\vec{v}_{i\perp} \approx \frac{1}{B_0}\vec{e_z} \times \nabla\phi + \frac{1}{eB_0 n_i}\vec{e_z} \times \nabla p_i - \frac{1}{B_0\Omega_i}\left(\frac{\partial}{\partial t} + \vec{v}_{i\perp}\cdot\nabla\right)\nabla_\perp\phi\,, \tag{13}$$

where $p_e = n_e T_e$ ($p_i = n_i T_i$) is the electron (ion) pressure, n_e (n_i) is the electron (ion) number density and $T_e(T_i)$ is the electron (ion) temperature.

We derive the governing nonlinear equations for long wavelength coupled electron-drift-dust ion-acoustic waves. Accordingly, we substitute Eq. (13) into the ion continuity equation and make use of a Boltzmann electron response, viz.

$$n_{e1} \approx n_{e0} \frac{e\phi}{T_e}, \tag{14}$$

which is valid for $|\partial/\partial t| \ll \nu_e \ll \Omega_e$ and $\nu_e|\partial/\partial t| \ll v_{Te}^2|\partial^2/\partial z^2|$, and assume quasineutrality $n_{e1} \approx n_{i1}$ to obtain the generalized Hasegawa-Mima equation [*Shukla et al.* 1991]

$$\left(\frac{\partial}{\partial t} + \frac{1}{B_0}\vec{e}_z \times \nabla\phi \cdot \nabla\right)(\phi - \rho_s^2 \nabla_\perp^2 \phi) + v_{e*}\frac{\partial \phi}{\partial y} + \frac{n_{i0}}{n_{e0}}\frac{T_e}{e}\frac{\partial v_{iz}}{\partial z} = 0\,, \tag{15}$$

where $\rho_s = c_s/\Omega_i$ is the ion gyroradius at the electron temperature, $c_s = (n_{i0}T_e/n_{e0}m_i)^{1/2}$ is the dust ion-acoustic speed, $v_{e*} = \rho_s c_s/L_n$ is the electron diamagnetic drift velocity, and $L_n = -n_{i0}(\partial n_{i0}/\partial x)^{-1}$ is the ion density gradient scale length.

The parallel component of the ion fluid velocity perturbation reads

$$\left(\frac{\partial}{\partial t} + \frac{1}{B_0}\vec{e}_z \times \nabla\phi \cdot \nabla\right) v_{iz} = -\frac{e}{m_i}(1+\tau)\frac{\partial \phi}{\partial z}\,, \tag{16}$$

where $\tau = 3n_{e0}T_i/n_{i0}T_e$. Equations (15) and (16) govern the dynamics of coupled electron-drift and dust ion-acoustic waves in a nonuniform dusty magnetoplasma with $n_{e0} \neq n_{i0}$.

In order to study the linear properties of coupled electron-drift and dust ion-acoustic waves, we invoke the standard WKB approximation and assume that the wavelengths of perturbations under consideration are much smaller than $|L_n|$. Supposing that the perturbed quantities ϕ and v_{iz} vary as $\exp[i(\vec{k}\vec{r} - \omega t)]$, we Fourier transform Eqs. (15) and (16) and combine the resultant equations to obtain the dispersion relation

$$\omega^2 + \omega\omega_{e*} - \omega_s^2(1+\tau) = 0\,, \tag{17}$$

where $\omega_{e*} = k_y v_{e*}/(1+k_\perp^2\rho_s^2)$ is the electron drift wave frequency, and $\omega_s = k_z c_s/(1+k_\perp^2\rho_s^2)^{1/2}$ is the dust ion-acoustic wave frequency in a magnetized dusty plasma. The solution of Eq. (17) is

$$\omega = -\frac{\omega_{e*}}{2} \pm \frac{1}{2}\left[\omega_{e*}^2 + 4\omega_s^2(1+\tau)\right]^{1/2}. \tag{18}$$

We now present possible stationary solutions of the nonlinear equations (15) and (16). We introduce a new variable $\xi = y + \alpha z - ut$, where α and u are constants, and assume that the wave potential and the parallel fluid velocity perturbations are functions of x and ξ only. Thus, Eqs. (15) and (16) can be written in the form

$$D_\xi\left(\phi - \nabla_\perp^2\phi\right) - \frac{v_{e*}}{u}\frac{\partial\phi}{\partial\xi} - \frac{n_{i0}}{n_{e0}}\frac{T_e}{e}\,\alpha\,\frac{\partial v_{iz}}{\partial \xi} = 0, \tag{19}$$

and

$$D_\xi v_{iz} = \frac{e\alpha}{m_i u}\,(1+\tau)\,\frac{\partial\phi}{\partial\xi}, \tag{20}$$

where $D_\xi = \partial_\xi - (1/uB_0)(\partial_x\phi\partial_\xi - \partial_\xi\phi\partial_x)$. Equation (20) is satisfied by

$$v_{iz} = \frac{e\alpha}{m_i u}\,(1+\tau)\,\phi, \tag{21}$$

which can be inserted into (19) to obtain

$$D_\xi\left(\nabla_\perp^2\phi + \frac{\delta}{\rho_s^2}\phi\right) = 0\,, \tag{22}$$

where

$$\delta = 1 - \frac{v_{e*}}{u} - \left(\frac{c_s^2\alpha^2}{u^2}\right)^2 (1+\tau)\,. \tag{23}$$

Equation (22) is satisfied by the Ansatz

$$\nabla_\perp^2\phi = C_1\phi + C_2 x\,, \tag{24}$$

where the constants C_1 and C_2 are related by $C_1 + (1/uB_0)C_2 = \delta/\rho_s^2$. It turns out that Eq. (24) admits a dipolar vortex solution, the form of which in the outer region is

$$\phi_{out} = \Phi_m K_1(\lambda_3 r)\cos\theta\,, \tag{25}$$

where Φ_m is a constant and $\lambda_3 = \delta/\rho_s^2$. In the inner region, the solution reads

$$\phi_{in} = [\Phi_i J_1(\lambda_4 r) + C_{2i}\frac{r}{\lambda_4^2}]\cos\theta\,, \tag{26}$$

where Φ_i is a constant, $C_{2i} = (\lambda_3^2+\lambda_4^2)uB_0$, and the constant λ_4 has to be determined from the transcendental equation

$$\frac{K_2(\lambda_3 R)}{\lambda_3 K_1(\lambda_3 R)} = -\frac{J_2(\lambda_4 R)}{\lambda_4 J_1(\lambda_4 R)}\,. \tag{27}$$

Furthermore, when $\delta = 0$ or

$$u = \frac{v_{e*}}{2} \pm \frac{1}{2}\left[v_{e*}^2 + 4c_s^2\alpha^2(1+\tau)\right]^{1/2}\,. \tag{28}$$

We observe that Eq. (22) is solved by the Ansatz [*Petviashvili and Pokhotelov* 1992]

$$\nabla_\perp^2\phi = \frac{4AK^2}{a^2}\exp\left[-\frac{2}{A}(\phi - uB_0 x)\right]\,, \tag{29}$$

where A, K and a are arbitrary constants. The solution of Eq. (29) is given by

$$\phi = uB_0 x + A\ln\left\{2\left[\cosh(Kx) + \left(1-\frac{1}{a^2}\right)\cos(Ky)\right]\right\}\,. \tag{30}$$

For $a^2 > 1$ the vortex profile, given by Eq. (30), resembles the Kelvin-Stuart "cat's eyes" that are chains of vortices.

4 Electromagnetic Vortices

4.1 Governing Equations

Here we present a detailed derivation of the nonlinear equations that describe the electromagnetic perturbations in magnetized dusty plasmas with complex geometries, in the regime when the dust dynamics may be neglected. The special cases of these equations that are of particular practical interest have been analyzed in sections 2 and 3.

We study nonlinear processes with low characteristic frequency (compared to the ion cyclotron and magnetosonic frequencies), whose spatial scale ranges between the ion inertial length and the electron collisionless skin depth. The plasma consists of electrons, singly charged positive ions and dust grains, whose unperturbed densities $n_e^{(0)}(x)$, $n_i^{(0)}(x)$ and $n_d^{(0)}(x)$ are weakly x-dependent, $n_\alpha^{(0)}(x) = n_{\alpha,0} + \delta n_{\alpha,0}(x)$. Here $\alpha = e,\ i,\ d$ and $\delta n_{\alpha,0}(x) \ll n_{\alpha,0} = \text{const}$. The density of dust grains is assumed to be large enough, so that their mutual distance is small compared to the ion gyroradius. In the frequency range of interest the dust grains may be regarded as infinitely heavy point charges. The plasma is immersed in a weakly inhomogeneous magnetic field $\vec{B}^{(0)}(x)$ which is directed predominantly along the z axis, $\vec{B}^{(0)}(x) = \vec{e}_z B_0 + \delta\vec{B}_0(x)$, with $|\delta\vec{B}_0(x)| \ll |B_0| = \text{const}$. In the unperturbed state there also exist sheared electron and ion flows, as well as a sheared magnetic field, all along the y axis.

We consider warm electrons and cold two-dimensional ions, $T_e \gg T_i$ and $v_{z,i} = 0$, where T_e, T_i and $v_{z,i}$ are the electron and ion temperatures, and the z-component of the ion fluid velocity, respectively. In other words, we restrict our analysis to the characteristic wavenumber and frequency regimes in which the ion acoustic and finite ion Larmor radius effects may be neglected. We study electromagnetic perturbations which are elongated along the magnetic field lines, $|\vec{B}\cdot\nabla| \ll |\vec{B}\times\nabla|$, and whose perpendicular (to the external magnetic field $\vec{B}^{(0)}$) wavelength is much shorter than the inhomogeneity scalelengths of the background quantities. Furthermore, in the small β limit [where $\beta \equiv 2n_0T_e/(c^2\epsilon_0 B^2) \ll 1$ is the ratio of the thermodynamic and magnetic pressures], the compressional component of the magnetic field perturbation decouples from the torsional, and it will not be regarded here. Under these conditions, the plasma electrons are appropriately described by the drift-kinetic equation, which fully accounts for the electron kinetic effects associated with their motion along perturbed magnetic field lines. Owing to the small mass of electrons, we disregard their finite Larmor radius and polarization-drift effects, and use the following simple form of the drift-kinetic equation

$$\left(\frac{\partial}{\partial t} + \vec{v}_{GC}\cdot\nabla - \frac{eE_\parallel}{m_e}\frac{\partial}{\partial v_\parallel}\right) f\left(v_\parallel\right) = 0, \tag{31}$$

where $v_\parallel = \vec{v}\cdot\vec{B}/|\vec{B}|$ and $E_\parallel = \vec{E}\cdot\vec{B}/|\vec{B}|$ are the electron velocity, respectively. electric field components parallel to the local magnetic field, $\vec{v}_{GC} = v_\parallel\vec{B}/|\vec{B}| + \vec{E}\times\vec{B}/B^2$

is the velocity of the guiding center, and $f\left(v_\parallel\right)$ is the electron distribution function integrated for the velocity components perpendicular to the local (perturbed) magnetic field. We also assume small deviations of the magnetic field and the number densities from the constant values $B_0\vec{e}_z$ and $n_{\alpha,0}$

$$|\delta\vec{B}| \equiv |\vec{B} - \vec{e}_z B_0| \ll B_0, \tag{32}$$

$$\delta n_\alpha \equiv n_\alpha - n_{\alpha,0} \ll n_{\alpha,0}, \quad \alpha = e,\, i,\, d., \tag{33}$$

and express the electric and magnetic fields in terms of the electrostatic and vector potentials

$$\vec{E} = -\nabla\phi - \vec{e}_z \frac{\partial A_z}{\partial t}, \quad \vec{B} = \vec{e}_z B_0 - \vec{e}_z \times \nabla A_z. \tag{34}$$

Note that in the above equation, the vector potential A_z stays for the sum of the the zero-order sheared magnetic field $\delta\vec{B}_0(x)$ and the torsional perturbation. To the leading order in the small parameters defined in Eqs. (33), (32), the electron drift-kinetic equation simplifies to

$$\left[\frac{\partial}{\partial t} + \frac{1}{B_0}(\vec{e}_z \times \nabla\phi)\cdot\nabla\right] f + \left[\frac{\partial}{\partial z} - \frac{1}{B_0}(\vec{e}_z \times \nabla A_z)\cdot\nabla\right] v_\parallel f$$

$$+\left\{\left[\frac{\partial}{\partial t} + \frac{1}{B_0}(\vec{e}_z \times \nabla\phi)\cdot\nabla\right] A_z + \frac{\partial\phi}{\partial z}\right\} \frac{e}{m_e}\frac{\partial f}{\partial v_\parallel} = 0. \tag{35}$$

The electron fluid phenomena are described by the electron continuity and parallel momentum equations, that can be easily obtained as the first two moments of the drift-kinetic equation (35)

$$\left[\frac{\partial}{\partial t} + \frac{1}{B_0}(\vec{e}_z \times \nabla\phi)\cdot\nabla\right] n_e - \left[\frac{\partial}{\partial z} - \frac{1}{B_0}(\vec{e}_z \times \nabla A_z)\cdot\nabla\right] \frac{j_\parallel}{e} = 0, \tag{36}$$

and

$$\left[\frac{\partial}{\partial t} + \frac{1}{B_0}(\vec{e}_z \times \nabla\phi)\cdot\nabla\right]\left(A_z + \frac{m_e j_\parallel}{e n_{e,0}^2}\right) -$$

$$\left(\vec{e}_z \times \nabla\frac{n_e T_e}{e n_{e,0}}\right)\cdot\nabla A_z + \frac{\partial}{\partial z}\left(\phi - \frac{n_e T_e}{e n_{e,0}}\right) = 0, \tag{37}$$

where the hydrodynamic quantities are defined as

$$n_e = \int_{-\infty}^{\infty} dv_\parallel f\left(v_\parallel\right), \quad j_\parallel = -e\int_{-\infty}^{\infty} dv_\parallel v_\parallel f\left(v_\parallel\right),$$

$$T_e = \frac{m_e}{n_e}\int_{-\infty}^{\infty} dv_\parallel v_\parallel^2 f\left(v_\parallel\right). \tag{38}$$

Assuming the isothermal electrons, $T_e = T_{e,0} = const.$, using the parallel component of the Ampere's law, $j_{\|} = c^2\epsilon_0\nabla_\perp^2 A_z.$, and making the appropriate linear combination, the fluid equations (36), (37) may be rewritten in the following symmetric form

$$\left\{\frac{\partial}{\partial t} \pm v_{T,e}\frac{\partial}{\partial z} + \frac{1}{B_0}\left[\vec{e}_z \times \nabla\left(\phi \pm v_{T,e}A_z\right)\right]\cdot\nabla\right\}$$

$$\left[\left(1 - d^2\nabla_\perp^2\right) v_{T,e}A_z \pm \frac{n_e T_e}{e n_{e,0}}\right] + v_{T,e}\frac{\partial}{\partial z}\left(\phi \pm v_{T,e}A_z\right) = 0, \tag{39}$$

where $d = c/\omega_{p,e}$ is the electron collisionless skin depth, and $\omega_{p,e} = \left[e^2 n_{e,0}/\left(m_e\epsilon_0\right)\right]^{1/2}$ is the electron plasma frequency.

For cold and two-dimensional ions, with the accuracy to the first order in the small parameter $(1/\Omega_i)(d/dt)$, where $\Omega_i = eB_0/m_i$ is the ion cyclotron frequency, the ion hydrodynamic velocity can be expressed from the perpendicular ion momentum equation, as the sum of the $\vec{E} \times \vec{B}$ and polarization drifts

$$\vec{v}_{\perp,i} = \frac{1}{B_0}\left\{\vec{e}_z \times \nabla\phi - \frac{1}{\Omega_i}\left[\frac{\partial}{\partial t} + \frac{1}{B_0}\left(\vec{e}_z \times \nabla\phi\right)\cdot\nabla\right]\nabla\phi\right\}. \tag{40}$$

Substituting the above expression into the ion continuity equation and using Eq. (33), we readily obtain

$$\left[\frac{\partial}{\partial t} + \frac{1}{B_0}\left(\vec{e}_z \times \nabla\phi\right)\cdot\nabla\right]\left(\frac{n_i}{n_{i,0}} - \rho_{s,i}^2\nabla_\perp^2\frac{e\phi}{T_{e,0}}\right) = 0, \tag{41}$$

where $\rho_{s,i} = c_s/\Omega_i$ is the ion inertial length and $c_s = \left(T_{e,0}/m_i\right)^{1/2}$ is the ion sound speed.

The system of equations is closed by the quasineutrality

$$n_i^{(0)}(x) = n_e^{(0)}(x) + z_d n_d^{(0)}(x), \quad \delta n_i = \delta n_e. \tag{42}$$

Here $n_\alpha^{(0)}(x)$, $\alpha = i, e, d$ are respectively the unperturbed number densities of ions, electrons and dust grains, δn_i and δn_e are the perturbations of the ion and electron densities, and z_d is the dust charge.

It is worth noting that Eqs. (36), (37) are identical to the basic equations of Refs. [*Pokhotelov et al.* 1999] and [*Schep et al.* 1994], which were obtained in the fluid approximation, using in the parallel electron momentum equation both the $\vec{E} \times \vec{B}$ and diamagnetic convective terms, along with the Braginskii's collisionless stress tensor contribution. Such approach permitted the authors to study the combined effects of inertia and compressibility on the nonlinear electron dynamics. In the dust-free limit it was demonstrated that these equations are of the non-canonical Hamiltonian type and thus posses a wealth of stationary solutions [*Kuvshinov et al.* 1998].

4.2 Nonlinear Stationary Solutions

In the strongly nonlinear regime studied in this paper, we seek a solution which is time stationary and z-independent in the reference frame which is tilted by a (small) pitch angle θ ($\tan\theta = u_y/u_z$) relative to the z-axes, and moving along the y axes with the velocity u_y, i. e. which depends only on the coordinate x and the quantity $y' \equiv y - u_y t + (u_y/u_z)z$. Then, using

$$\frac{\partial}{\partial t} = -u_y \frac{\partial}{\partial y}, \quad \frac{\partial}{\partial z} = \frac{u_y}{u_z}\frac{\partial}{\partial y}, \tag{43}$$

we may integrate equations (39) and (41) one time, yielding

$$\frac{n_e T_e}{e n_{e,0}} \pm \left(1 - d^2\nabla_\perp^2\right)\Psi = 2\,\mathcal{F}_\pm\left(\Phi \pm \Psi\right), \tag{44}$$

$$\frac{n_i T_e}{e n_{i,0}} - \rho_{s,i}^2 \nabla_\perp^2 \Phi = \mathcal{G}\left(\Phi\right), \tag{45}$$

where we used the notations $\Phi = \phi - B_0 u_y x$, $\Psi = v_{T,e}[A_z - B_0(u_y/u_z)x]$, while $\mathcal{F}_\pm$ and $\mathcal{G}$ are arbitrary functions of their arguments, that need to be determined from the appropriate boundary and continuity conditions. Our problem is further simplified by eliminating the density perturbation δn from Eqs. (44) and (45). Thus we arrive at the following simple equations

$$\left(1 - d^2\nabla_\perp^2\right)\Psi = \mathcal{F}_+\left(\Phi + \Psi\right) - \mathcal{F}_-\left(\Phi - \Psi\right) \tag{46}$$

$$\frac{T_e}{e}\frac{n_e^{(0)} - n_i^{(0)}}{n_{e,0}} + \frac{n_{i,0}}{n_{e,0}}\rho_{s,i}^2\nabla_\perp^2\Phi = \mathcal{F}_+\left(\Phi + \Psi\right) + \mathcal{F}_-\left(\Phi - \Psi\right) - \frac{n_{i,0}}{n_{e,0}}\mathcal{G}\left(\Phi\right), \tag{47}$$

which will be studied in details in the rest of this paper.

4.3 Dipolar Vortex

The dipoles in dusty plasma were sketched in [*Pokhotelov et al.* 1999] and here we present the detailed calculations. Dipolar vortices are obtained in plasmas with linear density profiles,

$$n_i^{(0)}(x) = n_{i,0}\left(1 - \frac{x}{L_i}\right), \quad n_e^{(0)}(x) = n_{e,0}\left(1 - \frac{x}{L_e}\right), \tag{48}$$

in the absence of both the sheared magnetic field, $\delta\vec{B}_0(x) = 0$, and the sheared electron and ion flows. Then, for the solutions that are localized in the x, y plane, $\phi, A_z \to 0$ when $x \to \infty$, the functions $\mathcal{F}_\pm$ and $\mathcal{G}$ asymptotically must be linear

$$\mathcal{F}_\pm\left(\xi\right) = F_{\pm 0} + F_{\pm 1}\cdot\xi, \quad \mathcal{G}\left(\xi\right) = G_0 + G_1\cdot\xi, \tag{49}$$

We will seek the *Ansatz* solution for which the functions $\mathcal{F}_\pm$ and $\mathcal{G}$ are linear on the entire x, y plane, but with different coefficients inside and outside of the arbitrary circle with the radius r_0. This circle is is usually referred to as *the vortex core.* In the outer region, $r > r_0$, the coefficients of the linear functions are determined so that the secularly growing terms in Eqs. (44), (45) cancel out, which yields

$$F_{\pm 1}^{(out)} = \frac{v_* u_z \pm v_{T,e} u_y}{2u_y (u_z \pm v_{T,e})}, \quad G_1^{(out)} = \frac{L_e}{L_i}\frac{v_*}{u_y}, \tag{50}$$

where v_* is the zero-order electron diamagnetic drift velocity, $v_* = T_e/(eB_0 L_e)$. The localized outer solution (i. e. the one that is valid for $r > r_0$) may be expressed in terms of the modified Bessel functions of the first order, K_1, as

$$\phi^{(out)} = \cos\theta \left[\beta_{1,1} K_1(\rho_1 r) + \beta_{1,2} K_1(\rho_2 r)\right], \tag{51}$$

where $\theta = \arctan(y'/x)$ and $r = (x^2 + y'^2)^{1/2}$. The characteristic e-folding numbers $\rho_{1,2}$ are determined from

$$\rho_1^2 + \rho_2^2 = -\frac{1}{\rho_{s,i}^2}\frac{v_*}{u_y}\left(\frac{L_e}{L_i} - \frac{n_{e,0}}{n_{i,0}}\right) - \frac{1}{d^2}\frac{u_z^2 (v_* - u_y)}{u_y \left(u_z^2 - v_{T,e}^2\right)}\left(1 - \frac{n_{e,0}}{n_{i,0}}\frac{c_A^2}{u_z^2}\right), \tag{52}$$

$$\rho_1^2 \rho_2^2 = \frac{1}{d^2 \rho_{s,i}^2}\frac{v_*}{u_y}\left(\frac{L_e}{L_i} - \frac{n_{e,0}}{n_{i,0}}\right)\frac{u_z^2 (v_* - u_y)}{u_y \left(u_z^2 - v_{T,e}^2\right)}, \tag{53}$$

where $c_A \equiv c(\Omega_i/\omega_{p,e})(m_i/m_e)^{1/2}$. It is important to note that in the absence of dust we have $L_e = L_i$ and $n_{e,0} = n_{i,0}$, and thus one of the e-folding numbers must be equal to zero. As the consequence, the electromagnetic dipoles in electron-ion plasmas (i. e. those which are expressed in terms of two linearly independent Bessel functions) behave as $\sim \cos\theta/r$ for $r > r_0$, and thus possess infinite energy. Such perturbation with a large energy content are difficult to create experimentally, and it may be expected that under realistic conditions the electromagnetic dipoles in electron-ion plasmas have short lifetimes. The presence of dust drastically changes the spatial behavior of the dipolar vortices, by breaking the anti-symmetry of the zero-order electron and ion currents, which leads to the localization of the magnetic field perturbations.

Within the vortex core, $r < r_0$, using Eq. (49) we may decouple Eqs. (46), (47), to obtain a simple fourth-order linear equation

$$\left(\nabla_\perp^2 + \kappa_1^2\right)\left(\nabla_\perp^2 + \kappa_2^2\right)(\phi - aB_0 u_y x) = 0. \tag{54}$$

The wavenumbers $\kappa_{1,2}$ and the parameter a are related to the slopes $F_{\pm 1}^{(in)}$ and $G_1^{(in)}$ through

$$\kappa_1^2 + \kappa_2^2 = \frac{n_{e,0}}{\rho_{s,i}^2 n_{i,0}}\left(\frac{n_{i,0}}{n_{e,0}} G_1^{(in)} - F_{+1}^{(in)} - F_{-1}^{(in)}\right) + \frac{1}{d^2}\left(F_{+1}^{(in)} + F_{-1}^{(in)} - 1\right), \tag{55}$$

$$\kappa_1^2\kappa_2^2 = \frac{1}{d^2}\frac{n_{e,0}}{\rho_{s,i}^2 n_{i,0}}\left[\left(F_{+1}^{(in)} + F_{-1}^{(in)} - 1\right)\left(\frac{n_{i,0}}{n_{e,0}}G_1^{(in)} - F_{+1}^{(in)} - F_{-1}^{(in)}\right)\right.$$
$$\left. + \left(F_{+1}^{(in)} - F_{-1}^{(in)}\right)^2\right] \tag{56}$$

$$a = 1 - \frac{1}{d^2}\frac{n_{e,0}}{\rho_{s,i}^2 n_{i,0}}\frac{v_*}{u_y}\left(F_{+1}^{(in)} + F_{-1}^{(in)} - 1\right)\left(\frac{n_{i,0}}{n_{e,0}}\frac{L_e}{L_i} - 1\right). \tag{57}$$

The *inside* solution is then readily found as a linear combination of two Bessel functions of the first order J_1

$$\phi^{(in)} = \cos\theta\left[\alpha_{1,1}J_1(\kappa_1 r) + \alpha_{1,2}J_1(\kappa_2 r) + aB_0u_y r\right], \tag{58}$$

A well-behaved and localized solution is possible also if one of the 'inside' wavenumbers is imaginary, e. g. for $\kappa_2^2 < 0$, when the Bessel function $J_1(\kappa_2 r)$ needs to be substituted with the modified Bessel function $I_1(|\kappa_2|r)$.

At the edge of the vortex core $r = r_0$ we require that the functions $\mathcal{F}_\pm$ and $\mathcal{G}$, potentials ϕ and A_z, and their radial derivatives $(\partial/\partial r)\phi$ and $(\partial/\partial r)A_z$ are continuous. After a close inspection, we find that the function $\mathcal{G}$ must have the same form both out- and inside the core, $G_1^{(out)} = G_1^{(in)}$, while the remaining continuity conditions may be satisfied in two different ways, which correspond to two nonlinear modes.

The dipole of the first kind is obtained when only the $\mathcal{F}_+$ (or $\mathcal{F}_-$) function is part-by-part linear, and consequently its argument $\Phi + \Psi$ (or $\Phi - \Psi$) must be equal to zero for $r = r_0$. After a somewhat lengthy, but straightforward algebra, eliminating the amplitudes $\alpha_{i,j}$ and $\beta_{i,j}$ from the appropriate continuity equations, we obtain the nonlinear dispersion relation $\mathcal{D}_{1,1}(\kappa_1, \kappa_2) = 0$, where

$$\mathcal{D}_{1,1}(\kappa_1, \kappa_2) \equiv$$

$$\begin{vmatrix} \mathcal{K}_{1,1} - \mathcal{K}'_{1,1} & \mathcal{K}_{1,2} - \mathcal{K}'_{1,2} & \mathcal{J}_{1,1} - \mathcal{J}'_{1,1} & \mathcal{J}_{1,2} - \mathcal{J}'_{1,2} \\ q_1\left(\mathcal{K}_{1,1} - \mathcal{K}'_{1,1}\right) & q_2\left(\mathcal{K}_{1,2} - \mathcal{K}'_{1,2}\right) & p_1\left(\mathcal{J}_{1,1} - \mathcal{J}'_{1,1}\right) & p_2\left(\mathcal{J}_{1,2} - \mathcal{J}'_{1,2}\right) \\ (b - q_1 a)\mathcal{K}'_{1,1} & (b - q_2 a)\mathcal{K}'_{1,2} & (b - p_1 a)\mathcal{J}'_{1,1} & (b - p_2 a)\mathcal{J}'_{1,2} \\ w_1(1 + q_1 a)\mathcal{K}_{1,1} & w_1(1 + q_2 a)\mathcal{K}_{1,2} & w_2(1 + p_1 a)\mathcal{J}_{1,1} & w_2(1 + p_2 a)\mathcal{J}_{1,2} \end{vmatrix}. \tag{59}$$

Here we used the notation $\mathcal{J}_{i,j} = J_i(\kappa_j r_0)$, $\mathcal{K}_{i,j} = K_i(\rho_j r_0)$, the primes denote the operator $(r_0/i)(\partial/\partial r_0)$, a is defined in Eq. (57), and

$$p_j = \frac{F_{+1}^{(in)} - F_{-1}^{(in)}}{1 - F_{+1}^{(in)} - F_{-1}^{(in)} + d^2\kappa_j^2}, \quad q_j = \frac{F_{+1}^{(out)} - F_{-1}^{(out)}}{1 - F_{+1}^{(out)} - F_{-1}^{(out)} - d^2\rho_j^2}$$

$$b = \frac{v_{T,e}}{u_z} + (a - 1)\frac{F_{+1}^{(in)} - F_{-1}^{(in)}}{1 - F_{+1}^{(in)} - F_{-1}^{(in)}},$$

$$w_1 = 1 + \frac{v_{T,e}}{u_z} - a - b, \quad w_2 = 1 + \frac{v_{T,e}}{u_z}.$$

This dispersion relation is solved numerically, and the magnetic part of the dipole is shown in Fig. 1 a).

The second nonlinear dipole mode is obtained when both functions $\mathcal{F}_+$ and $\mathcal{F}_-$ are part-by-part linear, and their respective arguments $\Phi + \Psi$ and $\Phi - \Psi$ both reduce to zero at the core edge, $r = r_0$. Following the similar procedure as above, we obtain the dispersion relation $\mathcal{D}_{1,1}(\kappa_1, \kappa_2) = 0$, $\mathcal{D}_{1,2}(\kappa_1, \kappa_2) = 0$ from which the wavenumbers κ_1 and κ_2 may be determined. The determinant $\mathcal{D}_{1,1}$ is defined in Eq. (59), and $\mathcal{D}_{1,2}(\kappa_1, \kappa_2) = 0$, where

$$\mathcal{D}_{1,2}(\kappa_1, \kappa_2) \equiv$$

$$\begin{vmatrix} \mathcal{K}_{1,1} - \mathcal{K}'_{1,1} & \mathcal{K}_{1,2} - \mathcal{K}'_{1,2} & \mathcal{J}_{1,1} - \mathcal{J}'_{1,1} & \mathcal{J}_{1,2} - \mathcal{J}'_{1,2} \\ q_1\left(\mathcal{K}_{1,1} - \mathcal{K}'_{1,1}\right) & q_2\left(\mathcal{K}_{1,2} - \mathcal{K}'_{1,2}\right) & p_1\left(\mathcal{J}_{1,1} - \mathcal{J}'_{1,1}\right) & p_2\left(\mathcal{J}_{1,2} - \mathcal{J}'_{1,2}\right) \\ (b - q_1 a)\,\mathcal{K}'_{1,1} & (b - q_2 a)\,\mathcal{K}'_{1,2} & (b - p_1 a)\,\mathcal{J}'_{1,1} & (b - p_2 a)\,\mathcal{J}'_{1,2} \\ \left(q_1 - \frac{v_{T,e}}{u_z}\right)\mathcal{K}_{1,1} & \left(q_2 - \frac{v_{T,e}}{u_z}\right)\mathcal{K}_{1,2} & 0 & 0 \end{vmatrix}. \quad (60)$$

The corresponding dipole is plotted in Fig. 1 b), with the same values of parameters as in the previous case. It features a better spatial localization, larger amplitude and opposite polarity than the dipole of the first kind.

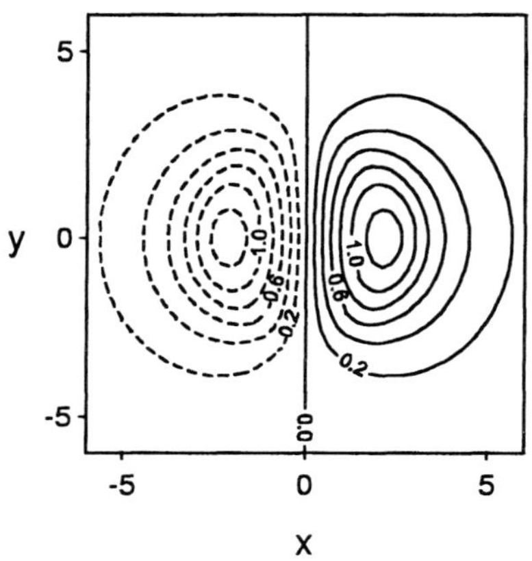

a)

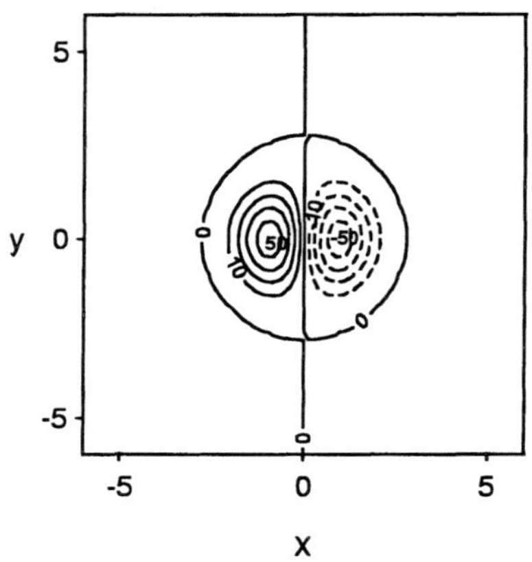

b)

Figure 1: Vector potentials A_z of the dipolar vortex of the first kind, a) and of the second kind, b). The plasma parameters are: $\rho_{s,i}/d = .75$, $n_{i,0}/n_{e,0} = 1.25$, $L_e/L_i = -2$, $v_{T,e}/u_z = .2$, $v_*/u_y = .8$, $r_0/d = 3$, the distance is normalized to d and the vector potential to $B_0 u_y r_0 / v_{T,e}$

4.4 Tripolar Vortex

Tripolar vortices arise in plasmas with parabolic density profiles in the presence of sheared magnetic fields and perpendicular shear flows [*Jovanović et al.* 1998]. Without loss of generality, we solve our equations in the Galilean system in which $\partial/\partial t = \partial/\partial z = 0$, and adopt the origin $x = 0$ so that the plasma background quantities are given by

$$\phi^{(0)}(x) = \frac{T_e}{e}\frac{x^2}{dL_\phi}, \quad v_{T,e}A_z^{(0)}(x) = \frac{T_e}{e}\frac{x^2}{dL_\psi},$$

$$n_\alpha^{(0)}(x) = n_{\alpha,0}\left(1 + \frac{x^2}{dL_\alpha}\right), \quad \alpha = e, i, d. \tag{61}$$

The functions $\mathcal{F}_\pm$ and $\mathcal{G}$ in the equations (46), (47) may be adopted as linear, in the form Eq. (49), only for the parabolic profiles of the background quantities whose maxima (minima) coincide, as in Eq. (61). Otherwise, $\mathcal{F}_\pm$ and $\mathcal{G}$ are asymptotically nonlinear, which prevents the excitation of tripoles, and other types of vortex structures, such as vortex chains, may become dominant instead. In the external region, $r > r_0$, same as before the parameters of the linear functions $\mathcal{F}_\pm$ and $\mathcal{G}$ are determined from the asymptotic considerations, yielding

$$F_{\pm 0}^{(out)} = \mp\frac{T_e}{e}\frac{d}{L_\psi}, \quad F_{\pm 1}^{(out)} = \frac{1}{2}\left(1 \pm \frac{L_\phi}{L_\psi}\right)^{-1}\left(\frac{L_\phi}{L_e} \pm \frac{L_\phi}{L_\psi}\right), \tag{62}$$

$$G_0^{(out)} = -2\frac{T_e}{e}\frac{\rho_{s,i}^2}{dL_\phi}, \quad G_1^{(out)} = \frac{L_\phi}{L_i}, \tag{63}$$

This gives the same e-folding numbers $\rho_{1,2}$ as those calculated for the dipoles, Eqs. (52), (53), if we perform the substitutions $v_{T,e}/u_z \to L_\phi/L_\psi$ and $v_*/u_y \to L_\phi/L_e$. Likewise, within the vortex core, the equations are decoupled as

$$\left(\nabla_\perp^2 + \kappa_1^2\right)\left(\nabla_\perp^2 + \kappa_2^2\right)\left(\phi + a\frac{T_e}{e}\frac{x^2}{dL_\phi} + c\right) = 0, \tag{64}$$

where the wavenumbers $\kappa_{1,2}$ and the parameter a are given by Eqs. (55) - (57), and

$$c = \frac{1}{d^2}\frac{n_{e,0}}{\rho_{s,i}^2 n_{i,0}}\left\{\frac{2T_e}{edL_e}\left[d^2 - \frac{\kappa_1^2+\kappa_2^2}{\kappa_1^2\kappa_2^2}\left(F_{+1}^{(in)} + F_{-1}^{(in)} - 1\right)\right]\right.$$
$$\times\left(1 - \frac{n_{i,0}}{n_{e,0}}\frac{L_e}{L_i}\right)\left(F_{+1}^{(in)} + F_{-1}^{(in)} - 1\right)\left(\frac{n_{i,0}}{n_{e,0}}G_0^{(in)} - F_{+0}^{(in)} - F_{-0}^{(in)}\right)$$
$$\left. + \left(F_{+1}^{(in)} + F_{-1}^{(in)}\right)\left(F_{+0}^{(in)} + F_{-0}^{(in)}\right)\right\}. \tag{65}$$

Now we can easily write down the solution as the sum of the zeroth and second cylindrical harmonics, in the form

$$\phi^{(out)} = \beta_{0,1} K_0 (\rho_1 r) + \beta_{0,2} K_0 (\rho_2 r) + \cos 2\theta \left[\beta_{2,1} K_2 (\rho_1 r) + \beta_{2,2} K_2 (\rho_2 r)\right], \tag{66}$$

$$\phi^{(in)} = \alpha_{0,1} J_0 (\kappa_1 r) + \alpha_{0,2} J_0 (\kappa_2 r) - a \frac{T_e}{e} \frac{x^2}{2dL_\phi}$$

$$-c + \cos 2\theta \left[\alpha_{2,1} J_2 (\kappa_1 r) + \alpha_{2,2} J_2 (\kappa_2 r) - a \frac{T_e}{e} \frac{x^2}{2dL_\phi}\right]. \tag{67}$$

Our solution (66), (67) must satisfy the standard physical boundary conditions, defined in the preceding Subsection. The resulting nonlinear dispersion relation appears to be fully determined by the second cylindrical harmonic only, and same as before, we have $G_{0,1}^{(out)} = G_{0,1}^{(in)}$. After some algebra, we obtain two types of nonlinear dispersion relations which are identical as those for the dipoles, Eqs. (59), (60), in which we introduced the substitutions $v_{T,e}/u_z \to L_\phi/L_\psi$, $v_*/u_y \to L_\phi/L_e$, $\mathcal{K}_{1,j} \to \mathcal{K}_{2,j}$ and $\mathcal{J}_{1,j} \to \mathcal{J}_{2,j}$. The solutions are shown in Figs. 3 and 4, obtained with the same choice of parameters as for the dipoles. Similarly to the dipole case, the solution of the second kind (with both functions $\mathcal{F}_\pm$ being part-by-part linear) has a considerably larger amplitude, smaller spatial extend and opposite polarity from that with only one nonlinear function. However, they may be less stable, due to the more complicated radial dependence.

5 Conclusions

We have studied the formation of coherent nonlinear vortex structures in dusty plasmas, in the characteristic frequency range below the ion cyclotron and magnetosonic frequencies, and above the dust-acoustic and dust charging frequencies, regarding the dust grains as immobile and with constant charges.

Strictly two-dimensional electrostatic perturbations in nonuniform multi-component dusty plasmas in a homogeneous magnetic field were considered in section 2. It has been found that the presence of static dust grains produces aperiodic oscillations associated with the finite frequency convective cell mode (Shukla-Varma mode), whose real part of the frequency is directly proportional to the gradient of the equilibrium dust number density. The nonlinear coupling of weakly interacting finite-frequency convective motions in nonuniform magnetized dusty plasmas is governed by the generalized Navier-Stokes equation. The stationary solution of the latter is represented as coherent dipolar vortex whose translational speed and the radius are arbitrary. In section 3 we have presented governing equations for nonlinearly coupled electron-drift and dust ion-acoustic waves. Stationary solutions of

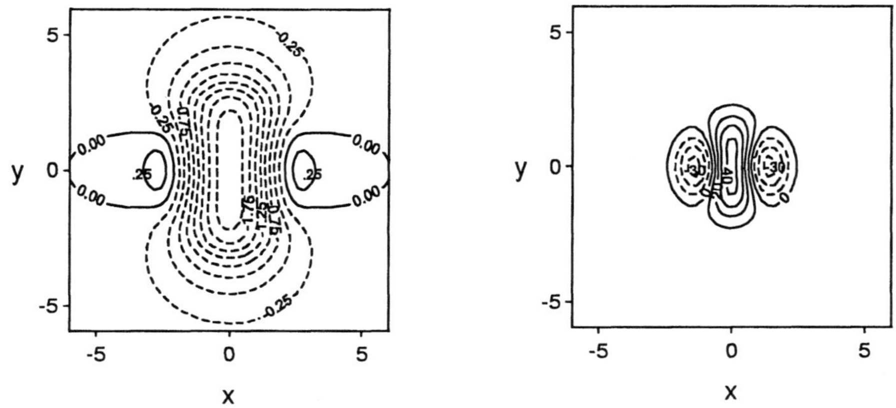

a) b)

Figure 2: Vector potentials A_z of the tripolar vortices of the first kind, a) and of the second kind, b). The plasma parameters are: $L_\phi/L_\psi = 0.2$, $L_\phi/L_e = 0.8$, and the others are same as in Fig. 1.

these equations have been found in the form of a dipolar vortex and a vortex street. The speeds of these coherent structures are found to be well defined in terms of the plasma parameters.

Nonlinear coherent structures were found also for the fully electromagnetic perturbations, for which the plasma fluid velocity is coupled with the torsion of the magnetic field lines. Depending on the profiles of the unperturbed plasma density, and the magnetic and velocity fields, we found two types of vortices. In the absence of magnetic and velocity shears, the solution has the form of a dipole that propagates relative to the surrounding plasma, while in the case of parabolic density profiles a coherent localized nonlinear solution is constructed in the form of a tripole, since stable simple monopolar vortices do not exist. The tripoles exist only for the parabolic profiles of the background quantities whose maxima (minima) coincide, and otherwise different types of vortex structures may become dominant instead, such as vortex chains.

The dust component drastically changes the character of the nonlinear phenomena studied in this paper, compared to their electron-ion plasma counterparts. Namely, in the presence of heavy dust grains, the electron and ion diamagnetic currents do not cancel out, which limits the energy of vortex structures and provides a good spatial localization of the magnetic field perturbation. Consequently, the dusty vortices are expected to have a much better stability and longer lifetime, compared to their ordinary-plasma counterparts.

Acknowledgements

The authors acknowledge the support of the International Institute for Space Science of Bern (Switzerland) for the team "Complex Plasmas in Space: Application to Solar System and Near Earth Space". The research was partially supported also by the European Commission (Brussels) through the contract No. HPRN-CT-2000-00140. One of the authors (D. J.) acknowledges the financial support of the Deutsche Forschungsgemeinschaft for a three months' stay at the University of Bayreuth, F. R. Germany.

References

Jovanović D., Pécseli H. L., Rasmussen J. Juul and Thomsen K. J. 1987, *J. Plasma Phys.* **37**, 81.

Jovanović D., Pegoraro F. and Rasmussen J. Juul 1998, *J. Plasma Phys.* **60**, 383.

Jovanović D. and Pegoraro F. 2000, *Phys. Rev. Lett.* **84**, 95.

Kuvshinov B. N., Lakhin V. P., Pegoraro F. and Schep T. J. 1998, *J. Plasma Phys.* **59**, 727.

Larichev V. D. and Reznik G. M. 1976, *Doklady Akad. Nauk. SSSR* **231**, 1077.

Petviashvili V. I. and Pokhotelov O. A. 1992, *Solitary Waves in the Plasmas and in the Atmosphere* (Gordon and Breach, London).

Pokhotelov O. A., Onischenko O. G., Shukla P. K. and Stenflo L. 1999, *J. Geophys. Res.* **104**, 19797.

Schep T. J., Pegoraro F. and Kuvshinov B. N. 1994, *Phys. Plasmas* **1**, 2843.

Shukla P. K., Yu. M. Yu and Bharuthram R. 1991, *J. Geophys. Res.* **96**, 21.

Shukla P. K. and Varma R. K. 1993, *Phys. Fluids* **5**, 236.

Shukla P. K., Mendis D. A. and Chow V. W. 1996, *The Physics of Dusty Plasmas*, (World Sci., River Edge, N. J.).

Spitzer L. S. 1978, *Physical Processes in the Interstellar Medium*, (Willey–Interscience).

Stenflo L. 1987, *Phys. Fluids* **30**, 3297.

Dust Agglomeration and Gravitation-Like Instabilities in Laboratory and Astrophysical Plasmas

R. Bingham
Rutherford Appleton Laboratory, Didcot, OX11 0QX , U.K.
and
University of Strathclyde, Glasgow, G1 1XQ, Scotland

D. P. Resendes
GOLP, Instituto Superior Técnico,
1096 Lisboa Codex, Portugal

V. N. Tsytovich
General Physics Institute, Russian Academy of Science,
Moscow, Russia

Abstract

Dust growth by agglomeration due to a new mechanism of dust attraction is shown to compete in space plasmas with the standard accretion mechanism by neutral particles. We demonstrate that dust growth by accretion of neutrals and plasma particles is aided by a pressure force due to a shadowing effect first proposed [Bingham *et al* 2001a] to explain rapid growth of dust in laboratory discharges. Consequences of this new force promoting dust growth by agglomeration in astrophysics are discussed including the role of a new gravitational-like instability which produces clumping enhancing the rate of gravitational collapse.

1 Introduction

The subject of dust formation in space and astrophysical plasmas is important in diverse environments such as dust molecular clouds, proto-planetary nebulae, stellar outbursts, and supernova explosions. The formation of dust proceeds the formation of stellar objects and planets. And the ultimate fate of most of the dust is to be consumed in the process of making stars, planets and smaller bodies such as comets and asteroids.

In all these environments the dust particles interact with both neutral and plasma particles as well as with (ultraviolet) radiation and cosmic rays. The conventional view of grain growth is one based on accretion where condensation nuclei containing a small number of atoms formed in the process of adiabatic expansion bound together by the Van der Waals and chemical forces grow by atoms sticking and attaching themselves to the surface [Watson and Salpeter 1972, Spitzer 1978]. This first stage of Brownian motion-driven dust growth was considered in detail both theoretically and numerically [Weidenschilling and Cuzzi 1993, Kempf *et al* 1999, Meaking 1991] and confirmed recently by micro-gravity experiments [Blum *et al* 2000]. This model has also been used to explain the slow growth of dust grains to a size of about $0.1\mu m$ in $10^6 - 10^9$ years. The problem in producing larger dust grains by accretion is due to a number of effects including, the sticking probability which reduces with increasing size, the ambient density and temperature of neutrals and plasma as well as the temperature of the dust grain itself. High grain temperatures prevent further growth. Observations [Winters *et al* 2000] of dust reveal typical sizes of the order of $10\mu m - 1000\mu m$. And even more importantly recent observations [Kempf *et al* 1999] of circumstellar dust shells around long-period variables [Winters *et al* 2000] demonstrate dust formation taking place on a time scale of 10 years. The winds of these C-stars are so effective at producing dust that the circumstellar dust shell totally obscures the central object. Multiple dust shells are also observed forming the circumstellar structure. Given the rate of production and multiple shell like structures we propose a new process of dust formation in astrophysical plasmas namely agglomeration. An important feature of the light curve from supernova SN1987A is a break at about 500 to 600 days coincident with an increase in the infrared flux and an increase in absorption of red-shifted emission lines [Chevalier 1992]. These observations confirm the formation of dust within the supernova eject after about 500 days. The dust in interstellar space can be studied using telescopes. Small particles of solid matter tend to polarize light reflected and scattered by their surfaces. Particles of about a micron in size appear brighter in forward-scattered light than in back-scattered light.

Dust agglomeration has been shown to be important in laboratory etching experiments [Garscadden *et al* 1994] where growth of dust is extremely rapid and is due in the final stages to dust-dust attraction by a plasma or neutral bombardment force known as the shadow force [Tsytovich *et al* 1996a, Tsytovich *et al* 1996b]. Since small grains of dust are attaching themselves together the rate of growth to large sized grains of order 10μm or larger is extremely rapid of order of hours in laboratory experiments and as we will demonstrate it is of order 10 years in C-stars [Bingham *et al* 2001a, Bingham *et al* 2001b].

Although the process of charged dust agglomeration is observed in most plasma etching experiments and inferred to be important in astrophysical plasmas, the experimental and theoretical understanding is poor. The only hypothesis which is proposed to date is the possibility that the dust particles in a plasma can for some reason exist with both positive and negative charge and hence attract each other. In the absence of ionizing radiation dust in a plasma is usually negatively charged due to the greater mobility of the electrons. The possibility of finding dust particles with either charge could occur if the average charge on the dust particles is close to zero and fluctuations will create charges of either sign. Alternatively secondary electron emission by dust particles could create simultaneously dust particles with charges of both signs. This effect appears only for large values of secondary emission yields. In low temperature plasma etching experiments with $T_e < 3\ eV$ secondary emission is small. We will concentrate on an alternative model of agglomeration where secondary emission is unimportant. We also consider the case of like charges with large charges $Q = -Z_D\, e$, where Z_D is the number of charges of order $10^4 - -10^5$. Experimentally charged dust particles are observed to grow to visible sized objects out of a plasma system to sizes of order 100 μm.

The main question that arises is why agglomeration is observed in spite of the presence of large charged dust particles? Here we will show that the flux of electrons, ions and neutrals on the dust particles can create an attractive force between the dust particles which under certain conditions will be larger than the repulsive force due to the Coulomb interaction between them. We restrict the analysis to the simplest case of spherical dust particles of equal radius a with $a \ll d_i$, where d_i is the ion Debye screening radius. The model that we outline was developed by Bingham and Tsytovich [Bingham *et al* 2001b] and is associated with ion and neutral drag forces commonly known as the shadow force [Bingham *et al* 2001b].

This rapid formation of large dust grains through agglomeration can lead to a new type of instability formed as a result of the net attractive force existing between the dust grains. By replacing the force of gravity by this force we derive a linear dispersion relation describing the growth rate of a gravitation like instability leading to clumping of the dust structures with a typical length for collapse smaller than the Jeans length for gravitational collapse. We find that the formation of clump like structures proceeds faster than gravitational collapse. The consequence is that this type of collapse is perhaps necessary before the onset of gravitational collapse resulting in a self similar or scale invariant process describing self-organization of dust structures.

The shadow force that is responsible for this phase of collapse is due to an anisotropic pressure of neutrals and plasma particles on the surface of a dust grain the shadow or shielding by a neighbouring grain. The shielding effect results in a reduction in pressure between the grains forcing them to coalesce or agglomerate. The force produced as a result of the pressure differential is proportional to the square of the ratio of dust size to separation distance. At large separation distances r this force is larger than the Van der Waals force which is proportional to r^{-8}.

In this article we estimate the rate of dust growth due to agglomeration and compare it with the normal accretion model. We will also demonstrate that it can lead to a rapid gravitational like collapse with a length much smaller than the Jeans length for gravitational collapse. The result is the formation of dust clumps.

2 Dust Growth

Our understanding of dust growth has benefited greatly from laboratory experiments [Garscadden *et al* 1994, Boufendi and Bouchoule 1994] carried out within the last 10 years. These experiments have been successfully explained by a combination of a series of stages of dust growth. In the first stage nucleation occurs with the chemical formation of atom clusters this stage is well documented for both laboratory and space scenarios. It is understood that the Van der Waals and chemical binding forces play a key role in forming these nm sized molecular clusters. One advantage in space is that gases from outbursts or winds from stars, supernova explosions or jets are adiabatically cooled to very low temperatures forming atom clusters when the attaching (sticking) coefficient is of the order of the de-attaching (evaporation) coefficient for neutral gas atoms. Gas jet experiments also confirm the formation of atom clusters. In previous calculations of dust growth in space these atom clusters accrete single neutral atoms through collisions. Neutral gas accretion requires dust temperatures of order $10 - 80$K, the dust can easily maintain these temperatures by radiation cooling. In molecular clouds dust at this temperature has been detected, however, for the dust to grow to the size of several microns [Spitzer 1978] takes about $10^6 - 10^9$ years by the process of accretion. This may be satisfactory to explain the observations of dust in molecular clouds but in outbursts, stellar winds and supernova explosions the dust formation is extremely rapid. The main difference between these regions of dust formation and molecular clouds is the presence of plasma which is known to assist in the process of dust growth. Laboratory etching experiments are carried out in plasmas where the temperature can be as high as 3×10^4K and the formation of atom clusters up to 10nm are still observed, these conditions are closer to those of outbursts, stellar winds and supernova explosions than to molecular clouds. The main difference being the absence of 3×10^4K plasma in the molecular clouds (the degree of ionization by cosmic rays or radioactivity is less than 10^{-7}). The exception is at boundaries illuminated by UV radiation resulting in rapid growth in several stages, where the presence of a plasma changes the growth processes considerably.

The second stage observed in laboratory experiments carried out in the presence of a plasma is rapid dust agglomeration up to grain sizes of order $0.1 - 1.0\mu m$. One of the main differences introduced by the plasma is the presence of charge on the grains which normally acquire a negative charge in the absence of UV radiation due to the greater electron mobility. Estimates show that for grains of 10nm in size charges of $10 - 50$ electron charges are possible on isolated grains. At the agglomeration stage the dust density is high which has the effect of decreasing the charge on the grains to about $3 - 20$ electrons. The shadow force produced by the pressure anisotropy can easily overcome the Coulomb repulsion between the grains leading to agglomeration of more and more grains forming larger and lower dust densities. The growth through agglomeration has two effects; to reduce the dust density and increase the charge on the dust to about 10^3 electrons which is large enough to overcome the shadow force through Coulomb repulsion slowing the growth [Boufendi and Bouchoule 1994].

The third stage of growth is the plasma deposition or plasma accretion stage where ions are attracted to the charged dust, increasing the size of dust to about 100μm or larger. During this stage irregular shaped dust soon becomes spherical due to a spherical converging flow of ions. The main difference between neutral accretion and ion accretion or deposition is that the ions are accelerated by the field produced by the charged dust to energies of the order of the electron temperature. The potential of the charged dust grain is of the order of the electron thermal energy in some cases this can be as high as 3×10^4K. This energy which is imparted to the ions causes ion implantation. The ion can be embedded deep within the dust in contrast to neutral particles which only sit on the surface. The presence of the plasma inevitably produces a continuous flux of ions which enhances the growth phase. For low levels of ionization the flux of neutrals can be much larger than the flux of ions. Contrary to ions the neutrals have a very short residence time on the dust unless the temperature of the dust is extremely low of order 10K [Spitzer 1978]. At higher temperatures greater than 100K the accretion of neutrals will not occur but accretion by ions will be rapid. In laboratory plasmas where the ion density is of order $10^9\mathrm{cm}^{-3}$ and electron ion temperature ratio $T_e/T_i \sim 10^2$, the ion deposition time for 50μm sized particles is about 30 minutes i.e. relatively fast, indicating that the ion implantation (bombardment) mechanism should always be considered for these conditions and needs to be considered for astrophysical conditions.

The attractive force for $a \ll d_i, r \ll d_i$ (r is the distance between the dust particles) has the same dependence on distance as the Coulomb repulsive force i.e. $\frac{1}{r^2}$. It has been shown by [Tsytovich *et al* 1996b] that for $r \gg d_i$ the screening $\sim e^{-\frac{r}{d_i}}$ significantly reduces the repulsive force while the attractive force due to bombardment are not affected by screening and dominate. There exists a minimum potential energy for $r \gg d_i$ forming a molecule from two dust particles. The main result of the present article is that for certain conditions the minimum value of r is of the order of a and the attractive force can dominate even for $r \ll d_i$. The only restriction made in the calculation of the attractive force [Tsytovich *et al* 1996b] is that $a \ll d_i$, but r is arbitrary. For $r \ll d_i$ we can neglect the screening of the Coulomb repulsive force and therefore also the effect of non-Debye screening (Which in fact differs from the usual Debye screening due to nonlinearities).

The mechanism of shadow attraction is very simple; the force of shadowing is proportional to the solid angle α^2/r^2 of the shadow and the pressure force $nT\sigma$, where σ is the effective surface area of dust shadowing. For ion flux shadowing it is known to be $1/\tau_i = T_e/T_i$ larger than the geometric cross-section πa^2.

The attractive force and the Coulomb repulsive force can be described by potentials U_a and U_c given by

$$U_c = \frac{Z_d^2 e^2}{r} \ ; \ U_a = -\eta_a \frac{a^2}{d_i^2} \frac{Z_d^2 e^2}{r}, \tag{1}$$

where the coefficient η_a consists of three parts

$$\eta_a = \eta_b + \eta_c + \eta_n \tag{2}$$

η_b is due to direct plasma particle bombardment, η_c is due to plasma particle scattering and η_n is due to neutral particle bombardment.

The nature of the attractive force is due to bombardment by charged plasma particles and neutrals. For a single dust particle in a plasma with an isotropic distribution of particles direct bombardment and deflection transfers no net momentum to an isolated dust particles and therefore no net force acts on a single dust particle. However, in the presence of another dust particle a shadow effect is created and the flux of plasma particles on the dust grains becomes asymmetrical and a net momentum transfer results, this is the origin of the asymmetrical bombardment force which is greater on the outer part of the line between two dust particles than on the inner part resulting in an attractive force. It is interesting to note that an early description of the gravitational forces between two bodies used a similar analogy but instead of a plasma the ether was the medium proposed to provide the bombardment force, the asymmetric absorption of the ether provided the net momentum transfer. The presence of the plasma creates a flux of charged particles which results in dust charging as well as electron and ion recombination on the surface resulting in deposition of plasma material on the dust particle. By using different plasmas and different dust materials particles can be created which could be called "designer dust". The momentum is mainly transferred from the ions because of their greater mass rather than the electrons, and the result depends on the ion attachment coefficient κ_i. A straightforward calculation gives η_b and η_c [Tsytovich *et al* 1996b]

$$\eta_b = \frac{\kappa_i}{2\sqrt{\pi}} \left(\frac{aT_i}{Z_d Z_i e^2}\right)^2 \int_{v_o}^{\infty} \left(1 + \frac{Z_i e|\phi_o|}{T_i y^2}\right)^{\frac{5}{2}} e^{-y^2} y^4 dy. \tag{3}$$

Here $v_o = \frac{a}{d_i}\sqrt{\frac{Z_i e|\phi_o|}{T_i}}$, the averaging is made on the Maxwellian ion distribution function with temperature $T_i, y = v/\sqrt{2}v_{Ti}; v_{T_i} = \sqrt{T_i/m_i}$, the ion charge is $Z_i e (e > 0)$ and $|\phi_o|$ is the floating potential of the dust particle,

$$\eta_c = \frac{\kappa_i}{4\sqrt{\pi}} \int_o^{\infty} \left(1 + \frac{Z_i e|\phi_o|}{T_i y^2}\right) \ell n\,[\beta]\, e^{-y^2} dy, \tag{4}$$

where

$$\beta = \left(d^2 + \frac{Z_d^2 Z_i^2 e^4}{4T_i^2 y^4}\right) \Big/ \left[a^2 \left(1 + \frac{Z_i e|\phi_o|}{T_i}\right) + \frac{Z_d^2 Z_i^2 e^4}{4T_i^2 y^4}\right].$$

For neutrals the simplest model is when the neutrals attach themselves onto the dust particle with attachment coefficient κ_n, and with detachment coefficient κ'_n. The temperature of neutrals before attachment is T_n and after thermalization on the dust surface the neutral temperature is the dust surface temperature of T_{ds}. Then

$$\eta_n = \frac{1}{8\left(\frac{e|\phi_o|}{T_e}\right)^2} \frac{T_n T_i}{T_e^2} \frac{n_n}{n_i} \left(\kappa_n - \kappa'_n \sqrt{\frac{T_{ds}}{T_n}}\right), \tag{5}$$

where n_n is the neutral density, n_i, is the ion density $\frac{n_i}{n_n}$ describes the degree of ionization; η_n is positive for $T_{ds} < T_n \frac{\kappa_n^2}{\kappa_n'^2}$. Attraction only occurs for the case where

$$T_{ds} < T_n \kappa_n^2 / \kappa_n'^2. \tag{6}$$

For $\kappa'_n < \kappa_n$ condition (6) can be satisfied even for $T_{ds} > T_n$. the expression (5) is valid for the case where the time of dust-dust agglomeration is much larger than the resident time of a neutral atom on the surface of the dust particle. In the opposite case $\kappa'_n = 0$ and forces are always attracting.

The expressions for η_a are written in the limit $d_i \ll d_e$, or $T_i \ll T_e$ which is the case in most etching experiments where usually $T_i/T_e \simeq 10^{-2}$. Often in etching experiments T_{ds} is slightly larger than T_n. The heat balance relations at the dust particle surface indicate that the difference between T_n and T_{as} should be rather small and a small excess of κ_n over κ'_n is sufficient to create the attraction.

It is obvious from expression (1) that the attraction is dominant only for $\eta_a \gg 1$ since $a \ll d_i$. From Eq. (3) and (4) η_b and η_c contain the ratio T_i/T_e to some high power and from Eq.(5) the factor n_n/n_i appears which in most experiments is of the order of $10^5 - 10^6$.

We can obtain two separate criteria for dust formation namely $\eta_n \ll \eta_b + \eta_c$ and $\eta_n \gg \eta_b + \eta_c$. For $\eta_n \ll \eta_b + \eta_c$ we have

$$\left(\frac{d_i}{a}\right)^2 > \left(\frac{d_i}{a}\right)^2_{critical} \equiv A^2 = \eta_b + \eta_c \tag{7}$$

and for For $\eta_n \gg \eta_b + \eta_c$ we have

$$\left(\frac{d_i}{a}\right)^2 > \left(\frac{d_i}{a}\right)^2_{critical} \equiv A^2 = \eta_n. \tag{8}$$

For the numerical calculations we introduce three dimensionless parameters assuming $|\phi_o| = \frac{Z_d e}{a}$

$$z = \frac{Z_d e}{a T_e} = \frac{e|\phi_o|}{T_e} \; ; \; \tau = \frac{T_i}{T_e} \; ; \; P = \frac{n_d Z_d}{n_e}. \tag{9}$$

The parameters P enters into the calculation through the quasi-neutrality condition

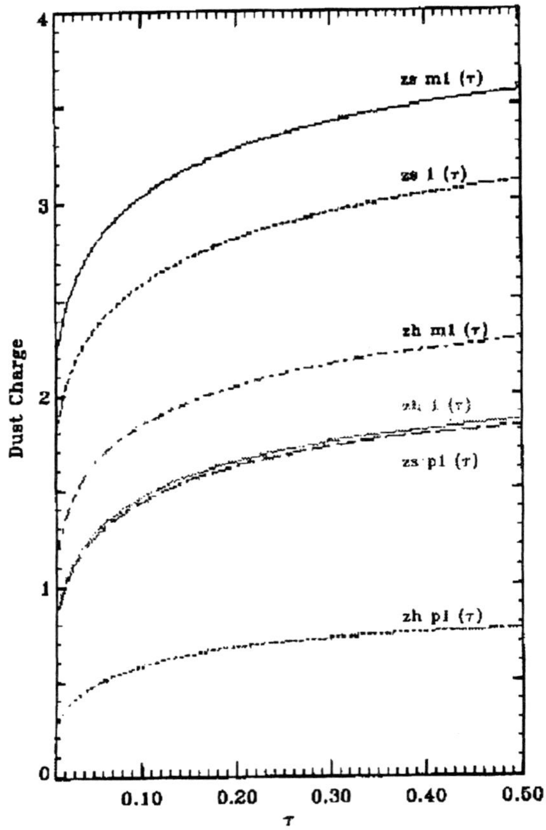

Figure 1: Dependence of dust charges on parameters T and P for Si^+ (three curves labelled with zs), subsequent labels on curves m1, 1, p1 (for example $zsm1(t)$) correspond to $P = 10^{-1}, P = 1, P = 10^1$

$$n_i = n_e + Z_d n_d = n_e(1 + P). \tag{10}$$

The equation for z corresponds to zero plasma particle current to the dust particle with ϕ_o the floating potential

$$\exp(-z) = \frac{1+P}{\sqrt{\mu\tau}}(\tau + z), \tag{11}$$

where $\mu = m_e/m_i$.

Calculations were performed for different ion species (hydrogen, argon, silicon and krypton) all singly charged $Z_d = 1$. Figure 1 shows z as a function of τ for different ions H^+ and Si^+ and for different dust charge states P.

Figure 2a,b shows the result of the critical dust size for agglomeration the value A as a function of τ for difference values of P. Criteria (8) contains a factor inversely

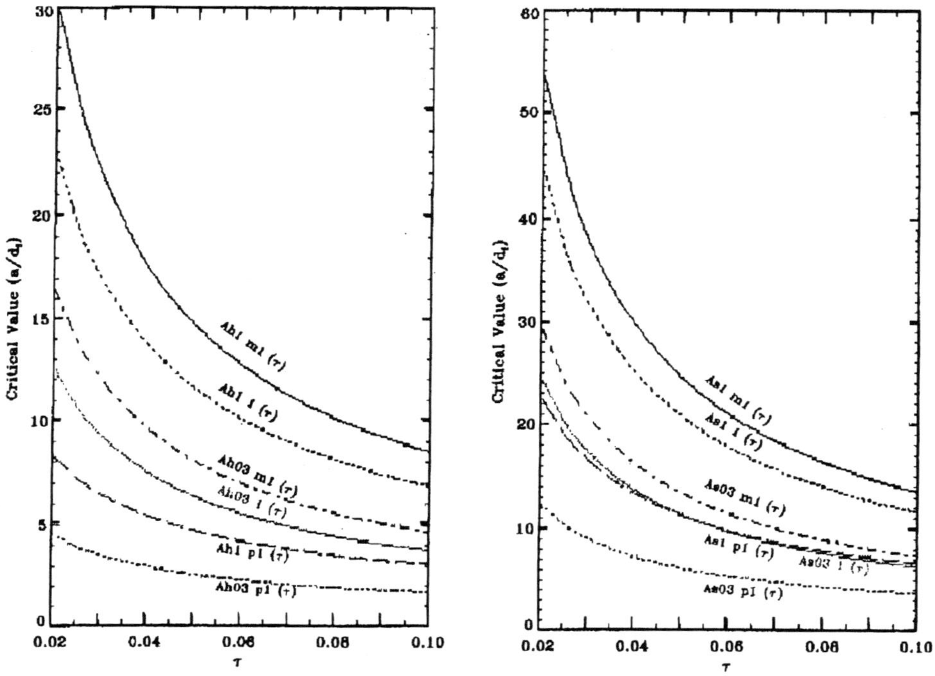

Figure 2: (a) Critical values for $A = a/d_i$ for ion bombardment attraction for H^+ as function of T, P and κ_i lables of the curves start as Ah (A for H^+), the subsequent labels 1, or 03 corresponds to $\kappa_i = 1$ or $\kappa_i = 0.3$ and the last labels m1, 1, p1 correspond to $P = 10^{-1}, 1, 10^1$. (b) The same but for Si^+. The labels differ only by using As for Si^+ instead of Ah for H^+.

proportional to the degree of ionization and is scaled proportional to n_n. In figure 3a,b the critical value A versus T is shown for different degrees of ionization.

All the results demonstrate that the criteria for agglomeration can be easily satisfied in existing experiments and can be used to design specific experiments for dust particle agglomeration.

The dust particles always have a surface temperature less than the neutral gas density for the case where the dust is cooled by radiation which is optically thin. This case corresponds to a situation with specially designed walls of the etching device and is a case often found in astrophysical dust-molecular clouds. Using the thermal balance equation we obtain the difference of the surface dust temperature and the neutral particle temperature.

$$\frac{T_n - T_{ds}}{T_n} = \frac{n_{cr}}{n_n} \frac{T_{ds}^4}{T_n^4}, \tag{12}$$

where the critical neutral density n_{cr} is

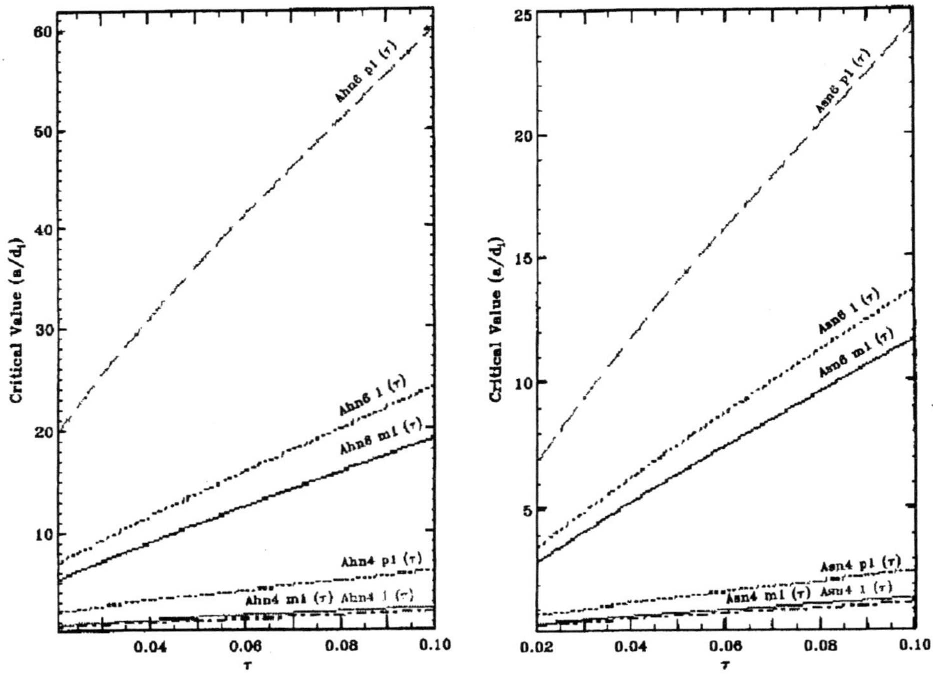

Figure 3: (a) Critical values for A for neutral bombardment attraction for H^+ as a function of T, $\kappa_n \frac{n_n}{n_i}$ (attached coefficient divided on degree of ionization) and P. Labels start with Ah (A for H^+), subsequent $n6, n4$ correspond to $\kappa_n \frac{n_n}{n_i} = 10^6$, $\kappa_n \frac{n_n}{n_i} = 10^4$ and the last labels m1, 1, p1 correspond to $P = 10^{-1}, 1, 10^1$. (b) The same but for Si^+. The labels As are used for A value of Si^+.

$$n_{or} = \mu\sigma\sqrt{\frac{\pi}{2} m_n T_n^5}. \tag{13}$$

Expressing the temperature in degrees Kelvin (K), σ in (13) has the value of the Stefan-Boltzmann constant $\sigma_{SB} = 5.67 \times 10^{-5} erg/cm^2 sec\,(K)^4$ divided by the three halves power of the Boltzmann constant $\kappa_B^{3/2}$, where $\kappa_B = 1.38 \times 10^{-16} erg/(K)$. Thus

$$\sigma = 3.9 \times 10^{19} cm^{-2} (K)^{-5/2} sec^{-1} erg^{-\frac{1}{2}}. \tag{14}$$

The difference between the temperatures is largest in the case where the neutral particle density is much less than the critical value and reaches a saturation value for the case where the neutral density becomes much larger then the critical value. In all cases the dust surface temperature is less than the temperature of neutrals and the force is attractive, but in the case where the density of neutrals becomes larger than the critical density the difference between temperatures becomes small and decreases inversely proportional to the neutral density. This leads to the phenomenon of saturation of the attraction forces which reach a value independent of the neutral density ($n \gg n_{cr}$) given by

$$F_a \to Fv_a = -1.178\frac{a^4}{r^2}T_n n_{cr}. \tag{15}$$

The expression for n_{cr} can be written in a form useful for practical estimations in etching devices:

$$n_{cr}^{14} = 0.832 \times \sqrt{A(T_n^{293})^5}, \tag{16}$$

where n_{cr}^{14} is the value in units $10^{14}cm^{-3}, T_n^{293}$ is the neutral particle temperature in degrees K in units of room temperature (293 K or 20 degrees Celsius). The numerical factor in (16) depends on the atomic number A. For silicon (Si) it is equal to 4.2, for argon (Ar) it is equal to 5.2, and for krypton (Kr) it is equal to 7.6. For the usual experimental conditions, the electron density is of the order of $2 \times 10^9 cm^{-3}$. For a degree of ionization of 10^6 we get $n_n = 2 \times 10^{15} cm^{-3}$ and this value is larger than n_{cr} for Ar, Kr, Si. This means that the radiative processes play an important role.

We use the equation of equilibrium for the dust charge for the case where the dust is charged by the plasma currents. Calculations then show that the charge is usually proportional to the size of the dust particle a. But for small dust sizes where agglomeration starts, the charge can be proportional to $\sqrt{a}$ as we will show later. In that case, the growth due to agglomeration exceeds the increase in equilibrium charge provided

$$P \equiv Z_d n_d \, / n_e \gg 1. \tag{17}$$

In most etching experiments $Z_d n_d / n_e \approx 10 - 30$ and thus (17) is fulfilled. The equilibrium dust charge depends on the actual local electron density, since negatively charged dust particles receive their charges from the surrounding electrons. For high dust density the electrons are depleted, resulting in the inequality (17). This depends on the rate of ionization, i.e., the production of new electrons, which finally recombine with ions on the dust particles. For $P \gg 1$ the average number of negative charges on dust particles exceeds the free electron charges in the volume.

As the dust particles grow by agglomeration, their number density decreases and correspondingly the density of free electrons can increase, the parameter P then decreases. There could exist two regimes where agglomeration takes place: (1) $z \ll 1$ and (2) z of the order of 1-2 where $z = Z_d e^2/aT_e$ is the dimensionless dust charge. When z is of the order of 1-2 the potential on the grain is the floating potential. The standard equation for the equilibrium charge has the form:

$$\exp(-z) = (1+P)\sqrt{\frac{m_e T_e}{m_i T_i}}\left(\frac{T_i}{T_e} + z\right). \tag{18}$$

Let us consider the possibility of agglomeration in both cases and show that in both cases, i.e., (1) small dust charges $z \ll 1$ and large dust densities $P \gg 1$, and (2) floating potential dust charges, $z \approx 1, P \approx 1$, agglomeration related to the dust cooling is possible. In both cases the agglomeration starts for dust particles exceeding some critical size ($a_{cr,1}$ for the small dust charges and $a_{cr,2}$ for floating potential dust charge). This means that there are two stages of agglomeration, in between which the dust particles can grow only by material deposition on the grains (condensation).

For the small charge case we find from (18):

$$z^2 = \frac{e^2}{aT_e}\sqrt{\frac{T_i m_i}{T_e M_e}\frac{n_e}{n_d}}, \tag{19}$$

from which it follows that $Z_d \propto \sqrt{a}$. Solution (19) is obtained for $T_i/T_e \ll z \ll 1$. Relation (17) is always satisfied if $z \ll 1$ for $\sqrt{m_i T_i/m_e T_e} \gg 1$; the relation $z \ll 1$ means

$$\frac{n_e}{n_d} \ll \frac{aT_e}{e^2}\sqrt{\frac{Pm_e T_e}{m_i T_i}}. \tag{20}$$

For example in those experiments [Boufendi and Bouchoule 1994] where detailed observations of the process of dust agglomeration for small dust charges was made, the condition (19) is satisfied.

We can introduce a "minimal" size of dust particles a_0 by the relation

$$a_0 = \frac{e^2}{T_e}. \tag{21}$$

This may be viewed as the size of a singly charged particle, whose "potential" e^2/a_0 equals the electron kinetic energy.

For typical experimental values $T_e \approx 3eV$ and $a_0 \approx 0.0001\mu m$ which is much less than the parameter range where the agglomeration of dust particles starts to be observed, $a \simeq 0.01\mu m$. We use (21) to write relation (15) in the form

$$\frac{n_e}{n_d} \ll \frac{a_o}{a}\sqrt{\frac{m_e T_e}{m_i T_i}}. \tag{22}$$

The relation showing that the dust particle carries more than one electron charge will contain the inverse ratio a_0/a:

$$\frac{n_e}{n_d} > \frac{a_0}{a}\sqrt{\frac{m_e T_e}{m_i T_i}}, \tag{23}$$

where $a \gg a_0$ there exists a broad range of dust densities where the dust particles should still have charges much larger than unit electron charge. The physical problem is why do they agglomerate in spite of their large electron repulsion? The answer we propose is that the attraction due to dust neutral atom bombardment can exceed the Coulomb repulsion. This condition gives a critical value of the dust size $a_{cr,1}$, which depends on dust densities. The attraction force increases rapidly with the size of the dust particles, which means that as soon as the size of the dust particles exceeds the critial value, the dust particles will agglomerate until condition $z \ll 1$ is violated or the ratio of electron to dust density increases.

The condition for agglomeration of dust particles with small charges is of the form

$$a > a_{cr,1} \tag{24}$$

and

$$n_{cr}a_{cr,1}^3 = \frac{N_e}{n_d}\frac{T_e}{T_n}\sqrt{\frac{m_i T_i}{m_e T_e}}. \tag{25}$$

Relation (23) can be considered as determining the size of dust particles for a given ratio of electron to dust densities or the critical ratio of densities for given sizes. The condition that the dust particles with large charges will agglomerate can be found by comparison of (24) with (21):

$$\frac{T_n}{T_e}n_{cr}a^3 > \frac{a_0}{a}. \tag{26}$$

The actual value of n_{cr} depends strongly on T_n and the small changes in T_n can substantially change $a_{cr,1}$. To avoid this first stage of agglomeration one can increase the neutral temperature to move the agglomeration to larger sized particles, or increase the electron density by additional external ionization. The exclusion of the regime $z \ll 1$ can be reached by violating condition (21). One can also operate in the nonsaturated regime $n_n \ll n_{cr}$ and decrease the attraction by decreasing n_n. There are also other possibilities to change the agglomeration, e.g., by having absorbing or totally reflecting walls, by having the thermal radiation of dust particles return to the plasma volume and by decreasing the radiative losses. In experiments condition $z \ll 1$ cannot be maintained as the particles are growing. The next stage should be the dust growth by plasma material deposition. the presence of this stage is manifested by the observations that irregular shaped particles embedded in the discharge grow in size and can soon become almost spherical. The observations of [Boufendi and Bouchoule 1994] show that the first stage of agglomeration is seen at $a \approx 0.07 - 0.1\mu m$ when deposition stage takes place.

No detailed experimental investigations of time sequences of the second stage of agglomeration exist, but it is certainly known that this stage exists in many experiments and that the dust particles can grow up to sizes which can be detected by eye — they are usually in the shape of "cauliflowers". The present model is able to explain both stages.

For z of the order of 1-2 we get the condition for agglomeration due to neutral atom bombardment in the form

$$a_{cr,2}^2 a_0 n_{cr} = z^2\frac{T_e}{T_n}. \tag{27}$$

For the conditions of the experiment the second stage of agglomeration starts at $a > 1\mu m$.

The present model of dust agglomeration deals with the non-collective shadow effect of two grains. The collective shadow attraction due to the presence of a collection of dust grains is of the same order of magnitude although rapidly increasing with P.

Dust agglomeration can be increased or suppressed by introducing three main experimental processes

1. A neutral gas jet which can be cooled or heated as required. This can raise or lower the ratios T_{ds}/T_n and n_n/n_i which are necessary to control specific operations.
2. Substrate heating plate controls the temperature of the etched material.
3. A temperature controlled wall and radiation absorption material surrounding the plasma processing volume. this is used to heat or cool the dust particles, providing the optimum conditions for dust fabrication.

3 Physical Explanation of the Shadow Force

The shadow force is the main force which acts in the fourth stage of dust growth. The shadow force due to neutrals and ions caused by dust is given by a simple expression which can easily be obtained qualitatively from Eq. (1):

$$F = \frac{a^2}{r^2} nT\pi a^2 A, \tag{28}$$

where a^2/r^2 is the solid angle, a is the dust size and r is the distance separating the dust particles, nT is the pressure of either neutrals or ions we can neglect electrons because of their smaller mass and A is a numerical coefficient of order of unity determined by the coefficients defined in Eq (2). In the calculation we have assumed that the dust particles are of equal size.

The physics of the shadow force is very simple. For a single dust particle the flux of absorbed charged and neutral particles on the dust surface is symmetric resulting in no net momentum transfer. Another dust particle at distance r shadows the flux to the first dust particle with a solid angle $\approx a^2/r^2$. The net momentum transfer will be proportional to the solid angle, to the surface area of the dust particle πa^2 and to the pressure nT as written in Eq. (28). The numerical coefficient A depends on whether the distance between the dust particle is less or larger than the mean free path of dust/neutral or dust/ion collisons being in both cases of the order of 1 [Khodataev *et al* 2001]. In the case where the distance between dust particles is larger than the mean free path the physics of the force is slightly different, one dust particle creates a temperature gradient in the neutral gas component and another dust particle is attracted by the thermophoretic force due to this temperature gradient [Khodataev *et al* 2001]. The attraction exists if the dust surface temperature is less than the temperature of the neutral gas, which is almost always the case in optically thin dust clouds where the dust surface is cooled by radiation losses. The energy flux of ions on dust particles which heats the dust surface is usually less than the radiation losses by more than 15 orders of magnitude and the heating by other types of radiation is less than the cooling for optically thin clouds. The shadow force is inversely proportional to r^2 as for gravity and Coulomb forces, it depends on the neutral n_n or ion density n_i and their temperature T.

The four stages of dust growth observed in laboratory experiments in the presence of plasma will also be important in space plasmas. For example in carbon rich star outbursts [Winters *et al* 2000] dust formation takes about 10 years in regions where there is a mixture of neutrals and plasma at relatively high temperatures. The gas density in these stellar outbursts is as high as $10^{10}\mathrm{cm}^{-3}$ with temperatures large enough that ionization is also present. The dust shells in these carbon rich stars form within 5 stellar radii in regions where the temperature is about 1000K and it is expected that the four stages of dust formation should proceed including ion implantation and agglomeration. In a gas or plasma the forces transmitted to the dust grains are the random forces transferred to the small particles during Brownian motion. If we bias the collisions so that they are not isotropic for example through shielding, results in the shadow force given by Eq.(28).

In previous studies of dust growth in space, Eq.(28) is not considered only the accretion of neutrals is assumed and isotropic collisions with dust grains results in no net force between the grains. This is also true if the neutral atom attaches itself and then de-attaches with the same temperature. If the neutral atom during its contact with the surface has sufficient time to thermalize to the dust temperature and is emitted with the energy corresponding to the dust temperature then the factor $\left(1 - \sqrt{\frac{T_\mathrm{ds}}{T_\mathrm{n}}}\right)$ [Khodataev *et al* 2001] appears in the expression for the force given by Eq.(28), where T_ds is the dust surface temperature and T_n is the neutral temperature. The force vanishes for $T_\mathrm{ds} = T_\mathrm{n}$ and is largest for $T_\mathrm{ds} \ll T_\mathrm{n}$ a condition often found in space where the dust cools by emission of radiation at the rate $\pi a^2 \sigma_\mathrm{SB} T_\mathrm{ds}^4$, where σ_SB is Stefan-Boltzmann constant. In this case the dust cloud is optically thin for radiation from dust a situation found in dust-molecular clouds. In [Khodataev *et al* 2001] it was shown that in an optically thin dust cloud the dust temperature is always smaller than the gas temperature, the forces between dust particles due to gas particle bombardment are always attractive and that with an increase of the gas density the attraction force "saturate" at densities of about $10^{16}\mathrm{cm}^{-3}$ much larger than in any dust-molecular cloud.

The force given by Eq.(28) operates in the presence or absence of plasmas. In the presence of plasma the force due to the ion and neutral flux should be compared to the Coulomb force F_c between two dust grains given by

$$F_c = \frac{Z_d^2 e^2}{r^2} = \left(\frac{Z_d e^2}{aT_e}\right)^2 \frac{n_i T_e^2 a^2}{e^2 n_i r^2} = z^2 \frac{a^2}{r^2} n_i T_e 4\pi d_e^2, \tag{29}$$

where $z = \frac{Z_d e^2}{aT_e}$ is the dimensionless dust charge of order 1-2 and $d_e = \sqrt{\frac{T_e}{4\pi n_e e^2}}$ is the electron Debye radius. Comparison of Eqs.(28)-(29) demonstrates that the bombardment force due to ions is $\frac{4d_e^2 z^2}{a^2}$ less than the Coulomb repulsive force for $r \ll d_e$. For $r \gg d_e$ the Coulomb force is screened while the attractive shadow force is not, this can lead to a contraction of the dust cloud until the inter-dust distances are comparable to 10 times the Debye radius. For distances larger than the Debye radius the ion accretion force ([Tsytovich *et al* 1996a][Tsytovich *et al* 1996b]) can dominate Coulomb repulsion and should be added to the attraction due to gas particle bombardment but this contribution is usually small. For outbursts of carbon rich stars where $n_n \simeq 10^{10}\text{cm}^{-3}, n_i$ is of order n_n and $T_e \sim 1eV$, resulting in a Debye radius of between 100 and 1000μm, attraction can occur between two equally charged grains. Using $r \sim 4d_e$ where the repulsive force can be overcome by the attractive force forming an attractive potential well, with the potential given by

$$V \simeq F_c r \approx z^2 n_i d_e a^2 T_e. \tag{30}$$

We find that a "dust molecule" is formed if

$$T_d < T_e n_i d_e a^2 z^2, \tag{31}$$

where T_d is the dust kinetic temperature not the surface temperature. In environments where dust is cooled by emission of radiation condition (31) can easily be satisfied. For $n_i \sim 10^{10}\text{cm}^{-3}, d_e \sim 10^{-2}\text{cm}$ i.e. $r \sim 10^3\mu\text{m}$ and $a \sim 1\mu\text{m}$ Eq.(31) results in $T_d < T_e z^2 \sim 10T_e$ which is easily satisfied in carbon rich star outbursts [Winters *et al* 2000].

Laboratory experiments [Thomas and Morfill 1996] demonstrate the possibility of producing regular dust structures known as dust crystals which occur for $Z_d e^2 n_d^{\frac{1}{3}} > 170T_d$. In space this condition is more difficult to satisfy namely

$$T_d > T_e n_i a d_e^2 z^2 \left(n_d^{\frac{1}{3}} a\right)/170, \tag{32}$$

but it may be possible which has lead to suggestions that dust crystals could exist in space environments [Morfill 2000]. For $T_i \ll T_e$ numerical results [Bingham *et al* 2001a] show that the attraction force Eq.(28) can dominate the Coulomb force for $a > (8-60)\, d_i$ thus leading to dust agglomeration, d_i is the ion Debye radius.

In dust molecular clouds d_e is very large and the degree of ionization is very low, however, cosmic rays produce ionization such that $\frac{n_i}{n_d} \sim 10^{-6} - 10^{-8}$ which is only about one order of magnitude less than in laboratory experiments. In most laboratory experiments the neutral and ion shadow forces are of the same order of magnitude and give rise to the second and the fourth stages of agglomeration. In dust-molecular clouds if the dust is not radioactive the main attraction force is determined by the neutral flux and Eq.(28) can be used if the dust surface temperature is lower than the gas temperature. The phenomenon of dust agglomeration due to neutral particle flux bombardment was not previously considered for dust-molecular clouds. It can operate effectively for a range of temperatures above the critical temperature range of 10 – 80K for gas accretion. At larger temperatures this will serve as the main mechanism responsible for dust growth. An estimate of the agglomeration time is found by taking into account the relative number of particles in phase space with energies less than the attractive potential well

$$\frac{dm_{\mathrm{d}}}{dt} = 2m_{\mathrm{d}}v_{\mathrm{d}}n_{\mathrm{d}}^{\frac{1}{3}}\left(\frac{V}{T_{\mathrm{d}}}\right)^{\frac{3}{2}}; \quad v_{\mathrm{d}} = \sqrt{\frac{T_{\mathrm{d}}}{m_{\mathrm{d}}}}. \tag{33}$$

Since the force acting is in the direction separating the two dust particles we can use for the time scale the time for the dust to travel the inter-dust distance which is a direct consequence of (33) given by

$$\frac{1}{\tau_{\mathrm{aggl.}}} \simeq \sqrt{\frac{T_{\mathrm{d}}}{m_{\mathrm{d}}}}n_{\mathrm{d}}^{\frac{1}{3}}\left(\frac{T_{\mathrm{e}}n_{\mathrm{i}}a^2d_{\mathrm{e}}z^2}{T_{\mathrm{d}}}\right)^{3/2} \tag{34}$$

for $a \sim 1\mu$m, $T_{\mathrm{d}} \sim 100$K, $n_{\mathrm{d}} \sim 10^{-3}$cm^{-3} we estimate $\tau_{\mathrm{aggl.}}$ to be about 10^9 seconds $\simeq 30$ years.

4 Gravitational Like Instability

The shadow force also creates a gravitational like instability. The dispersion relation for the agglomeration instability for the dust component can easily be written using the dispersion relation for the Jeans instability by replacing the force of gravity by the shadow force resulting in

$$\omega^2 = k^2v_{\mathrm{s}}^2 - G_{\mathrm{eff}}n_{\mathrm{d}}m_{\mathrm{d}}, \tag{35}$$

where according to (28)

$$G_{\mathrm{eff}} = \frac{nT_{\mathrm{n}}\pi a^4}{m_{\mathrm{d}}^2} \tag{36}$$

and

$$v_{\mathrm{s}}^2 = \frac{n_{\mathrm{d}}T_{\mathrm{d}} + n_{\mathrm{n}}T_{\mathrm{n}}}{n_{\mathrm{d}}m_{\mathrm{d}} + n_{\mathrm{n}}m_{\mathrm{n}}} \sim \frac{n_{\mathrm{n}}T_{\mathrm{n}}}{n_{\mathrm{d}}m_{\mathrm{d}} + n_{\mathrm{n}}m_{\mathrm{n}}} \tag{37}$$

if $n_{\mathrm{n}}T_{\mathrm{n}} \gg n_{\mathrm{d}}T_{\mathrm{d}}$. Although the attraction forces are acting only on dust particles the expression for v_{s} will correspond to the sound speed for frequencies much less than the dust-neutral collision frequency [Tsytovich 1997].

The corresponding Jeans type length is determined by the relation

$$L = \frac{2\pi v_{\rm s}}{\sqrt{G_{\rm eff} n_{\rm d} m_{\rm d}}} \simeq \frac{2\sqrt{\pi}}{n_{\rm d} a^2 \sqrt{(1 + n_{\rm n} m_{\rm n}/n_{\rm d} m_{\rm d})}}. \tag{38}$$

For $a \sim 1\mu$m, $n_{\rm d} \sim 10^{-3}$cm^{-3} we have $L \sim 10^{11}$cm, resulting in dust clumps of order $10^{11} - 10^{12}$cm in size.

These clumps are much smaller than the Jeans length for gravitational collapse and will proceed stellar and planetary formation. It is envisaged that clumping could proceed through this mechanism forming larger and larger scales all the way to the Jeans length scales, at each scale the structures are similar and scale invariant. These structures are a result of self-organization found in complex physical systems, each scale could collapse forming structures from asteroids in size all the way up to star formation which could be enhanced by additional compression produced by shock waves. The characteristic time for the Jeans like instability to develop is given by $\tau_{\rm ag.inst}$ i.e. the agglomeration instability time scale

$$\tau_{\rm ag.inst} \simeq \sqrt{\frac{m_{\rm d}}{n_{\rm n} T_{\rm n} \pi a^4 n_{\rm d}}}, \tag{39}$$

which for $n_{\rm d} \sim 10^{-3}{\rm cm}^{-3}, n_{\rm n} \sim 10^3, T_{\rm n} \sim 50{\rm K}, a \sim 1 - 3\mu{\rm m}$ is

$$\tau_{\rm ag.inst} \sim 10 - 100 \text{ years}. \tag{40}$$

This is 3 orders of magnitude faster than the Jeans gravitational collapse time and the Jeans length is 10^3 orders of magnitude larger. The result is a clumpy structure which is often observed in dust-molecular clouds.

5 Conclusions

Recent experiments [Blum *et al* 2000] demonstrated dust agglomeration under micro-gravity conditions. It should be emphasized that the terminology for dust growth varies from paper to paper. In this paper we discuss growth of large dust grains in a high density regime whereas in [Blum *et al* 2000] the growth is in a lower density regime where the term agglomeration is used for the case where Van der Waals force at small distances overcomes the repulsion causing the sticking of dust particles. At higher densities our effect dominates through Eq.(28) which is larger than the Van der Waals force which varies as $1/r^8$. The agglomeration discussed in [Blum *et al* 2000] is in our terminology the first chemical stage of dust clumping forming clumps containing many molecules (second stage).

We emphasized here the possibility of the existence of a fourth stage similar to that in laboratory experiments, even in a completely neutral gas the shadow force operates producing dust particles of the size 10μm or larger. In many cases the presence of ionization can also be important for dust growth. There is the possibility to check our mechanism in micro-gravity experiments. The experiments already performed deal with very low neutral gas pressure where according to our estimate the potential V given by Eq.(30) is less than that due to the Van der Waals force. To check our results it is desirable to work with larger sized dust particles and with a larger neutral gas pressure. In present experiments the attraction force Eq.(28) can help dust clumping at large distances, where the force given by Eq.(28) over-comes the Van der Waals force.

In this article we have described a new force of attraction giving rise to rapid dust agglomeration in astrophysical plasmas, the rate of dust formation is much more rapid than that due to the accretion of neutral particles. This new force arises due to a shadowing of the dust particles in either a neutral or plasma atmosphere. The force of attraction that results from the shadow effect gives rise to the formation of smaller scale structurs on faster time scales than that due to the standard Jeans instability. The structures that are then formed could easily be responsible for the formation of planetesimals.

Acknowledgements

This research was supported by the European Commission through contract No. HPRN-CT-2000-00140 for carrying out the task of the Human Potential Research Training Networks "Complex Plasmas: The Science of Laboratory Colloidal Plasmas and Mesospheric Charged Aerosols" as well as the International Space Science Institute, Bern, for the international team "Complex Plasmas in Space: Application to Solar System and Near Earth Space".

References

Bingham R. and Tsytovich V.N. 2001, Astron. and Astrophys. **376**, L43 – L47.

Bingham R. and Tsytovich V.N. 2001, IEEE Trans. Plasma Sci. **29**, 158.

Blum J. et al. 2000, Phys. Rev. Lett. **85**, 2426.

Boufendi L. and Bouchoule A. 1994, Plasma Sources Sci. Technology **3**, 62.

Chevalier R.A. 1992, Nature **355**, 691.

Garscadden A., Ganguly B.N., Healand P.D. and Williams J. 1994, Plasma Sources Sci. Technology **3**, 239.

Kempf S., Pfalzner S. and Henning T. 1999, Icarus **141**, 388.

Khodataev Ya.K., Morfill G.E., and Tsytovich V.N. 2001, J. Plasma Phys., **65**, 257.

Meaking P. 1991, Rev. Geophys. **29**, 317.

Morfill G.E. 2000, Private communication.

Spitzer L. Jr. 1978, Physical Processes in the Interstellar Medium, John Wiley and Sons, New York.

Thomas H.M. and Morfill G.E. 1996, Nature **379**, 806.

Tsytovich V.N., Khodataev Y. and Bingham R. 1996a, Comments on Plasma Physics and Controlled Fusion **17**, 249.

Tsytovich V.N., Khodataev Y., Bingham R. and Tarakanov V. 1996b, in Advances in Dusty Plasma, World Scientific, Singapore, eds. P.K.Shukla, D.A. Mendis and T. Desai, page 212.

Tsytovich, V.N. 1997, Physics Uspekhi **40**, 53.

Watson W.D. and Salpeter E.E. 1972, Ap.J. **174**, 321.

Weidenschilling S.J. and Cuzzi J.N. 1993, Protostars and Planets III, edited by E Levy and J I Lunine (Univ. Arizona Press, Tucson 1993) page 1031.

Winters J.M., Keady J.J., Gauger A. and Sada P.V. 2000, Astron. Astrophys. **359**, 651.

Some Exotic Phenomena in Dusty Plasmas

J. T. Mendonça[1], P. K. Shukla[2], R. Bingham[3],
D. P. Resendes[1] and A. Guerreiro[1]

[1]GoLP, Instituto Superior Técnico,
1049-001 Lisboa, Portugal

[2] Institut für Theoretische Physik IV,
Fakultät für Physik und Astronomie, Ruhr-Universität Bochum,
D-44780 Bochum, Germany

[3] Rutherford Appleton Laboratory, Chilton, Didcot,
Oxon OX11 OQX, U.K.

Abstract

Three examples of exotic phenomena in dusty plasmas are presented. These are the Ising model for an ensemble of elongated charged dust rods, the Casimir effect between two conducting plates, or between two spherical dust grains in a plasma, and the chargeon concept which results from the quantization of very low-frequency dust charge oscillations in a dusty plasma. First, we consider an array of elongated charged dust rods in a plasma. This is a physical system that allows for the Ising model to be directly tested by experiments, contrary to the usual situation where the Ising model is nothing more than an useful but abstract theoretical model for phase transitions. The magnetization of the usual Ising model is here replaced by the dust particle alignment with respect to ion flows, and the role of the temperature of the system is played by the energy of the electrostatic turbulence in the discharge. Second, the analogy between the Casimir force acting on two parallel plates in vacuum and a similar force, of classical nature, acting on two parallel plates in a turbulent plasma is also explored. Like the vacuum case, the force is attractive in a plasma, but its magnitude depends now on the frequency spectra as well as the associated energy densities of the turbulent fluctuations. The Casimir interaction between two nearby dust grains is also examined. Finally, we consider the field quantization of low-frequency oscillations emerging from the dust charge fluctuations in the high fugacity limit. When the surface charge oscillations of the dust grains dominate over the density oscillations of the plasma particles, the usual phonon concept ceases to hold, and it is replaced by the chargeon one, which corresponds to the minimum quantum variations of the dust charge.

1 Introduction

Dusty plasma physics is becoming one of the most exciting areas of plasma physics [Shukla and Mamun 2002]. New theoretical concepts and new observations in laboratory and in space have led to surprising discoveries, such as the plasma crystals [Chu and I 1994; Thomas *et al.* 1994], the dust acoustic wave [Rao *et al.* 1990; Barkan *et al.* 1995], or the dust acoustic Mach cones [Havnes *et al.* 1995; Samsonov *et al.* 1999]. Furthermore, the well known concepts from other areas of physics could be adopted to this new field, not in a simple and mechanical way, but revealing interesting new aspects. As an example, we can refer to the change in character of the Brownian motion of charged dust particles, where the plasma field fluctuations replace the molecular collisions of the traditional Brownian models, and charge fluctuations introduce a new dimension on the stochastic dynamics [Mendonça *et al.* 1997]. A dust particle in a plasma can also be seen as a kind of gigantic atomic system, where the electron orbitals are replaced by the polarization plasma cloud that are responsible for the electrostatic shielding of a dust particle [Mendonça *et al.* 2002].

In recent years, we have explored a number of concepts which can be seen as exotic in the context of current research in this field, but which can eventually be relevant to future experiments and interpretation of observations. These are, in particular, the Ising model adopted to chains of interacting elongated dust rods [Mendonça *et al.* 2001c], the Casimir force between nearby electric plates or spherical grains in a plasma [Mendonça *et al.* 2001a], and the concept of chargeon which we propose as a substitute for the phonon, when applied to ultra low-frequency oscillations of dusty plasmas [Mendonça *et al.* 2001b]. These three examples of exotic phenomena in dusty plasmas will be reviewed here.

The Ising model is an idealized model for the magnetization, which has been highly successful in stimulating theoretical work over the last few decades, but has no simple direct application to physical reality [Binney *et al.* 1995]. Here, we introduce a physical system which can be directly described by the Ising model: it is based on the presence of arrays of elongated charged dust rods in a plasma sheath where they are aligned by ion flows [Molotkov *et al.* 2000]. Hence, elongated charged dust rods introduce new degrees of freedom [Mahmoodi *et al.* 2000], which become the analog of the spin states of the atom. Such analogy will be explored here remembering that the elongated shape of the dust particles is an essential ingredient of our system.

Our second example of an exotic phenomena in dusty plasma physics is associated with the Casimir effect. It was first shown by Casimir [1948] that two parallel conducting plates in vacuum can be attracted to each other by a very small force [Casimir and Polder 1948; Lifshitz 1956]. This attractive force is a very peculiar, and conceptually important, manifestation of the electromagnetic field fluctuations of vacuum. It results from the fact that, in the region between the two plates, the allowed spectrum of fluctuations is reduced because of the existence of the boundary conditions, implying that the fluctuating electric field is equal to zero at the surface of the plates. The existence of the Casimir effect has been experimentally confirmed by Abrikosova and Deryaguin [1953], and it is still under investigation [Lamoreaux 1998; Mohideen and Roy 1998]. Recent review papers describe the essential features of the immense amount of research related with this topic [Elisalde and Romeo 1991; Pulnien *et al.* 1986; Kardar and Golestanian 1999]. In its more fundamental aspects, this problem is related with the calculation of the energy content of the vacuum. But other problems related with material media in thermal equilibrium with radiation have also been considered. In particular, extensions to thermal fluctuations in super-fluid films, liquid crystals and biological media should also be mentioned [Kardar and Golestanian 1999]. The case of a two-dimensional Wigner crystal was also recently considered [Lau *et al.* 2000]. A closely related work [Ford and Svaiter 2000] has discussed the possibility of focusing the vacuum fluctuations by a parabolic mirror. In this article, we show how to extend the previously mentioned works to include a turbulent plasma, thereby generalizing in two aspects the previous theoretical models of the Casimir effect: i) by considering the case of a plasma; and ii) by considering the plasma in an arbitrary turbulent state far away from thermal equilibrium.

The above mentioned generalizations have relevance to dusty plasmas. The two main geometric model used to discuss the Casimir effect will be considered: the interaction between two parallel conducting plates, and the interaction of two spherical dust grains suspended in the plasma. The first model is mainly used as a paradigm of interacting forces between two macroscopic objects due to the background fluctuations. It can nevertheless become interesting and even of some experimental relevance [Dubinov *et al.* 2000], especially if the conducting plates are replaced by two pieces of a conducting mesh. It is found that two parallel thin foils or meshes have the tendency to attract each other when they are immersed in a plasma. This new Casimir force, which adds to the well known electrostatic interactions, results from the difference in content of the plasma turbulent spectrum inside and outside the volume delimited by the parallel plates. Another way of describing it is saying that the difference on the plasmon, phonon or photon pressure, on the inner and outer surfaces of the plates, produces an attractive force that adds to the force due to their electrostatic interactions.

The second geometric configuration, corresponding to two spherical grains, can be of more practical relevance, especially because it can be applied to describe the interaction between two dust grains in a plasma. It is well known that there are several interacting forces between two nearby dust particles in a plasma, including monopole and dipole interactions [Resendes *et al.* 1998], wakefield effects [Nambu *et al.* 1995, Shukla and Rao 1996] and shadowing effects [Tsytovich et al. 1998, Lampe *et al.* 2000]. The present work will show that in plasmas there always exists an additional force due to the existence of plasma potential fluctuations, and it can be directly associated with the Casimir effect.

Finally, let us turn to our third exotic phenomena in dusty plasmas. It is well known that the acoustic type of plasma oscillations can change its character due to the dust contributions. In particular, dusty plasmas support a novel dust-acoustic wave (DAW) in the ultra low-frequency regime [Rao *et al.* 1990]. An important distinctive feature of dusty plasmas, when compared with the usual electron-ion plasma, is the existence of dust charge fluctuations [Varma *et al.* 1992; Melandsø *et al.* 1993]. When the contributions due to charge fluctuations are treated on an equal footing with the dust collective motion, dusty plasmas are shown to support a new kind of wave, the so called dust-Coulomb wave (DCW) [Rao 1999]. The essential difference between these two modes is that the restoring force for the DAW is provided by the plasma thermal pressure. In contrast, for the DCW this restoring force is due to the Coulomb pressure, which depends on the grain charge and size [Rao 2000]. The physical mechanisms responsible for these two modes are then quite distinct and, for a given wavelength, the frequency regime for the DAW is much larger than that for the DCW. Here, we describe the quantization procedure for the ultra-low frequency electrostatic field in a dusty plasma, in order to account for the grain charge fluctuations and to establish the main quantization differences between the DAW and the DCW. In the DAW we will simply obtain the modified phonons, as the elementary quantum excitations of the medium. On the other hand, for the DCW the elementary quantum excitations will change in character, in order to account for the oscillations of the dust charge, which can become dominant. In this case, we propose to call the resulting quasi-particles as the 'chargeons'. Of course, the quantization of any specific physical field will lead to specific quasi-particles, and we should not abuse of the names. But here the results are quite clear, the importance of the charge fluctuations is dominant and the resulting new concept seems quite natural and appealing.

2 The Ising Model

Let us consider a plasma created by a gas discharge in such a way that there is a vertical ion current $\vec{J_i} = q_i n_i \vec{u}$, where $\vec{u} = -u\vec{e_z}$ is the ion drift velocity. We first discuss the case of an isolated single cylindrical dust rods (with radius a and length l) that are suspended in such a plasma. Two possible equilibrium positions can be considered for the dust with respect to the ion flow: one corresponding to a parallel orientation with respect to the ion flow direction, and the other to a perpendicular orientation, as illustrated in figure 1. The oblique positions will be unstable to torque.

The number of ions that impact on the element of surface dS per unit time is $n_i(\vec{u} \cdot \vec{n})dS$, where $\vec{n}$ is the unit vector normal to the surface of the dust. It can easily be realized that the force exerted by the ion flow on the dust, on the parallel and perpendicular positions, is simply given by

$$\vec{F}_{\|} = \pi a^2 \frac{m_i}{q_i} \vec{J}_i = -\pi a^2 p_i \vec{e}_z \quad , \quad \vec{F}_{\perp} = \frac{2l}{\pi a} \vec{F}_{\|}, \tag{1}$$

where we have denoted $p_i = n_i m_i u$. This means that we will have $F_{\perp} > F_{\|}$ if the cylindrical dust rod is elongated enough, viz. $l > \pi a/2$. The equilibrium position of the centre of mass of the dust inside the discharge will then be slightly lower for the perpendicular equilibrium position. In order to determine this equilibrium position, we examine the equation of motion for the particle centre of mass

$$M \frac{d^2 \vec{R}}{dt^2} = Q\vec{E}(z) - Mg\vec{e}_z + \vec{F}_{\|,\perp}, \tag{2}$$

where M and Q are the mass and charge of the dust and g is the gravity constant.

We assume that the dust is made of a good conducting material and that it acquires a negative charge $Q < 0$ inside the plasma. The stationary vertical position of the dust centre of mass corresponds to a position where the static electric field pointing upwards in the discharge exactly compensates for the dust weight and for the additional ion force $\vec{F}_{\|,\perp}$. This means that we can characterize it by

$$|Q| E(z_{\|,\perp}) = Mg + F_{\|,\perp}. \tag{3}$$

For massive dusts such that $Mg \gg |F_{\|,\perp}|$, we have $z_{\|} \simeq z_{\perp}$. We can then develop the value of the electric field in series around $z = z_{\|}$ and derive from the equilibrium condition the following value for the shift in the centre of mass

$$\Delta z = z_{\perp} - z_{\|} = \frac{(F_{\perp} - F_{\|})}{|Q|(\partial E/\partial z)_{\|}}. \tag{4}$$

We now look at the total energy of the dust, for these two distinct equilibrium positions. This energy will have two terms, the gravitic and the electrostatic ones, according to $W(z) = -|Q|V(z) - Mgz$, where $V(z)$ is the electrostatic potential. By expanding the latter around $z_{\|}$, we can easily see that the energy difference between the two dust equilibrium positions is given by

$$\Delta W = F_{\|} \Delta z = \frac{F_{\|}}{|Q|(\partial E/\partial z)_{\|}} (F_{\perp} - F_{\|}). \tag{5}$$

Let us now associate the number $s = +1$ to the parallel equilibrium state, and the number $s = -1$ to the perpendicular one. The Hamiltonian of the isolated dust grain is then defined as

$$H_0(s) = -Bs, \tag{6}$$

where B is a constant to be defined later. It is clear that $H_0(+1) = -B$ and that $H_0(-1) = +B$. If we now identify the energy difference between these two states, ΔH_0 with the energy difference given by equation(5), we obtain then

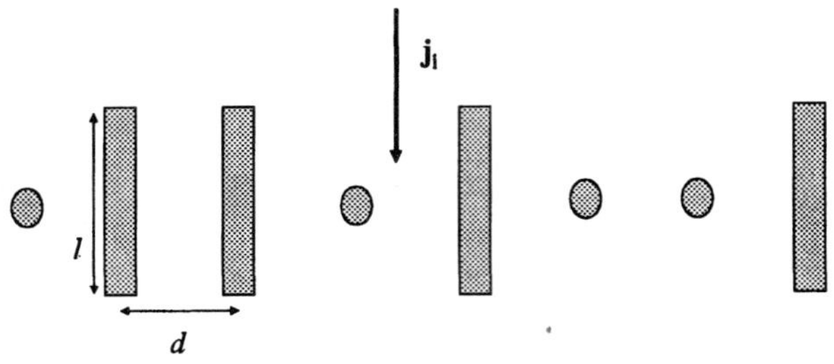

Figure 1: One dimensional chain of elongated dust particles in a plasma, where the two possible equilibrium states of the particles are illustrated.

$$B = \frac{l}{\pi a}(1 - \frac{\pi a}{2l})\frac{p_i^2}{|Q|(\partial E/\partial z)_{\|}} \tag{7}$$

If, instead of a single isolated dust particle we have an ensemble ($N \gg 1$) of dust particles in a discharge, with an average separation distance $d \gg$ than the electron Debye radius λ_{De} so that they can be regarded as independent, we can then characterize their equilibrium states by the total Hamiltonian

$$H_0 = -B\sum_{i=1}^{N} s_i, \tag{8}$$

where each number s_i can take the values ± 1.

Let us now consider the interaction between two adjacent dust particles. This is relevant when their mean distance d becomes comparable with λ_{De}. We can then have two distinct situations for a couple of nearby dust particles: the state (p) if they are parallel to each other, and the state (a) if they are perpendicular to each other. Notice that the state (p) occurs when both particles are characterized by variables s_i and s_j with the same sign (both positive or both negative), and the state (a) corresponds to opposite signs of these variables.

For each relative position between the two dust particles, we have a different value of the electrostatic energy. The explicit value of this interaction energy is very difficult to derive, because a debate is still going on, even for point particles, concerning the attraction or repulsion between them.

There will be a repulsive force acting on the dust particles if we consider the simple models for negative monopole charges [Mohideen *et al.* 1998], partially screened by the surrounding plasma. In this case, the interaction energy will be positive, and larger for the state (p) than for the state (a). In contrast, the two dust particles can eventually attract each other due to induced dipole interactions [Resendes *et al. 1998*], or due to wakefield effects [Nambu *et al.* 1995, Shukla and Rao 1996], or even due to shadowing mechanisms [Tsytovich *et al.* 1998, Lampe *et al.* 2000]. If the centre of mass of the dust particles is contained in a plane perpendicular to the ion flow, the wakefield mechanisms will be absent, but the dipole interactions will lead to a negative value of the interaction energy.

The interest of our present model is that it can be built up without knowing the exact nature of the dust-dust interaction. A different behavior will be, however, expected for the attractive and repulsive scenarios. Let us then assume that we have the following interaction energy W_p for the parallel configuration. By geometrical considerations we can write, for the other configuration

$$W_a \simeq \frac{2a}{\pi l} W_p. \tag{9}$$

We can now introduce an interaction Hamiltonian between the two nearby dust particles, labelled by i and j, of the form

$$H_{ij} = \frac{1}{2} J s_i s_j, \tag{10}$$

where we have $s_i = s_j$ for the parallel configuration, and $s_i = -s_j$ for the anti-parallel one. This means that, apart from an additive constant, we can write the energy of these configurations as $H_p = J/2 = W_p$, and $H_a = -J/2 = W_a$. This allows us to write

$$J = W_p - W_a = \left(1 - \frac{2a}{\pi l}\right) W_p. \tag{11}$$

If we neglect interactions between dust particles that are not immediate neighbours, we can establish the total Hamiltonian of the system of dusty particles in a plasma:

$$H = -B \sum_i s_i + \frac{1}{2} \sum_{i,j} J_{ij} s_i s_j, \tag{12}$$

where $J_{ij} = J$ if i and j represent neighbours and is equal to zero otherwise.

We realize that this is formally identical to the Hamiltonian of the usual Ising model, but that the physical content is quite different. Here, the usual magnetic field B (aligned along the z-axis) is replaced by the ion flow aligned along the same direction. The formal similarities and the physical differences become even more pronounced when we establish the partition function of the system

$$Z = \sum_{[s_i]} \exp \left[(B \sum_i s_i - \frac{1}{2} \sum_{i,j} J_{ij} s_i s_j) / \beta_{es} \right] , \tag{13}$$

where the sum over $[s_i]$ represents the sum over all the possible (spin) configurations, but where we have replaced the usual parameter $\beta = 1/k_B T$ by a new one, $\beta_{es} = N/W_{es}$, where W_{es} represents the energy density of the electrostatic fluctuations of the background plasma and N is the number of dust particles per unit volume. Such a replacement of β is due to the fact that we always have $W_{es} \gg N k_B T$, because the plasma discharges are never in thermal equilibrium but contain a high level of electrostatic turbulence (electron and ion plasma fluctuations). It is this electrostatic spectrum of fluctuations, and not the thermal fluctuations, that act on the charged dust rods and make them change their position and orientation with respect to the ion flow.

Following the usual Ising theory, we can then predict that the mean orientation of the one-dimensional dust chain will be given by

$$< s >= \sinh(B\beta_{es}) \left[\cosh^2(B\beta_{es}) - (1 - \exp\{-4|J|\beta_{es}\}) \right]^{-1/2} , \tag{14}$$

where we have used $J = -|J|$. This means that, with our model, we can have direct access to a real Ising system, and by this way we can compare the observations with the theoretical predictions. In particular, the main aspects can be experimentally tested: i) the role of the plasma fluctuations W_{es} on the dust rod alignment, ii) the influence of the ion flow, represented by B, and iii) the interaction potential between two nearby dust rods, represented by $|J|$.

The procedure can obviously be extended to two-dimensional arrays of elongated dust rods, where exact solutions are also available. In particular, it was demonstrated long time ago by Onsager [1944] that spontaneous magnetization (in our case, we should say spontaneous alignment) for $B = 0$ (in the absence of ion flows, if we have $\beta_{es} > \beta_c = 0.4407/\epsilon$. In our case, this spontaneous alignment would be observed for

$$W_{es} < 2.27N \left(1 - \frac{2a}{\pi l} \right) |W_p|, \tag{15}$$

where W_p is the coupling potential between two parallel plates. The above result remains valid for $W_{es} \gg B$, which requires very low ion flows. But exact results are also available for finite ion flows [Zamalodchikov 1989]. Equations (14) and (15) are good candidates for an experimental test of the present model.

3 Casimir Effect

3.1 Two Parallel Plates

Let us first study the influence of two parallel metallic plates, or meshes, on the energy spectrum of the field fluctuations associated with the state of turbulence in a plasma. This can be done by generalizing the usual formulation of the Casimir effect [Itzykson and Zuber 1980], in order to account for the plasma case.

We consider the field modes inside the volume $V = L^2 a$ delimited by the two square plates of side L, placed at a distance z inside the plasma. In the absence of the plates, the electrostatic, or electromagnetic, energy of the plasma fluctuations contained in this volume would be determined by

$$W_0 = V \int W(\vec{k}) \frac{d\vec{k}}{(2\pi)^3} = \int L^2 \frac{d\vec{k}_\parallel}{(2\pi)^2} \int_{-\infty}^{\infty} z \frac{dk_z}{2\pi} W(\vec{k}). \tag{16}$$

The energy density can be determined by

$$W(\vec{k}) = s\hbar\omega(\vec{k}) N(\vec{k}), \tag{17}$$

where s is the number of independent polarization states ($s = 1$ for electrostatic modes, and $s = 2$ for electromagnetic ones) and $N(\vec{k})$ is the number of quasi-particles (meaning plasmons or phonons, for electrostatic turbulence, and photons for electromagnetic one).

The quantity $\omega(\vec{k}) = \omega$ is determined by the dispersion relation of the various field modes that can be excited in the medium. In the simple case of vacuum, we would simply have $\omega(\vec{k}) = kc$, where $k = \sqrt{k_\parallel^2 + k_z^2}$. Returning to the general case of a turbulent medium, it is convenient to write equations (16) and (17) as

$$W_0 = s\hbar \int L^2 \frac{d\vec{k}_\parallel}{(2\pi)^2} \int_{\infty}^{\infty} z \frac{dk_\perp}{2\pi} \omega(\vec{k}_\parallel, k_z) N(\vec{k}_\parallel, k_z). \tag{18}$$

In contrast, when the two plates are immersed into the plasma medium, not all of the previously existing fluctuation field modes are possible. Owing to the new boundary conditions imposed by the plates, only those modes corresponding to an integer number of wavelengths in the direction Oz perpendicular to the plates will survive. This can be described by imposing

$$N(\vec{k}_\parallel, k_z) = \frac{\pi}{z} \sum_{n=-\infty}^{\infty} N(\vec{k}_\parallel, n) \delta(k_z - n\frac{\pi}{z}). \tag{19}$$

Inserting equation (19) into equation (18), we obtain for the energy of the field fluctuations inside the volume V in the presence of the metallic plates. The result is

$$W = \frac{s}{2}\hbar \sum_{n=-\infty}^{\infty} \int L^2 \frac{d\vec{k}_\parallel}{(2\pi)^2} \omega(\vec{k}_\parallel, n) N(\vec{k}_\parallel, n), \tag{20}$$

where we have introduced $\omega(\vec{k}_\parallel, n) = \omega(\vec{k}_\parallel, k_\perp = n\pi/z)$. Now, we assume the following symmetry condition for the wave spectrum

$$N(\vec{k}_\parallel, n) = N(\vec{k}_\parallel, -n). \tag{21}$$

Hence, we can write

$$W = \frac{s}{2}\hbar \int L^2 \frac{d\vec{k}_\parallel}{(2\pi)^2} \left\{ \omega(\vec{k}_\parallel, 0) N(\vec{k}_\parallel, 0) + 2\sum_{n=1}^{\infty} \omega(\vec{k}_\parallel, n) N(\vec{k}_\parallel, n) \right\}. \tag{22}$$

We are now in a position to establish the energy difference, per unit surface, between the two states of the medium (with and without the parallel plates). We have

$$E = \frac{1}{L^2}(W - W_0). \tag{23}$$

At this point we can assume that the turbulence is isotropic in the parallel direction. This means that we can use $d\vec{k}_\parallel = 2\pi k dk$, and assume that $N(\vec{k}_\parallel, n)$ is only dependent on the absolute value of the parallel wavevector $k = |\vec{k}_\parallel|$, and is insensitive to its direction. We can then write equation (22) in the form

$$W = \frac{s\hbar}{4\pi} L^2 \int_0^\infty kdk \left\{ \omega(k, 0) N(k, 0) + 2\sum_{n=1}^{\infty} \omega(k, n) N(k, n) \right\}. \tag{24}$$

A similar expression can be obtained for W_0. Using a real variable $n = (z/\pi)k_z$ in equation (18), we can write

$$W_0 = \frac{s\hbar}{2\pi} L^2 \int_0^\infty kdk \int_0^\infty \omega(k, n) N(k, n). \tag{25}$$

It is now convenient to define the quantity

$$f(n) = \int_0^\infty \omega(\xi, n) N(\xi, n) d\xi, \tag{26}$$

where ξ is a new variable determined by

$$\xi = \left(\frac{zk}{\pi}\right)^2. \tag{27}$$

Using the definition (26) into equation (24), we obtain

$$W = s\hbar \frac{\pi}{4z^2} L^2 \left\{ \frac{1}{2} f(0) + \sum_{n=1}^{\infty} f(n) \right\}. \tag{28}$$

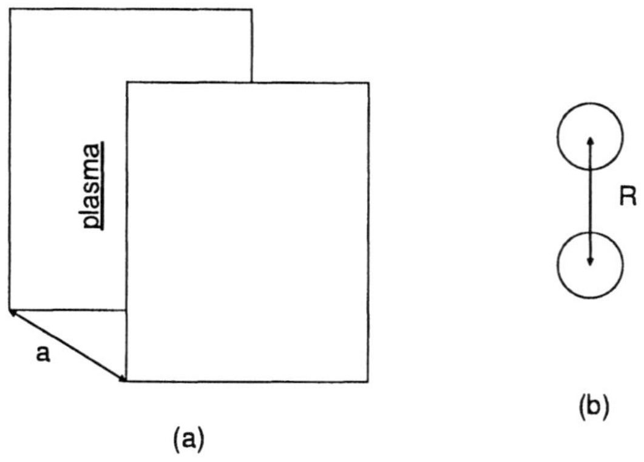

Figure 2: Two typical examples of Casimir force in a plasma: (a) two parallel plates or conducting grids; (b) two spherical dust grains.

Similarly, from equation (25), we obtain for the non-perturbed energy inside the same volume

$$W_0 = s\hbar \frac{\pi}{4z^2} L^2 \int_0^\infty f(n) dn. \tag{29}$$

We can now use the expressions (28) and (29) into equation (23), and calculate the energy difference per unit surface as

$$E = s\hbar \frac{\pi}{4z^2} \left[\frac{1}{2} f(0) + \sum_{n=1}^{m} f(n) - \int_0^\infty f(n) dn \right]. \tag{30}$$

Here, it should be noticed that the sum and the integral over the (discrete and continuous) variable n, can only be extended to some upper-limit m, above which we necessarily have $N(\xi, n \geq m) = 0$. For electrostatic waves this cut-off for large values of n is dictated by dispersion or damping. For electromagnetic waves, in contrast, the cut-off is imposed by the transparency of the plates for extremely high perpendicular wavenumbers. In the case of a conducting mesh, the cut-off wavenumber would be determined by the size of the mesh. This means that, according to equation (26), we will also have $f(n \geq m) = 0$.

In order to solve equation (30), we can now use the Euler-MacLaurin formula that establishes a relation between the sum and the integral over n [Arfken 1985]

$$\int_0^\infty f(n) dn = \frac{1}{2} f(0) + f(1) + f(2) + \ldots + \frac{1}{2} f(m) \tag{31}$$

$$-b_2[f'(m) - f'(0)] - b_4[f^{(3)}(m) - f^{(3)}(0)] - \dots,$$

where $b_n = B_n/n!$, and B_n are the Bernoulli numbers defined by

$$B_n \left[\frac{d^n}{dx^n} \left(\frac{x}{e^x - 1} \right) \right]_{x=0} . \tag{32}$$

The first few, that are important here, are

$$B_0 = 1 \quad , \quad B_2 = \frac{1}{6} \quad , \quad B_4 = -\frac{1}{30}. \tag{33}$$

Equation (30) can then be written as

$$E = s\hbar \frac{\pi}{4z^2} \left\{ b_2[f'(m) - f'(0)] + b_4[f^{(3)}(m) - f^{(3)}(0)] \right\} . \tag{34}$$

Equation (34) is a generalization of the Casimir energy for a turbulent plasma medium. It contains, as a special case, the vacuum fluctuations. The resulting force per unit surface between the two parallel plates, at a distance $z = a$ by x, will be given by the derivative of this energy difference at $x = a$

$$F = -\left(\frac{dE}{dz} \right)_{z=a} . \tag{35}$$

In the following, we apply our results to a few different examples that are relevant to vacuum and in a plasma.

3.2 Types of Turbulent Spectra

i) - Electromagnetic fluctuations in vacuum

In order to reduce equation (34) to this case we consider $s = 2$ and

$$N(\xi, n) = \frac{1}{2} \quad , \quad (n < m). \tag{36}$$

Here we can also express

$$\omega(\xi, n) = c|\vec{k}| = \frac{\pi}{z} c \sqrt{\xi + n^2}. \tag{37}$$

From equation (26), we have

$$f(n) = \frac{\pi}{z} c \int_{n^2}^{\infty} \sqrt{\xi'} d\xi', \tag{38}$$

where we have introduced $\xi' = \xi + n^2$. From the expression (??) we conclude that $f'(0) = 0$ and $f^{(3)}(0) = -4\pi c/z$. If we assume a smooth upper cut-off such that $f'(m) = f^{(3)}(m) = 0$, we recover from equation (35) the usual expression for the Casimir effect in vacuum

$$F = 3\hbar c b_4 \frac{\pi^2}{a^2} = -\frac{\pi^2}{240}\frac{\hbar c}{a^4}. \tag{39}$$

ii) - Ion-acoustic turbulence

In this case of electrostatic turbulence in a plasma, we only have one single polarization state: $s = 1$. In order to stress the analogies with the previous example, we will neglect the dispersion effects in the ion-acoustic dispersion relation (which are only significant in the low-frequency range where the fluctuations are usually negligible), and write

$$\omega(\xi, n) = |\vec{k}|v_s = \frac{\pi}{z}v_s\sqrt{\xi + n^2}, \tag{40}$$

where $v_s \simeq \sqrt{T_e/m_i}$ is the ion-acoustic or ion-sound speed. Now, instead of equation (38), we have

$$f(n) = \frac{\pi}{z}v_s\int_{n^2}^{\infty}\sqrt{\xi'}N(\xi', n)d\xi'. \tag{41}$$

From the expression (41) we can easily derive

$$f'(n) = -\frac{2\pi}{z}v_s n^2 g(n)N_0, \tag{42}$$

where we have denoted

$$N(\xi' = n^2, 0) = g(n)N_0. \tag{43}$$

The definition (43) introduces a form factor $g(n)$ which is nearly equal to one at very low-frequencies $g(0) \simeq 1$, and tends to zero at the end of the spectral range $g(m) = 0$. In the case of a smooth cut-off in the upper-frequency range, we still have $f^{(3}(0) \gg f^{(3)}(m)$, which leads to the following expression for the resulting Casimir force

$$F = 3\hbar v_s b_4 \frac{\pi^2}{a^2} = -\frac{\pi^2}{240}\frac{\hbar c}{a^4}N_0. \tag{44}$$

The expression (44) is similar to the previous result for electromagnetic fluctuations in vacuum, except that the speed of light c is now replaced by $v_s N_0$.

iii) - Electron plasma turbulence

Here, again, we have $s = 1$, but the dispersion relation is slightly more complicated than in the two previous examples

$$\omega(\xi, n) = \frac{\pi}{z}S_e\sqrt{\xi + n^2 + \xi_D}, \tag{45}$$

where $S_e = (3T_e/m_e)^{1/2}$, and $\xi_D = z^2\omega_p^2/\pi^2 S_e^2$. In this case, equation (41) is replaced by

$$f(n) = \frac{\pi}{z} S_e \int_{n^2+\xi_D}^{\infty} \sqrt{\xi'} N(\xi', n) d\xi'. \tag{46}$$

Taking

$$g(n) = \frac{1}{N_0} N(\xi' = n^2 + \xi_D, 0), \tag{47}$$

and, using the same arguments as before, we obtain an expression similar to equation (44), except that the ion-acoustic speed v_s replaced by the electron thermal speed S_e.

iv) - General expressions for the Casimir force

The above discussion allows us to establish the following general expression for the Casimir force

$$F = -\frac{s\pi^2}{240} \frac{\hbar v}{a^4} N_0, \tag{48}$$

which is valid for both the vacuum and the plasma case, and for electrostatic or electromagnetic fluctuations. For the electromagnetic vacuum, we use $v = c$, $s = 2$ and $N_0 = 1/2$, and recover the well known result of the quantum Casimir effect.

In the case of a plasma, we have to consider several kinds of turbulence. For transverse electromagnetic fluctuations, we have $v = c$, $s = 2$ as in vacuum, but the occupation number can now be very large $N_0 \gg 1$, thus leading to Casimir forces larger by many orders of magnitude than in vacuum. To be precise, N_0 is the occupation number for very low wavenumber photons, below $k = \pi/a$. In the case of electrostatic turbulence, we have $s = 1$, and $N_0 \gg 1$ is the occupation number for $k \to 0$. For the electron plasma wave turbulence, we have $v = S_e$, and for the ion-acoustic turbulence, we have $v = v_s$. The above results are already quite general, but they were derived under the assumption that the occupation numbers $N(\vec{k}_\parallel, k_z)$ do not change in the presence of the parallel plates. This means that, due to the appearance of the plates, only some of the modes will not be excited, viz. those which are not compatible with the new boundary conditions.

But we can push this generalization of the Casimir effect even further by noting that the presence of the plates can reduce the actual contents of the turbulent spectrum. In such a general case, equation (19) is not valid and we should write, for the energy difference (23)

$$E = s\hbar z \int \omega_k [N(\vec{k}) - N_0(\vec{k})] d\vec{k}, \tag{49}$$

where $N(\vec{k})$ is the spectrum of fluctuations in the presence of the plates and $N_0(\vec{k})$ is the spectrum in the absence of the plates (or outside the volume confined by them).

In order to illustrate equation (49), let us make the strong assumption that the turbulence spectrum disappears inside the confined volume for wavenumbers below $|k| = \pi/z$, even for the parallel propagation. This will mean that $N(\vec{k}) = 0$ for $|k| \leq \pi/z$, and that $N(\vec{k}) = N_0(\vec{k})$, for $|k| > \pi/z$. Equation (49) reduces to

$$E = -s\hbar z \int \omega_k N_0(\vec{k}) d\vec{k}, \tag{50}$$

which can be approximately written as

$$E \simeq -s\hbar z \bar{\omega} \bar{N}_0 V_k, \tag{51}$$

where $\bar{\omega}$ and $\bar{N}_0$ are averaged values of the frequency and the occupation number, and $V_k \simeq (\pi/z)^3$. A simple estimate gives $\bar{\omega} \simeq kv \simeq \pi v/z$. Equation (51) is then reduced to

$$E = -s\hbar v \bar{N}_0 \frac{\pi^4}{z^3}, \tag{52}$$

which leads to the following attractive force between the two plates at a distance $z = a$

$$F = -s\hbar v \bar{N}_0 \frac{\pi^4}{a^4}. \tag{53}$$

Apart from the numerical factors, equation (53) is very similar to equation (48), just showing the consistency of our approach, and opening the way to more sophisticated models.

3.3 Two Nearby Dust Grains

Let us consider two spherical dust grains, with diameters a_1 and a_2, immersed at a distance R in a plasma. We assume that this pair of grains is acted by a fluctuating electrostatic field of the form

$$\vec{E}(\vec{r}, t) = \vec{E}_0 \exp(i\vec{k} \cdot \vec{r} - i\omega t) + c.c., \tag{54}$$

where $\vec{E}_0 = E_0 \vec{k}/k$ and *c.c.* stands for the complex conjugate. This field induces two dipole momenta $\vec{d}_1(t) = \alpha_1 \vec{E}(\vec{r}_1, t)$ and $\vec{d}_2(t) = \alpha_2 \vec{E}(\vec{r}_2, t)$. If the dust grains are two conducting spheres, we can write $\alpha_{1,2} = -a_{1,2}^3$, if we neglect the contributions from the surrounding plasma, as some static theoretical models seem to suggest [Hamaguchi and Farouki 1994]. But a convincing evaluation of these quantities, for the dynamical case is still missing. For grains with a small diameter ($a_{1,2} \ll R$), we can easily establish

$$\vec{d_1} = \alpha_1 \vec{E_0} \exp(-i\omega t) + c.c., \tag{55}$$
$$\vec{d_2} = \alpha_2 \vec{E_0} \exp(ikR\cos\theta - i\omega t) + c.c.,$$

where θ is the angle between the wavevector $\vec{k}$ and the orientation of the two grains $\vec{R}$. In these expressions, we have assumed that the grain 1 was centered at $\vec{r} = 0$, which means that $\vec{R} = \vec{r}_2$. These two induced dipoles interact with each other and produce an interaction energy given by [Resendes *et al*, 1998, Resendes 2000]

$$W_{int} = \frac{\exp(-R/\lambda_D)}{R^3}\left[w(R)\vec{d_1}\cdot\vec{d_2} - 3w'(R)(\vec{d_1}\cdot\vec{n})(\vec{d_2}\cdot\vec{n})\right]. \tag{56}$$

For $\lambda_D \to \infty$ and $w(R) = w'(R) = 1$, the expression (56) recovers the usual result for the interaction between two dipole moments in vacuum. In a plasma, for $a, R \ll \lambda_D$, we can use

$$w(R) \simeq (1 - R/\lambda_D) \quad , \quad w'(R) \simeq (1 + R/\lambda_D). \tag{57}$$

Inserting equation (57) into equation (56) and averaging over a time interval much larger than $1/\omega$, we obtain

$$< W_{int} >= \frac{\exp(-R/\lambda_D)}{R^3}|E_0|^2\beta(\vec{k},\vec{R}), \tag{58}$$

where

$$\beta(\vec{k},\vec{R}) = 2\alpha_1\alpha_2\left[w(R) - 3w'(R)\frac{(\vec{k}\cdot\vec{R})^2}{k^2R^2}\right]\cos(kR). \tag{59}$$

Equation (59) reveals that the interaction energy can be significant if the distance between the two dust grains is much smaller than the wavelength, viz. $kR \ll 1$. For larger distances its value decreases typically as $1/R^3$ and oscillates due to the cosine appearing in the expression for $\beta(\vec{k},\vec{R})$.

The oscillating field defined by equation (54) could well be excited by an electrostatic probe. A more general situation corresponds to a turbulent plasma, where a large spectrum of fluctuations can be spontaneously excited in the discharge. The total electric field in the plasma can then be defined as

$$\vec{E}(\vec{r},t) = \int \vec{E}_k \exp(i\vec{k}\cdot\vec{r} - i\omega_k t)\frac{d\vec{k}}{(2\pi)^3}, \tag{60}$$

where the frequency $\omega = \omega_k$ is a function of $\vec{k}$ as established by the appropriate dispersion relation. In this more general case, we can similarly obtain for the total mean energy of the dipole-dipole interaction

$$W = < W_{int} > = \frac{\exp(-R/\lambda_D)}{R^3} \int |E_k|^2 \beta(\vec{k}, \vec{R}) \frac{d\vec{k}}{(2\pi)^3}. \tag{61}$$

The resulting interacting force between the two nearby dust spherical grains can now be determined by

$$\vec{F} = -\nabla W = -\int |E_k|^2 \vec{f}(\vec{k}, \vec{R}) \frac{d\vec{k}}{(2\pi)^3}, \tag{62}$$

where

$$\vec{f}(\vec{k}, \vec{R}) = \frac{\partial}{\partial \vec{R}} \left[\frac{\exp(-R/\lambda_D)}{R^3} \beta(\vec{k}, \vec{R}) \right]. \tag{63}$$

The expression (54) is the value of the Casimir force between two dust grains in a turbulent plasma. This force is not necessarily negative due to the oscillatory character of the coupling coefficient $\beta(\vec{k}, \vec{R})$, as clearly shown by equation (59) for the case of a single oscillating field. This qualitative aspect is not new in the theory of the Casimir force (see the review articles by Elisalde and Romeo [1991], Pulnien *et al.* [1986], Kardar and Golestanian [1999]), and it is one of its distinctive features, contrary to the van der Waals attractive force between two elementary atoms or molecules.

Two comments should be made here, before concluding. One is that, apart from the dipole interactions considered above, another interacting energy is usually present, resulting from the dust charge variation associated with the plasma potential fluctuations of the turbulent spectrum. This is particularly true for the lower frequency part of the turbulent spectrum, because, for field frequencies ω_k larger than the characteristic dust charge relaxation frequency [Shukla 2001], the average charge fluctuation would be equal to zero. This new interaction energy could be easily established but it will not be considered here because its value would always give a small correction to the usual and well known monopole and dipole charge interaction energies associated with the two nearby charged dust grains, and it would not be physically relevant.

Second, we have assumed that the correlation length of the turbulent spectrum l_{corr} is much larger than the distance R between the two dust grains, which means that both grains were acted by the same field. This is consistent with the fact that we are considering very small distances between the dust grains (even if larger than their diameter). But we could also have easily considered the opposite case of a very short correlation length, $l_{corr} \ll R$, where the field seen by one grain is not correlated with the field seen by the other. In this case, the dipole induced by the turbulent field on one grain would induce another dipole on the second one, and the resulting interaction energy will then fall more rapidly with distance, as $1/R^6$, which is typical of the usual van der Waals interaction.

4 Chargeons and Phonons

Let us consider the case of an isotropic dusty plasma, made of three different particle populations: electrons, ions and dust particles. We assume that the electron and ion fluids are in Boltzmann equilibrium in the plasma wave potential ϕ. Their number densities n_α, with $\alpha = e, i$, are therefore determined by

$$n_\alpha = n_{\alpha 0} \exp\left[\frac{q_\alpha \phi}{T_\alpha},\right] \tag{64}$$

where $n_{\alpha 0}$ is the equilibrium number density of the particle species $\alpha(= e, i)$, T_α is the temperature and q_α is the particle charge.

In contrast, the dust fluid is described by the mass and momentum conservation equations

$$\frac{\partial n_d}{\partial t} + \nabla \cdot (n_d \vec{v}_d) = 0, \tag{65}$$

and

$$\frac{\vec{v}_d}{\partial t} + \vec{v}_d \cdot \nabla \vec{v}_d = -\frac{q_d}{m_d} \nabla \phi - \frac{V_{td}^2}{2} \nabla \ln n_d, \tag{66}$$

where q_d and m_d are the charge and the mass of the dust particles, and n_d and $\vec{v}_d$ are their mean density and velocity, respectively. We have also denoted $V_{td}^2 = 2\gamma_d T_d / m_d$, where γ_d is the adiabatic index of the dust fluid and T_d is the dust temperature.

The plasma potential fluctuations are governed by Poisson's equation

$$\nabla^2 \phi = -\frac{\rho}{\epsilon_0} = -\frac{1}{\epsilon_0} \sum_{\alpha = e,i,d} q_\alpha n_\alpha. \tag{67}$$

For ultra-low frequency oscillations, it is possible to replace Poisson's equation by the quasi-neutrality condition, and use

$$\sum_\alpha q_\alpha n_\alpha = q_d n_d + q_i n_i - e n_e \simeq 0. \tag{68}$$

We now consider perturbations δq and δn in the grain charge and the number density, and accordingly write

$$q_d = \bar{q} + \delta q \quad , \quad n_d = \bar{n} + \delta n, \tag{69}$$

where $\bar{q}$ and $\bar{n}$ are the respective equilibrium quantities. In contrast to the ion charge, the grain charge perturbation (δq) and hence the grain surface potential perturbation ($\delta \psi$) are not zero, but are related to the plasma wave potential perturbation ($\delta \phi$) through the current balance equation which, to the lower order, yields [Rao 1999]

$$\delta \phi \simeq -\delta \psi \equiv -\frac{\delta q}{R}, \tag{70}$$

where R is the size (radius) of the dust grain. The physical implication of equation (70) is obvious: For ultra low-frequency fluctuations, the net grain surface associated with a volume element of the plasma will have to be equal and opposite to the plasma potential, in order to maintain the overall quasi-neutrality of the plasma.

Linearizing equations (65) and (66) with respect to the above perturbations, we obtain

$$\frac{\partial}{\partial t}\delta n + \bar{n}\nabla \cdot \vec{v}_d = 0, \tag{71}$$

and

$$\frac{\partial}{\partial t}\vec{v}_d = \frac{\bar{q}}{m_d R}\nabla \delta q - \frac{V_{td}^2}{2\bar{n}}\nabla \delta n, \tag{72}$$

which can be combined to obtain the governing equation relating the charge and the number density fluctuations. We have

$$\frac{\partial^2}{\partial t^2}\delta n + \frac{\bar{n}\bar{q}}{m_d R}\nabla^2 \delta q - \frac{V_{td}^2}{2\bar{n}}\nabla^2 \delta n = 0. \tag{73}$$

We now introduce the dimensionless parameter called fugacity of the dusty plasma [Rao 1999]

$$f = 4\pi \bar{n}\lambda_D^2 R, \tag{74}$$

where the dusty plasma Debye radius λ_D is determined from $\lambda_D^{-2} = \lambda_{De}^{-2} + \lambda_{Di}^{-2}$. Here, λ_{De} and λ_{Di} are the electron and ion Debye radii, respectively. In the high fugacity (dense) regime ($f \gg 1$), wherein the DCWs exist, the grain charge fluctuations are directly proportional to the dust number density fluctuations, and accordingly the quasi–neutrality condition (67) yields

$$\bar{q}\,\delta n + \bar{n}\,\delta q \approx 0, \tag{75}$$

which implies that an increase of the dust concentration is compensated by a decrease of the dust charge so as to maintain the charge neutrality. Thus, the bulk oscillations associated with the DCWs are coupled to the charge oscillations, which is unlike the case of DAWs which exist in the low fugacity (tenuous) regime when the grain charge is constant.

Using equation (75) into equation (71), we obtain the wave equation for the grain charge fluctuations in a dusty plasma

$$\frac{\partial^2}{\partial t^2}\delta q - C_q^2 \nabla^2 \delta q = 0, \tag{76}$$

where

$$C_q^2 = \frac{\bar{q}^2}{m_d R} + \frac{V_{td}^2}{2}. \tag{77}$$

For perturbations of the type $\exp[i(\vec{k}\cdot\vec{r}-\omega t)]$, we obtain from Eq. (76) the dispersion relation of the so-called dust-Coulomb waves

$$\frac{\omega^2}{k^2} = C_{DC}^2 + \frac{V_{td}^2}{2}, \tag{78}$$

where the characteristic phase speed of the DCWs is given by

$$C_{DC} = \frac{\bar{q}}{\sqrt{m_d R}}, \tag{79}$$

which depends only on the grain parameters, namely, the charge ($\bar{q}$), the mass (m_d) and the size (R).

The phase velocity of DCWs has two distinct terms: The first one is associated with the mean Coulomb charge $\bar{q}$ and has no counterpart in the usual dust acoustic waves (DAWs). The second one is due to the dust thermal motion and depends on the dust thermal velocity V_{td}. The latter is required to be typically much smaller than C_{DC} in order that the modes are not damped due to kinetic effects such as Landau damping. A more detailed derivation of the DCW dispersion relation is given by Rao [2000].

In order to quantize the charge field $\delta q(\vec{r}, t)$ associated with the DCWs, we consider its Fourier spectrum

$$\delta q(\vec{r}, t) = \int \delta q_k(t) e^{\vec{k}\cdot\vec{r}} \frac{d\vec{k}}{(2\pi)^3}. \tag{80}$$

Using equation (80) into equation (76), we reduce the equation for the amplitude of each mode $\delta q_k(t)$ to that of an harmonic oscillator

$$\frac{\partial^2}{\partial t^2}\delta q_k + \omega_k^2 \delta q_k = 0, \tag{81}$$

which implies that for each mode, $\delta q_k(t) \propto \exp(-i\omega_k t)$ with $\omega_k = kC_q$.

Consider now the energy associated with the dust-Coulomb waves per unit volume

$$W = \int W_k \frac{d\vec{k}}{(2\pi)^3}. \tag{82}$$

For each mode of the field, we have two distinct contributions for the total energy density

$$W_k = W_\epsilon(k) + W_p(k). \tag{83}$$

The first term in equation (83) is the electrostatic energy associated with the potential fluctuations at the surface of the dust grain, while the second term is due to the motion of the dust particles in the wave field. Roughly speaking, one can identify the first term as the potential energy and the second term as its kinetic counterpart.

For the electrostatic energy associated with the k-component of the dust-Coulomb spectrum, we write

$$W_\epsilon(k) = \frac{1}{4}\epsilon_0|E(k)|^2\frac{\partial}{\partial\omega}[\omega\chi(\omega)]_{\omega_k}, \tag{84}$$

where $\chi(\omega)$ is the susceptibility defined, according to equation (76), as

$$\chi(\omega) = 1 - \frac{k^2C_q^2}{\omega^2}. \tag{85}$$

Noting that $|E(k)|^2 = k^2|\delta\psi(k)|^2$, while from equation (69), $|\delta\psi(k)|^2 = |\delta q_k|^2/R^2$, equation (84) can be written in the form

$$W_\epsilon(k) = \frac{\epsilon_0}{2}\frac{k^2}{R^2}|\delta q_k|^2. \tag{86}$$

On the other hand, for the kinetic energy of the oscillating dust grains, we have

$$W_p(k) = \frac{1}{2}\bar{n}m_d v_d^2, \tag{87}$$

where the dust fluid velocity is calculated from the continuity equation and the quasi-neutrality condition, yielding

$$v_d = C_q\frac{\delta n}{\bar{n}} = -C_q\frac{\delta q}{\bar{q}}. \tag{88}$$

Using equation (88) into equation (87), we write the kinetic energy as

$$W_p(k) = \frac{1}{2}\bar{n}m_d\frac{C_q^2}{\bar{q}^2}|\delta q_k|^2. \tag{89}$$

Finally, combining equations (86) and (89), we write the total energy density per mode as

$$W_k = \frac{1}{2}\alpha_k\omega_k^2|\delta q_k|^2, \tag{90}$$

where

$$\alpha_k = \frac{\epsilon_0}{R^2C_q^2} + \frac{\bar{n}m_d}{\bar{q}^2k^2}. \tag{91}$$

We now follow the usual steps [Landau and Lifshitz 1995] for the quantization of the phonon fields by introducing a pair of canonical variables, Q_k and P_k, such that

$$Q_k = \sqrt{\alpha_k}(\delta q_k + \delta q_k^*),$$
$$P_k = \frac{dQ_k}{dt} = -i\omega_k\sqrt{\alpha_k}(\delta q_k - \delta q_k^*). \tag{92}$$

Thus, Eq. (90) can be rewritten in terms of the new variables as

$$W_k = \frac{1}{2}\left[P_k^2 + \omega_k^2 Q_k\right], \tag{93}$$

which is formally identical to the energy of an harmonic oscillator with unit mass and position Q_k. In order to quantize the charge field, we introduce the creation and destruction operators for the plane wave modes with a frequency $\omega_k = kC_q$ as

$$a_k^{\pm} = \frac{1}{\sqrt{2\hbar\omega_k}}\left[\omega_k Q_k \mp iP_k\right]\delta q, \tag{94}$$

where Q_k and P_k are now operators. The operators $a_k^{\pm}$ satisfy the following commutation relations which are typical of a boson field

$$[a_k^+, a_{k'}^-] = \delta_{kk'} \quad , \quad [a_k^-, a_{k'}^-] = [a_k^+, a_{k'}^+] = 0. \tag{95}$$

The Hamiltonian operator of such a field reads

$$\hat{H} = \int \hat{h}_k \frac{d\vec{k}}{(2\pi)^3}, \tag{96}$$

where

$$\hat{h}_k = \hbar\omega_k \left[a_k^+ a_k^- + \frac{1}{2}\right]. \tag{97}$$

The eigenstates of the system can thus be represented by

$$|k, n> = C_k \frac{1}{n!}|a_k^+|^n |k, 0>, \tag{98}$$

where C_k is a constant and the vector $|k, 0>$ represents the vacuum state for a given mode k. In order to characterize the chargeon concentration, we consider the chargeon occupation number

$$N_k = < a_k^+ a_k^- > = \frac{\alpha_k}{2\hbar}\omega_k |\delta q_k|^2. \tag{99}$$

The expression (99) shows that a single chargeon corresponds to a charge fluctuation δq_k, with an amplitude given by

$$|\delta q_k|_{min} = \left(\frac{2\hbar}{\alpha_k \omega_k}\right)^{1/2}, \tag{100}$$

which establishes the scale of the charge quantization in a dusty plasma.

5 Discussion

In this article, we have considered three exotic phenomena in dusty plasmas. First, we have shown that a chain of elongated dust rods in a plasma can adequately be described by the analogue of an Ising model, where ion flows play the role of the magnetic field and the plasma turbulent fluctuations play the role of the thermal fluctuations. With this work a new field of research is proposed in the area of dusty plasma physics and of dusty crystals, by suggesting that new types of experiments have to be conducted in the light of our analytical models and predictions.

We have also extended the theory of the Casimir interaction between two nearby macroscopic bodies in a medium to the case of a turbulent plasma. With this extension we have explored two new directions: i) a medium far away from thermal equilibrium was considered for the first time; ii) the case of a plasma was explicitly considered for the first time. We have discussed the two basic geometric configurations of the Casimir theory. The first consists of two parallel conducting plates suspended in a turbulent plasma. The second corresponds to two nearby spherical particles, which can be associated, in the case of a plasma, with two nearby dust grains. Our results point towards the existence of fluctuation induced forces between two nearby particles in a plasma, which have to be added to the well known electrostatic forces associated to their electric charges and their corresponding polarization clouds. The expressions derived in this work are valid for both the electrostatic and the electromagnetic turbulence spectra. They also include the vacuum force as a limiting case. Moreover, the usual Casimir effect between two parallel plates has been generalized for a turbulent plasma. Fluctuation-induced forces between two nearby spherical dust grains were also shown to exist. Future work will have to include the contribution of the induced charge fluctuations of the dust grains, as well as the finite propagation time of the interacting fields.

Finally, we have extended the concept of the phonon to the case of high-fugacity dusty plasmas, where grain charge fluctuations give rise to a normal mode, which is now referred to as the dust-Coulomb wave. While this mode exists in the ultra low-frequency regime, the elementary quantum excitation of the grain charge leads to the introduction of a new quasi-particle, which can be called the 'chargeon'. In many respects, the chargeon is similar to the phonon, but it describes a distinct physical entity which is unique to and relevant in the dense regime. We have limited our discussion to the case of high fugacity which typically, though not necessarily, corresponds to high dust concentration. But the approach followed here can easily be extended to the general case of arbitrary fugacity when dusty plasmas are known [Rao 1999, 2000] to support a unified mode combining both the DAW and CDW modes. We have also restricted ourselves to electrostatic fluctuations in the linear regime. On the other hand, quantization of the electromagnetic oscillations of an electron-ion plasma has been recently considered [Mendonça *et al.* 2000] by including the contributions due to the nonlinear terms. This suggests that a more general quantization procedure can be followed for the case when the electromagnetic as well as the nonlinear properties of a dusty plasma have to be self-consistently taken into account. It is hoped that the present work will stimulate and help to extend the physics of dusty plasmas into new areas of research.

Acknowledgements

The authors dedicate this paper to the memory of Nagesha N. Rao whose untimely and unfortunate death deprived many of a stimulating physicist and true friend.

This research was partially supported by the International Space Science Institute at Bern, Switzerland for the team "Dust Plasma Interaction in Space", as well as by the European Commission (Brussels) through the contract No. HPRN-CT-2000-00140 for carrying out the tasks of the Human Potential and Research Training Network entitled "Complex Plasmas in Space: The Science of Laboratory Colloidal Plasmas and Mesospheric Charged Aerosols".

References

Abrikosova I. I. and Deryaguin B. V. 1953, *Dokl. Akad. Nauk SSSR* **90**, 1055.

Arfken G. 1985, *Mathematical Methods for Physicists*, Academic Press, Orlando, third edition.

Barkan A., Merlino R.L. and D'Angelo N. 1995, *Phys. Plasmas* **2**, 3563.

Binney J. J., Dowrich N. J., Fisher A. J. and Newman M. E. J. 1995, *The theory of Critical Phenomena*, Oxford University Press, Oxford.

Casimir H. B. G. 1948, *Proc. Kon. Ned. Akad. Wet.* **51**, 793.

Casimir H. B. G. and Polder D. 1948, *Phys. Rev.* **73**, 360.

Chu, J. H. and I. L. 1994, *Phys. Rev. Lett.* **72**, 4009.

Dubinov A. E., Zhdanov V. S., Ignatov A M, Kornilov S. Yu, Sadovoi S. A. and Selemir V. D. 2000, in *Frontiers of Dusty Plasmas*, edited by Nakamura Y., Yokota T. and Shukla P. K., Elsevier Science, Amsterdam, pp. 389-392.

Elisalde E. and Romeo A. 1991, *Am. J. Phys.* **58**, 711.

Ford L. H. and Svaiter N. F. 2000, *Phys. Rev. A* **62**, 062105.

Hamaguchi S. and Farouki R.T. 1994, *Phys. Rev. E* **49**, 4430.

Havnes O., Li F., Melandso F., Aslaksen T., Hartquist T. W., Morfill G. E., Nitter T. and Tsytovich, V. N. 1995, *J. Geophys. Res.* **100**, 1731.

Itzykson C. and Zuber J. B. 1980, *Quantum Field Theory*, McGraw-Hill, New York.

Kardar M. and Golestanian R. 1999, *Rev. Mod. Phys.* **71**, 1233.

Lamoreaux S. K. 1998, *Phys. Rev. Lett.* **81**, 5475.

Lampe M, Joyce G., Ganguli G. and Gavrishchaka V. 2000, *Phys. Plasmas* **7**, 3851.

Landau L. and Lifshitz E. M. 1995, *Statistical Physics*, Butterworth-Heinemann, Oxford.

Lau A. W. C., Levine Dov. and Pincus P. 2000, *Phys. Rev. Lett.* **84**, 4116.

Lifshitz E. M. 1956, *Sov. Phys. JETP* **2**, 73.

Mahmoodi J., Shukla P. K., Tsintsadze N. L. and Tskhakaya D. D. 2000, *Phys. Rev. Lett.* **84**, 2626.

Melandsø, F., Aslaksen, T. and Havnes, O. 1993, *Planet. Space Sci.* **41**, 321.

Mendonça J. T., Bingham R., Shukla P. K. and Resendes D. P. 2001a, *Phys. Lett. A* **289**, 233.

Mendonça J. T., Martins A. M. and Guerreiro A. 2000, *Phys. Rev. E* **62**, 2989.

Mendonça J. T., Rao N. N. and Guerreiro A. 2001b, *Europhys. Lett.* **54**, 741.

Mendonça J. T., Shukla P. K. and Bingham R. 2001c, *Phys. Lett. A* **282**, 288.

Mendonça J. T., Shukla,P.K., Martins A. M. and Guerra R. 1997, *Phys. Plasmas* **4**, 674.

Mendonça J. T., Tsintsadze N. L. and Guerreiro A. 2002, *Europhys. Lett.* **57**, 362.

Mohideen U., Rahman H. U., Smith M. A., Rosenberg M. and Mendis D. A. 1998, *Phys. Rev. Lett.*, **81**, 349.

Mohideen U. and Roy A. 1998, *Phys. Rev. Lett.* **81**, 4549.

Molotkov V. I., Nefedov A. P., Putsyl'nik M. Y., Torchinsky V. M., Fortov V. E. and Khrapak A. G. 2000, *JETP Lett.* **71**, 102.

Nambu M., Vladimirov S. V. and Shukla P. K. 1995, *Phys. Lett. A* **203**, 40.

Onsager L. 1944, *Phys. Rev.* **65**, 117.

Pulnien G., Mueller B. and Greiner W. 1986, *Phys. Rep.* **134**, 87.

Rao N. N. 1999, *Phys. Plasmas* **6**, 4414.

Rao N. N. 2000, *Phys. Plasmas* **7**, 3214.

Rao N. N., Shukla P.K. and Yu M.Y. 1990, *Planet. Space Sci.* **38**, 543.

Resendes D.P. 2000, *Phys. Rev. E* **61**, 793.

Resendes D. P., Mendonça J. T. and Shukla P. K. 1998, *Phys. Lett. A* **239**, 181.

Samsonov D., Goree J., Ma Z. W., Bhattacharjee A., Thomas H. M. and Morfill G. E. 1999, *Phys. Rev. Lett.* **83**, 3649.

Shukla P. K. and Rao N. N. 1996, *Phys. Plasmas* **3**, 1770.

Shukla P. K. 2001, *Phys. Plasmas* **8**, 1793.

Shukla P. K. and Mamun A. A. 2002, *Introduction to Dusty Plasma Physics*, Institute of Physics Publishing, Bristol.

Thomas H. M., Morfill G . E., Demmel V., Goree J., Feuerbacher B. and Mohlmann D. 1994, *Phys. Rev. Lett.* **73**, 652.

Tsytovich V. N., Khodataev Ya. K., Bingham R., Morfill G. E. and Resendes D. P. 1998, *Comments on Plasma Phys. and Contr. Fusion*, **18**, 345.

Varma R. K., Shukla P. K. and Krishan V. 1993, *Phys. Rev. E* **47**, 3612.

Zamalodchikov A. B. 1989, *Int. J. Mod. Phys.* **A4**, 4235.

INDEX

A

B

C

D

E

F

G

H

I